T0310420

Anechoic and Reverberation Chambers

Anechoic and Reverberation Chambers

Theory, Design, and Measurements

Qian Xu
College of Electronic and Information Engineering
Nanjing University of Aeronautics and Astronautics
China

Yi Huang
Department of Electrical Engineering and Electronics
The University of Liverpool
UK

This edition first published 2019

© 2019 John Wiley & Sons Ltd

The right of Qian Xu and Yi Huang to be identified as the authors of this work has been asserted in accordance with law.

Registered Offices

John Wiley & Sons, Inc., 111 River Street, Hoboken, NJ 07030, USA

John Wiley & Sons Ltd, The Atrium, Southern Gate, Chichester, West Sussex, PO19 8SQ, UK

Editorial Office

The Atrium, Southern Gate, Chichester, West Sussex, PO19 8SQ, UK

For details of our global editorial offices, customer services, and more information about Wiley products visit us at www.wiley.com.

Wiley also publishes its books in a variety of electronic formats and by print-on-demand. Some content that appears in standard print versions of this book may not be available in other formats.

MATLAB® is a trademark of The MathWorks, Inc. and is used with permission. The MathWorks does not warrant the accuracy of the text or exercises in this book. This work's use or discussion of MATLAB® software or related products does not constitute endorsement or sponsorship by The MathWorks of a particular pedagogical approach or particular use of the MATLAB® software.

Library of Congress Cataloging-in-Publication data applied for

ISBN: 9781119361688

Cover Design by Wiley

Set in 10/12pt Warnock by SPi Global, Pondicherry, India

Printed in Singapore by C.O.S. Printers Pte Ltd

10 9 8 7 6 5 4 3 2 1

To our families

Contents

About the Authors

Dr. Qian Xu received the BEng and MEng degrees from the Department of Electronics and Information, Northwestern Polytechnical University, Xi'an, China, in 2007 and 2010 respectively, and received the PhD degree in electrical engineering at the University of Liverpool (UoL), UK, in 2016. He is now an Associate Professor at the College of Electronic and Information Engineering, Nanjing University of Aeronautics and Astronautics, China. He worked as a Radio Frequency (RF) engineer in Nanjing, China in 2011, an Application Engineer in CST (Computer Simulation Technology) Company, Shanghai, China in 2012, and a Research Assistant at UoL in 2016.

His work at the UoL was sponsored by Rainford EMC Systems Ltd. (now part of Microwave Vision Group) and Centre for Global Eco-Innovation. He finished his PhD in three years and has authored/co-authored more than 70 papers. Three of his papers received the Best Student Paper Award (third prize, first author) at LAPC (Loughborough Antenna & Propagation Conference) 2014, Best Non-Student Paper Award (first prize, coauthor) at LAPC 2015 and Best Non-Student Paper Award (third prize, coauthor) at LAPC 2016 respectively. He is a reviewer of Institute of Electrical and Electronics Engineers (IEEE) Transactions on Electromagnetic Compatibility (EMC) and IEEE Antennas and Wireless Propagation Letters. He developed new reverberation chamber and anechoic chamber measurement control systems at UoL. He also proposed a series of new measurement methods in reverberation chambers and quantified the stirring efficiency by using the total scattering cross section. He has developed a new software package for the Rainford EMC Systems which accelerates the anechoic chamber design greatly.

Prof. Yi Huang received his BSc in Physics (Wuhan, China), MSc (Eng) in Microwave Engineering (Nanjing, China) and DPhil in Communications from the University of Oxford, United Kingdom, in 1994. He has been conducting research in the areas of radio communications, applied electromagnetics, radar, and antennas since 1986. He has been interested in reverberation chamber theory and applications since 1991. His experience includes three years spent with NRIET (China) as a Radar Engineer and various periods with the Universities of Birmingham, Oxford, and Essex as a member of research staff. He worked as a Research Fellow at British Telecom Labs in 1994 and then joined the Department of Electrical Engineering & Electronics, University of Liverpool (UoL), UK, as a Lecturer in 1995, where he is now the Chair in Wireless Engineering, the Head of High Frequency Engineering Group, and the Deputy Head of the Department.

Dr. Huang has published over 300 refereed research papers in international journals and conference proceedings and is the principal author of the popular book *Antennas: From Theory to Practice* (John Wiley & Sons, 2008). He has received many research grants from research councils, government agencies, charity, EU, and industry; acted as a consultant to various companies; served on a number of national and international technical committees (such as the UK Location & Timing KTN, IET, EPSRC, and European ACE); and has been an Editor, Associate Editor or Guest Editor of four international journals. He has been a keynote/invited speaker and organiser of many conferences and workshops (e.g. IEEE iWAT, WiCom, LAPC, UCMMT, and Oxford International Engineering Programmes). He is at present the Editor-in-Chief of *Wireless Engineering and Technology*, an Associate Editor of *IEEE Antennas and Wireless Propagation Letters*, the Rep for the UK and Ireland to European Association of Antennas and Propagation (EurAAP), a Senior Member of IEEE, and a Fellow of IET.

About the Contributors

Dr. Tian-Hong Loh
National Physical Laboratory, Engineering, Materials & Electrical Science Department, 5G & Future Communications Technology Group, Teddington, United Kingdom

Prof. Xiaoming Chen
School of Electronic and Information Engineering, Xi'an Jiaotong University, Xi'an, China

Acknowledgements

The authors would like to express their gratitude to the following people who have either directly or indirectly contributed to the production of this book. Their time and efforts will never be forgotten, without them, the book would not have been possible.

John Noonan – Microwave Vision Group, UK
Paul Duxbury – Microwave Vision Group, UK
David Seabury – MPE Company, UK
Dr. Tian-Hong Loh – National Physical Laboratory, UK
Dr. Jiafeng Zhou – University of Liverpool, UK
Prof. Xu Zhu – Harbin Institute of Technology, China
Prof. Xiaoming Chen – Xi'an Jiaotong University, China
Dr. Lei Xing – Nanjing University of Aeronautics and Astronautics, China
Dr. Zhihao Tian – University of Liverpool, UK
Dr. Chong Li – University of Glasgow, UK
Dr. Stephen J. Boyes – Defense Science & Technology Laboratory, UK
Prof. Andy Marvin – University of York, UK
Dr. Matt Fulton – University of Liverpool, UK
Doug McInnes – University of Liverpool, UK
Sandra Grayson – Wiley, UK
Ashmita Thomas – Wiley, UK
Karthika Sridharan – SPi Global, India

Chapter 8 is contributed by Prof. Xiaoming Chen, Yuxin Ren and Zhihua Zhang; Chapter 9 is contributed by Dr. Tian-Hong Loh and Wanquan Qi.

We would also like to thank our colleagues, friends and students who have contributed to this book in one way or another: Prof. Yongjiu Zhao, Dr. Yonggang Zhou, Dr. Zhenyi Niu, Dr. Hongwei Deng, Dr. Min Wang, Prof. Changqing Gu, Dr. Zhuo Li, Dr. Hengyi Sun, Dr. Xinlei Chen, Dr. Weiqiang Liu, Dr. Junqing Zhang, Dr. Long Shi, Dr. Chaoyun Song, Dr. Sheng Yuan, Dr. Neda Khiabani, Dr. Ping Cao, Dr. Zhouxiang Fei, Dr. Saqer Alja'afreh, Dr. Jingwei Zhang, Dr. Manoj Stanely, Dr. Qianyun Zhang, Tianyuan Jia and Letian Wen. We have enjoyed working with you all and appreciate your ideas, support and good sense of humour.

Part of the work in this book was supported by Project 139 of Centre for Global Eco-Innovation and Rainford EMC Systems (Microwave Vision Group), UK, National Natural Science Foundation of China (61701224, 61601219) and Natural Science Foundation of Jiangsu Province (BK20160804).

Further we would like to thank the International Electrotechnical Commission (IEC) for permission to reproduce information from its International Standards. All such extracts are copyright of IEC, Geneva, Switzerland. All rights reserved. Further information on the IEC is available from *www.iec.ch*. IEC has no responsibility for the placement and context in which the extracts and contents are reproduced by the author, nor is IEC in any way responsible for the other content or accuracy therein.

Acronyms

AAC	average absorption coefficient
AC	anechoic chamber
ACF	auto-correlation function
ACS	absorption cross section
AF	antenna factor
AR	axial ratio
ASCII	American Standard Code for Information Interchange
AUT	antenna under test
AVF	antenna validation factor
BEM	boundary element method
BER	bit error rate
BFS	breadth-first search
BSE	base station emulator
CAD	computer-aided design
CDF	cumulative distribution function
CEM	computational electromagnetics
CISPR	International Special Committee on Radio Interference (Comité International Spécial des Perturbations Radioélectriques)
CLF	chamber loading factor
CLT	central limit theorem
COM	component object model
CPU	central processing unit
CV	coefficient of variance
CVF	chamber validation factor
DG	diversity gain
DoF	degrees of freedom
DUT	device under test
EM	electromagnetic
EMC	electromagnetic compatibility
EUT	equipment under test
FACET	fast anechoic chamber evaluation tool
FD	frequency domain
FDFD	frequency domain finite-difference
FDTD	finite-difference time domain
FEM	finite element method
FFT	fast Fourier transform
FIT	finite integration technique
FITD	finite integration time domain

FMM	fast multipole method
FT	Fourier transform
FU	field uniformity
GA	genetic algorithm
GEV	generalised extreme value
GO	geometric optics
GPU	graphics processing unit
GTEM	gigahertz transverse electromagnetic
GUI	graphical user interface
IE	integral equation
IEC	International Electrotechnical Commission
IEEE	Institute of Electrical and Electronics Engineers
IFFT	inverse fast Fourier transform
IFT	inverse Fourier transform
i.i.d.	independent and identically distributed
IP	intellectual property
LoS	line-of-sight
LPDA	log-periodic dipole array
LUF	lowest usable frequency
MIMO	multiple-input multiple-output
MLE	maximum likelihood estimation
MLFMM	multilevel fast multipole method
MoM	moment method or method of moment
NLoS	none-line-of-sight
NPL	National Physical Laboratory
NSA	normalised site attenuation
NUAA	Nanjing University of Aeronautics and Astronautics
OOP	object-oriented programming
OTA	over-the-air
OUT	object under test
PBC	periodic boundary condition
PC	personal computer
PDF	probability density function
PEC	perfect electric conductor
PILA	planar inverted-L antenna
PMC	perfect magnetic conductor
PML	perfectly matched layer
PO	physical optics
PVC	polyvinyl chloride
P2A	point-to-area
P2L	point-to-line
P2P	point-to-point
RAM	radio absorbing material
RC	reverberation chamber
RCS	radar cross section
RF	radio frequency
RIMP	rich isotropic multipath environment

Rx	receiving
SA	site attenuation
SBR	shooting bouncing ray
SE	shielding effectiveness
SF	spreading factor
SNR	signal-to-noise ratio
SR	stirring ratio
SSH	scalar spherical harmonics
STD	standard deviation
STL	stereolithography
SVSWR	site voltage-standing-wave ratio
TCS	transmission cross section
TD	time domain
TE	transverse electric
TEM	transverse electromagnetic
TG	time gating
TIS	total isotropic sensitivity
TLM	transmission line matrix
TM	transverse magnetic
TRP	total radiated power
TSCS	total scattering cross section
Tx	transmitting
UoL	University of Liverpool
UK	United Kingdom
USRP	universal software radio peripheral
VNA	vector network analyser
VSH	vector spherical harmonics
WV	working volume

1

Introduction

1.1 Background

Anechoic chambers (ACs) and reverberation chambers (RCs) are two very different types of indoor measurement facilities and have been widely used in acoustics as well as in electromagnetics. It is interesting to note that these chambers share similar phenomena, physical quantities, and mathematical expressions in some ways. This book is about ACs and RCs in electromagnetics. Inside an AC, electromagnetic (EM) waves are absorbed by the absorbing materials at the boundary, while inside an RC, EM waves are reflected by the conducting reflector at the boundary. Over the years, these two different chambers have found some common or complimentary applications in antennas, electromagnetic compatibility (EMC), and radio communication measurements. Each has its advantages and disadvantages. Thus, it makes a perfect sense to bring these two different chambers into one book. They are like two sides of one coin: one is based on deterministic theory and the other is based on statistical theory; people working on RCs can be inspired by those working on ACs, and vice versa. Dual quantities can also be found in absorbing and scattering phenomena. This book is aimed at providing a clear and systematic approach to their design, measurement, and applications. Some latest developments are also included. In this chapter, we present an overview of both chambers while more details are provided in later chapters.

1.1.1 Anechoic Chambers

An ideal AC is a room designed to emulate free space – no radio waves are reflected from the walls, ceiling, and floor. The reason for using an AC is well-known: an ideal free space is required for EM measurement in an indoor environment that is not affected by the weather and interference outside the chamber, thus repeatable results can be obtained. A typical AC is given in Figure 1.1a and a typical measurement scenario with an aircraft is shown in Figure 1.1b.

In practice, because no ACs can absorb EM waves perfectly and reflections always exist, the performance of an AC needs to be characterised to show how close it is to the ideal free space. Thus, how to design an AC effectively and efficiently becomes an important issue. A problem is how to optimise the performance of such a chamber for a given chamber size using the least amount of radio absorbing materials (RAMs) to minimise the cost and maximise the test volume (i.e. the equipment under the test area). The cost of the RAM depends on its size and type. How to choose the RAMs and arrange them properly is another key problem. Currently, the design of the chamber depends on the designer's experience and sometimes a trial-and-error approach or a large safe margin has to be adopted. Intuitively, a large space with high-performance absorbing materials leads to a good AC, but to quantify the chamber performance a well-defined and accurate mathematical model needs to be created. Thus, a scientific and objective way to find the best solution is required. An analytical solution is

Anechoic and Reverberation Chambers: Theory, Design, and Measurements, First Edition. Qian Xu and Yi Huang.
© 2019 John Wiley & Sons Ltd. Published 2019 by John Wiley & Sons Ltd.

(a) (b)

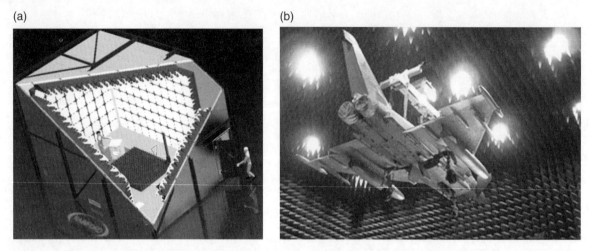

Figure 1.1 Anechoic chamber: (a) 3D model with a cutting plane and (b) measurement with an aircraft inside an AC (pictures from Rainford EMC Systems, Microwave Vision Group).

almost impossible for such a complex system, which offers an opportunity to bring computational electromagnetics (CEMs) and real engineering problems together.

If an efficient computer-aided design (CAD) tool was available to predict the performance of an AC, the designer could design the chamber better, faster, and more accurately with the help of computers, not just relying on experience.

The figures of merit used to characterise the chamber performance in practice are site attenuation (SA) for a full AC (all walls are covered with RAMs) and normalised site attenuation (NSA) for a semi-AC (no RAMs on the floor), field uniformity (FU), and site voltage-standing-wave ratio (SVSWR) [1, 2]. The procedures to measure these figures of merit and acceptable limits are given in relevant standards [1, 2].

It is well-known that the performance of ACs is closely related to the reflectivity of RAMs and how to arrange them [3, 4]. The first patented absorber was used to improve the front-to-back ratio of an antenna in 1936 [5]. During World War II (1939–1945), –20 dB (near normal incident angles) in the frequency range of about 2–15 GHz was obtained as the well-known Jauman absorber [6]. During the war years, Neher [7] demonstrated that the reflection from a long pyramidal shaped structure was much smaller than the reflection from a panel of the same absorber. This demonstrated the important role of geometry in the reflection reduction of RAMs. The first commercially available absorber started in 1953. In the early 1950s, 'dark-rooms' were built at a number of government and commercial organisations [8–10]; at that time, a typical level of reflected signal at *S* band was about 20 dB below the level of the direct signal. In the late 1950s, a new generation of broadband absorbers was able to produce a reflection coefficient of about –40 dB for near-normal incident angles. In the 1960s, by using ferrite underlayers, the thickness of the absorber was reduced greatly at low frequencies and the tapered chamber was developed, which showed a better performance than the rectangular chamber [10, 11]. The normal reflection coefficient at high frequencies achieved –60 dB. Nowadays, by combing the ferrite tiles and the pyramid absorbers, the reflection coefficient can achieve –25 dB at 30 MHz and –51 dB at 18 GHz (http://www.mvg-world.com/en/system/files/fiche_uh_absorbers_hypyr-loss_en_bd_oct_25th.pdf). More details will be discussed in the following chapters.

Three basic types of AC are used in practice, as shown in Figure 1.2: the rectangular chamber (Figure 1.2a), the tapered chamber (Figure 1.2b), and the compact chamber (Figure 1.2c). Test regions are marked with a circle, waves propagate along the lines ideally and absorbers are plotted as small triangles. In practice, because of the reflection and scattering of the RAMs, and because extraneous signals exist, the field in the test region is not uniform. The tapered chamber normally can provide a better FU than the rectangular chamber at lower frequencies, but the SA of a tapered chamber does not follow the Friis free-space transmission formula because of the multiple reflections from the tapered walls [3]. This should be noted for some special measurements such as using the three-antenna method to measure the gain of antennas. A compact chamber can be used to illuminate a large object with plane wave at higher frequencies because the object under test needs to be placed at the far-field region. When the frequency is high, the far-field condition cannot be satisfied without the use of a reflector. A parabolic reflector is normally used to generate a plane wave at higher frequencies, as shown in Figure 1.2c.

How to obtain an optimised AC has been investigated for many years. A well-known book was written by L. H. Hemming in 2002 [3] that provided an overview of this topic, including RAM characteristics, ACs of different shapes, and measurements in ACs. Geometric optics (GO) was mentioned as a general method to analyse the AC, but the calculation was done by hand and how to implement it using a computer was not given. In a recent book [12], B. K. Chuang reviewed the GO method for AC design in one chapter. Although CEMs have evolved over the years, compared with other CEM methods the GO method is still the most robust and efficient in AC analysis. The most attractive advantage is that no detailed information of the material properties (permittivity, conductivity, and permeability) needs to be known; only the reflection coefficient is enough to describe the RAMs. The simulation time is also short with an acceptable error.

In this book we present a systematic solution for AC design, from theory to measurement. The solution proposed in this book is meant to be general and useful for all types of ACs, that is, not limited to specific shapes; it is also possible to use this solution to explore new chamber shapes with special requirements.

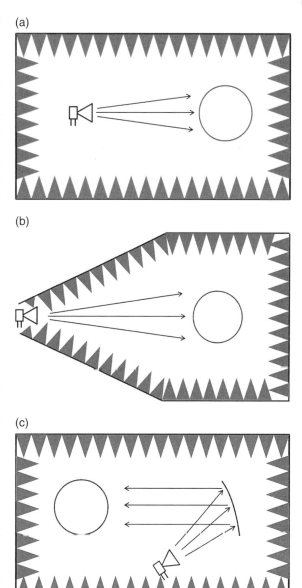

Figure 1.2 Three types of ACs: (a) rectangular chamber, (b) tapered chamber, and (c) compact chamber.

1.1.2 Reverberation Chambers

Unlike an AC, an RC is an electrically large conducting-screened room with electrically large stirrers used to stir the field inside the chamber (https://en.wikipedia.org/wiki/Electromagnetic_reverberation_chamber). The RC is also known in the literature as a reverberating chamber, a reverb, a mode-stirred chamber or a mode-tuned chamber. In this book the term 'reverberation chamber' is used as it is now widely used and accepted. The EM field inside the chamber is expected to be statistically uniform and isotropic. Two RCs are shown in Figure 1.3. In Figure 1.3a, two stirrers at the corner of the RC are used while in Figure 1.3b, one stirrer is employed near the middle of the RC.

(a)

(b)

Figure 1.3 Reverberation chambers: (a) RC at the University of Liverpool, UK, width 3.6 m, length 5.8 m, height 4 m, and (b) RC at the National Physical Laboratory, UK, width 5.8 m, length 6.5 m, height 3.5 m.

The first RC was probably proposed by H. A. Mendes in 1968 for EMC measurements [13] and was then adopted by American military standard MIL-STD-1377. An international standard on using an RC for EMC testing was published in 2003 and revised in 2011 [14]. Over the years, many researchers have worked on it and made significant advancements: P. Corona improved our understanding of RCs for EMC measurements [15] and D. A. Hill proposed a plane-wave integral representation of fields in the RC [16]. Like the Friis transmission equation in free space, Hill's equation reveals the transmission law in a multipath environment [16]. Systematic theory has been found to describe the fields in an ideal RC [16, 17]. It should also be noted that, when the RC is not working in the over-mode condition (i.e. not electrically large), the statistical behaviour deviates from the expected distribution functions. Thus, there is a blurred region from deterministic behaviour to statistical behaviour [17]. In practical engineering, we try to avoid working in this region as an RC of different shapes may have different behaviour, and it would be difficult to have a general theory to fit the measurement results in this region. At a lower frequency (i.e. the chambers are electrically small cavities), deterministic theory can be used, at a higher frequency (i.e. the chambers are electrically large cavities) Hill's theory can be applied.

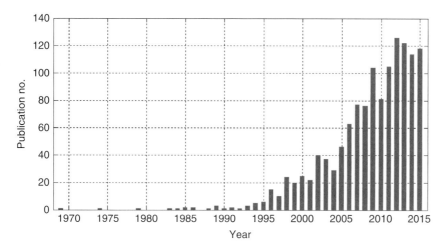

Figure 1.4 Number of publications per year (data from IEEE Xplore, key word: reverberation chamber).

The number of RC papers published over the past 45 years is shown in Figure 1.4. The numbers are from IEEE database. It is interesting to see that the number increased significantly in the 2000s, but the initial study on RCs is from 1968! The reason seems to be that the RC is no longer a specialised facility for EMC tests; it has gradually become a test facility for antenna measurement, bio-electromagnetics, material measurement, radio channel emulation, etc.

Like an AC, there are also many useful parameters to characterise an RC, such as K-factor, independent sample, correlation coefficient, FU, stirrer efficiency, enhanced backscatter constant, total scattering cross section, etc. Currently there is no unique parameter to summarise all these effects, and the relationships between some of them are still ambiguous. Because of the complexity of the statistical electromagnetics, there are still things we do not know and the application of the RC is expanding thanks to the researchers on this topic all over the world.

The stirring technique is a lasting topic in RC research. It is well known that to achieve a statistical EM field in a cavity, some kind of stirring mechanism needs to be involved to stir the field inside. From the integral expression of the electric field (integrate source with Green's function), to stir the electric field we can change the boundary conditions or change the source. The boundary conditions can be changed by (i) altering the internal structure, for example using asymmetric stirrers [14], helical stirring [18] or a carrousel stirrer [19–21], and (ii) changing the boundary structure, for example by a wall vibrating boundary [22–25], an oscillating stirrer [26], a sliding wall [27] or reactively loaded antennas [28]. If the source is changed, one can change the position or orientation of the source [29–31] or use multiple source excitations [32]. Frequency stir is an effective and efficient method: by mixing the results from different frequencies separated by more than the coherent bandwidth [33, 34], a similar effect can be obtained. It should be noted that not all stirring methods are applicable to all applications. For example, if the Q factor of an RC needs to be kept constant (Q factors can also be treated as random variables [35]) during the stirring process, the sliding wall method may not be suitable as it changes the volume of the RC when stirring; if a device under test has a very narrow frequency band response, frequency stir should be very carefully applied, as frequency stir normally presumes other factors unchanged in the stirring bandwidth. If it is wider than the working bandwidth of the device, the response could be smoothed out.

Unlike ACs, the performance of an RC is not sensitive to the shape of the cavity when it is electrically large, as the mode number is not sensitive to the shape but the volume of the cavity. Rectangular shape is the most popular as it is easy to build, but this does not mean that it is the best shape from an electrical point of view. Other shapes of RC are also possible, such as triangular [36] and non-parallel walls [37, 38], which could provide better performance in some cases. However, considering the fact that one can add scatterers or panels into a rectangular RC to change its internal shape, and it is not easy to extend a special shape (e.g. triangular shape) into a rectangular shape after the RC is built, rectangular shape is still the favourite choice for its reconfigurability and generality.

In the RC part of this book we present fundamental theory, typical measurements, and design principles of RCs. We have tried to minimise the mathematical derivations and make the book easy to use in practice. The book not only includes knowledge that is already known, but also presents information that is relatively new.

1.1.3 Relationship between Anechoic Chambers and Reverberation Chambers

There have been some discussions on the relations between ACs and RCs. People get inspiration on RCs from ACs and vice versa. The RC can be considered as an opposite environment to the AC. The two types of chamber can be related and compared as follows.

1) *Different philosophies behind the two chambers.* The RC takes advantage of multipath waves while the AC tries to eliminate them.
2) *The AC is a deterministic environment while the RC is a statistical environment.* Correspondingly, the AC can be used to verify conclusions from CEMs, and the RC can be used to verify results from statistical electromagnetics. There can be statistical variables in deterministic theory, and there are deterministic quantities in statistical theory. For both chambers, we find uncertainties in certainties and find certainties in uncertainties.
3) From the communication channel point of view, the AC can be considered as an ideal Gaussian channel, while the RC can be considered as a Rayleigh channel or Rician channel. This is very useful when one wants to emulate the channel for a communication system to measure the bit error rate (BER), total isotropic sensitivity (TIS), channel capacity, etc.
4) Some physical quantities are very difficult to measure in one kind of chamber, but are easy to measure in another kind of chamber. An AC is very good for measurements with directional variables such as radiation pattern, antenna gain, and scattering cross section, while an RC is good at measurements with assembled variables such as antenna radiation efficiency, TIS, total radiated power, average absorption cross section, and shielding effectiveness.
5) Dual physical quantities exist in absorbing and scattering phenomenon. In the RC, the vector superposition of the random scattered field of many stirrer positions tends to be zero, as if it is absorbed. This provides insight in the RC design: if a stirrer leads a better random scattered field than another it means the performance of the stirrer is better. This is discussed in detail in the book.

1.2 Organisation of this Book

The AC has been used in the radio frequency (RF) and microwave industry for many years. However, design guidelines are mostly based on the experience accumulated over the years. There is one book related to AC design, published a few years ago [4], but the CEM algorithms have been developed greatly in the last 10 years or so. The industry has also moved on; different companies need to share information without disclosing sensitive data. This book provides the latest systematic solutions for AC design using state-of-the-art CEM algorithms. By using CAD, chamber designers can now optimise the chamber (structure, absorber layout, antenna positions) to maximise the performance while minimise the cost. This book will provide guidelines on this and show real design examples verified by measurements.

As a very different chamber, the RC has been used in EMC measurements and tests for a long time, but recent advances show that RCs can be used in many other applications and could be even better than ACs in some applications. There are a couple of books relating to RCs [16, 17] but the emphasis of these books is on EM theory and EMC measurement protocols. In recent years many new applications of RCs have been developed [39, 40]. This book covers a series of the latest measurement methods in RCs. New understandings of RCs from the time domain are also included, providing new points of view which cannot be seen from only the frequency domain.

This book covers the most recent advances in AC and RC designs and measurements. It will be interesting to show that these two types of chambers are closely related, the design of the RC can be inspired by the design of the AC, and there exist dual quantities between random scattering and absorption. The book is organised as follows:

Chapter 2. *Theory for Anechoic Chamber Design.* This chapter details the theory for AC design without considering how to realise it. Basic knowledge on absorbing materials is given. CEM algorithms are reviewed and discussed. Two forms of GO methods are introduced and it is shown that one is easier to use than the other in software realisation.

Chapter 3. *Computer-aided Anechoic Chamber Design.* This chapter focuses on how to realise the hybrid geometric optics–finite element method (GO-FEM) in AC design. Details on algorithm implementation are presented. It is shown how an AC design problem can be solved by using a CEM model step by step. This chapter mixes computer graphics and electromagnetics. Acceleration strategies are also given, and the reader could benefit from the use of computer graphics and graphics processing unit (GPU) computing.

Chapter 4. *Anechoic Chamber Design Examples and Verifications.* This chapter explains the figure of merits of AC performance: NSA, SVSWR, and FU. Procedures on how to measure the figures of merit are given and physical understandings are also addressed. Practical design examples are given together with simulation and measurement results.

Chapter 5. *Fundamentals of the Reverberation Chamber.* This chapter introduces the basic theory of RCs, and definitions of figures of merit, such as lowest usable frequency (LUF), working volume (WV), FU, and stirrer efficiency, are explained. Discussions on the CAD of RCs are also given. Unlike AC design, currently there is no mature software tool for RC design, but the design process can be aided by using a computer.

Chapter 6. *The Design of a Reverberation Chamber.* This chapter focuses on the design guidelines and time domain behaviour of the chamber; it is shown that some difficulties in the frequency domain measurement can be resolved from the time domain measurement. The stirrer efficiency is defined by using the total scattering cross section of stirrers. The theoretical limit of the performance of stirrers (which is a longstanding problem) can be obtained from the time domain information. The time domain understanding can also be applied to the RC design.

Chapter 7. *Applications in the Reverberation Chamber.* This chapter summarises a range of measurements inside an RC, including radiated immunity, radiated emission, antenna measurement (S parameters, efficiency, diversity gain, and radiation pattern), material measurement, shielding effectiveness measurement, channel emulation, and volume measurement. Theories, measurement procedures, and data processing are also explained.

Chapter 8. *Measurement Uncertainty in the Reverberation Chamber.* RC measurement data are usually analysed from a statistical point of view, this chapter investigates the measurement uncertainty in the RC. This chapter is authored by Xiaoming Chen, Yuxin Ren, and Zhihua Zhang.

Chapter 9. *Inter-Comparison Between Antenna Radiation Efficiency Measurements Performed in an Anechoic Chamber and in a Reverberation Chamber.* To have an in-depth understanding of both ACs and RCs, this chapter compares measurements of antenna efficiency in ACs and RCs at the National Physical Laboratory in the UK. This chapter is authored by Tian-Hong Loh and Wanquan Qi.

Chapter 10. *Discussion on Future Applications.* This chapter predicts possible future applications and highlights some unsolved problems which could serve as a good starting point for researchers.

Appendices. In the appendices, some relevant detailed information is provided which includes code snippets, reference values, report template, and frequently used statistics.

References

1 CISPR 16-1-4 (2012). *Specification for Radio Disturbance and Immunity Measuring Apparatus and Methods – Part 1–4: Radio Disturbance and Immunity Measuring Apparatus – Antennas and Test Sites for Radiated Disturbance Measurements*, 3.1e. IEC Standard.

2 IEC 61000-4-3 (2008). *Electromagnetic Compatibility (EMC) – Part 4–3: Testing and Measurement Techniques – Radiated, Radio-Frequency, Electromagnetic Field Immunity Test*, 3.1e. IEC Standard.

3 Emerson, W. H. (1973). Electromagnetic wave absorbers and anechoic chambers through the years. *IEEE Transactions on Antennas and Propagations* 21 (4): 484–490.

4 Hemming, L. H. (2002). *Electromagnetic Anechoic Chambers: A Fundamental Design and Specification Guide*. New York, NY: Wiley-IEEE Press.

5 Naamlooze Vennootschap Machmerieen, French Patent 802 728, 1936.

6 Du Toit, L. J. (1994). The design of Jauman absorbers. *IEEE Antennas and Propagation Magazine* 36 (6): 17–25.

7 Neher, L.K. (1953). Nonreflecting background for testing microwave equipment. US Patent 2656 535.

8 Simmons, A. J. and Emerson, W. H. (1953). *Anechoic Chamber for Microwaves*. Tele-Tech., vol. 12 (7).

9 Simmons, A.J. and Emerson, W.H. (1953). An anechoic chamber making use of a new broadband absorber material. 1958 IRE International Convention Record, New York, NY, USA. pp. 34–41.

10 Emerson,W.H. (1967). Anechoic chamber. US Patent 3308 463.

11 King, H., Shimabukuro, F., and Wong, J. (1967). Characteristics of a tapered anechoic chamber. *IEEE Transactions on Antennas and Propagation* 15 (3): 488–490.

12 Chen, Z. N., Liu, D., Nakano, H. et al. (2016). *Handbook of Antenna Technologies*. Springer Reference.

13 Mendes, H.A. (1968). A new approach to electromagnetic field-strength measurements in shielded enclosures. Wescon Technical Papers, Wescon Electronic Show and Convention, Los Angeles.

14 IEC 61000-4-21 (2011). *Electromagnetic Compatibility (EMC) – Part 4–21: Testing and Measurement Techniques – Reverberation Chamber Test Methods*, 2.0e. International Electrotechnical Commission.

15 Migliaccio, M., Gradoni, G., and Arnaut, L. R. (2016). Electromagnetic reverberation: the legacy of Paolo corona. *IEEE Transactions on Electromagnetic Compatibility* 58 (3): 643–652.

16 Hill, D. A. (2009). *Electromagnetic Fields in Cavities: Deterministic and Statistical Theories*. Wiley-IEEE Press.

17 Demoulin, B. and Besnier, P. (2011). *Electromagnetic Reverberation Chambers*. Wiley.

18 Arnaut, L. R., Moglie, F., Bastianelli, L., and Primiani, V. M. (2017). Helical stirring for enhanced low-frequency performance of reverberation chambers. *IEEE Transactions on Electromagnetic Compatibility* 59 (4): 1016–1026.

19 Wellander, N., Lundén, O., and Bäckström, M. (2007). Experimental investigation and mathematical modeling of design parameters for efficient stirrers in mode-stirred reverberation chamber. *IEEE Transactions on Electromagnetic Compatibility* 49 (1): 94–103.

20 Lundén, O., Wellander, N. and Bäckström, M. (2010). Stirrer blade separation experiment in reverberation chambers. Proceedings of International Symposium on Electromagnetic Compatibility, Fort Lauderdale, FL. pp. 526–529.

21 Fedeli, D., Iualè, M., Primiani, V.M., and Moglie, F. (2012). Experimental and numerical analysis of a carousel stirrer for reverberation chambers. Proceedings of International Symposium on Electromagnetic Compatibility, Pittsburgh, PA. pp. 228–233.

22 Leferink, F. (1998). Test Chamber. Patent NL1010745.

23 Leferink, F. and van Etten, W.C. (2000). Optimal utilization of a reverberation chamber, Euro EMC 2000, Symposium on EMC, Brugge. pp. 201–206.

24 Leferink, F. and van Etten, W.C. (2001). Generating an EMC test field using a vibrating intrinsic reverberation chamber. EMC Society Newsletter, Spring. pp. 19–25.

25 Leferink, F. (2008). In-situ high field strength testing using a transportable reverberation chamber. Asia-Pacific Symposium on Electromagnetic Compatibility and 19th International Zurich Symposium on Electromagnetic Compatibility, Singapore. pp. 379–382.

26 Jensen, P.T., Mynster, A.P., and Behnke, R. B. (2014). Practical industrial EUT testing in reverb chamber: Experiences, findings and practical observations on high amplitude immunity testing of industrial equipment in reverberation chamber. Proceedings of EMC Europe 2014 International Symposium Electromagnetic Compatibility, Gothenburg. pp. 274–279.

27 Johnston, R. H. and McRory, J. G. (1998). An improved small antenna radiation-efficiency measurement method. *IEEE Antennas and Propagation Magazine* 40 (5): 40–48.

28 Voges, E. and Eisenburger, T. (2007). Electrical mode stirring in reverberating chambers by reactively loaded antennas. *IEEE Transactions on Electromagnetic Compatibility* 49 (4): 756–761.

29 Huang, Y. and Edwards, D.J. (1992). A novel reverberating chamber: source-stirred chamber. IEEE 8th International Conference on Electromagnetic Compatibility, Edinburgh. pp. 120–124.

30 Huang, Y. (1993). The Investigation of Chambers for Electromagnetic Systems, DPhil Thesis, University of Oxford.

31 Cerri, G., Primiani, V. M., Pennesi, S., and Russo, P. (2005). Source stirring mode for reverberation chambers. *IEEE Transactions on Electromagnetic Compatibility* 47 (4): 815–823.

32 Kildal, P.-S. and Carlsson, C. (2002). Detection of a polarization imbalance in reverberation chambers and how to remove it by polarization stirring when measuring antenna efficiencies. *Microwave and Optical Technology Letters* 34 (2): 145–149.

33 Loughry, T. A. (1991). *Frequency Stirring: An Alternate Approach to Mechanical Mode-Stirring for the Conduct of Electromagnetic Susceptibility Testing*, PL-TR-91-1036. Phillips Laboratory.

34 Hill, D. A. (1994). Electronic mode stirring for reverberation chambers. *IEEE Transactions on Electromagnetic Compatibility* 36 (4): 294–299.

35 Arnaut, L. R. and Gradoni, G. (2013). Probability distribution of the quality factor of a mode-stirred reverberation chamber. *IEEE Transactions on Electromagnetic Compatibility* 55 (1). 35–44.

36 Huang, Y. (1999). Triangular screen chambers for EMC tests. *Measurement Science and Technology* 10: 121–124.

37 Sato, K. and Koyasu, M. (1959). On the new reverberation chamber with nonparallel walls (studies on the measurement of absorption coefficient by the reverberation chamber method, II). *Journal of the Physical Society of Japan* 14: 670–677.

38 Leferink, F. (1998). High field strength in a large volume: the intrinsic reverberation chamber. IEEE EMC Symposium. International Symposium on Electromagnetic Compatibility. Symposium Record, Denver, CO. 1. pp. 24–27.

39 Kildal, P.-S. (2015). *Foundations of Antenna Engineering: A Unified Approach for Line-Of-Sight and Multipath*. Artech House.

40 Boyes, S. and Huang, Y. (2016). *Reverberation Chambers: Theory and Applications to EMC and Antenna Measurements*. Wiley.

2

Theory for Anechoic Chamber Design

2.1 Introduction

The design of an anechoic chamber (AC) is a systematic problem. The designer needs to consider a lot of things, such as electromagnetic performance, structures, ventilation, lighting, access, power lines, fireproofing, cost, etc. From the perspective of electromagnetics, a series of questions need to be answered: what kind of absorber to use, how to arrange them, and how to estimate the performance of the chamber. In this chapter, the problems are categorised into two levels: the micro level and the macro level. The micro level problems describe the phenomena up to few wavelengths while the electrically large problems belong to the macro level.

At the micro level, we focus on the radio absorbing material (RAM) but not the overall structure of the AC. Basic knowledge of RAMs is introduced first in this chapter and followed by the RAM measurement. At the macro level, the theory used for the AC design is reviewed – geometric optics (GO). GO is normally related with ray tracing or the imaging method. From an analytical point of view, GO is an intuitive theory and easy to understand. However, to implement it in practice, the problem is not just an electromagnetic problem but also related to computer graphics. The realisation of GO and its applications will be presented in the next chapter. In this chapter, we focus on the theoretical part of GO and do not consider the complex realisation.

2.2 Absorbing Material Basics

2.2.1 General Knowledge

There are two common absorbers: ferrite absorbers and dielectric absorbers [1]. Dielectric absorbers are normally used at microwave frequency range (GHz) and ferrite absorbers are frequently used at a lower frequency range (MHz). It is interesting to review the development history of absorbing materials. The early dielectric absorbers were made from matted horse hair impregnated with conductive carbon solution [1, 2]. Now the absorbing materials have been made fire-resistant to pass the NRL 8093 tests [3], and thus they are much safer than before.

In the 1960s and 1970s, many conductive and magnetic materials were used for absorption, including carbon, metals, and conducting polymers, and absorbers were made from foams, netlike structures, or honeycomb and coated with a paint containing particulate or fibrous carbon, evaporated metal, nickel chromium alloy, or even plasma [4]. In the 1980s, computers were used to calculate the reflectivity from the RAM, and the scattering of these materials was analysed based on the Floquet theorem [4, 5]. Materials have been extended to use carbon or graphite, carbonyl iron, and ferrites, and artificial dielectrics have also been used by adding inclusions such as rods, wires, discs, and spheres [6]. Since the 1990s, conducting polymers and composite materials have been used along with coated fibres and fabrics, and dynamic RAMs have also been proposed where the resonant

Anechoic and Reverberation Chambers: Theory, Design, and Measurements, First Edition. Qian Xu and Yi Huang.
© 2019 John Wiley & Sons Ltd. Published 2019 by John Wiley & Sons Ltd.

frequency of the absorber can be tuned through the variation of resistive and capacitive elements in the absorber [7]. Nowadays, various types of RAM can be made with designed properties to be fire retardant, waterproof, flexible, have high power capacity, etc. Table 2.1 gives a summary of typical characteristics of RAMs used in ACs.

The three types of absorbers that are most frequently used are presented in Figure 2.1 although arbitrary shapes could be possible. The pyramidal shape (Figure 2.1a) offers a tapering change of impedance matching from free-space impedance (377 Ohms) to the bottom of the absorber [13–16]. The peak of the pyramid can be twisted at 45°, which can increase the performance further. The tapered matching configuration allows good absorption over a very wide frequency range, but the height of the pyramid must be comparable to or larger than the wavelength. A typical value of 50 dB absorption at normal incidence can be achieved at high frequencies (GHz). The wedge absorber (Figure 2.1b) can give better performance at a wide-angle incident when the incident wave comes from the directions parallel to the ridge of the absorber [1, 17–23]. This type of absorber is normally used to reduce large angle reflections in a compact chamber or radar cross section (RCS) chamber; a typical absorption of 47 dB at normal incidence can be achieved at high frequencies. The convoluted absorber (Figure 2.1c) is primarily useful in millimetre wave bands [24, 25]. A typical absorption of 50 dB at normal incidence at 30 GHz can be achieved. Ferrite absorbers (Figure 2.2a) normally have very high permeability, different from dielectric absorbers (the loss is due to the imaginary part of permittivity), ferrite absorbers attenuate electromagnetic waves due to the imaginary part of permeability [22, 26–31] by combining the ferrite tiles and

Table 2.1 Typical RAMs.

Material/structure	EM property	Typical characteristics
Ferrite	High magnetic conductivity [8]	Thin and heavy, typically used in 30 MHz–1 GHz, reflection coefficient < −10 dB
Fibrous carbon, foam with graphite	Electrically conductive [9]	Tapered structure, light weight, typically used in >1 GHz, reflection coefficient can be < −50 dB
Conductive polymer, evaporated metal	Electrically conductive [10]	Easy to conform to curved structures, the performance is comparable with the foam structure
Honeycomb structure	Electrically conductive [11]	Light weight, high stiffness-to-weight ratios, used in >1 GHz, reflection coefficient can be < −15 dB
Artificial structure	Resonant structure [7]	Active component, relatively narrow band but tunable
Metamaterial	Controllable permittivity and permeability [12]	Normally narrow band, high degrees of freedom

(a)   (b)   (c)

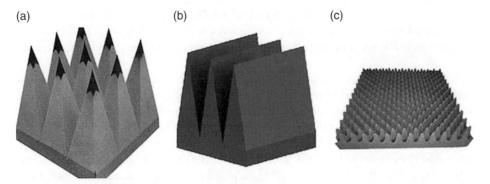

Figure 2.1 Typical dielectric absorbers: (a) pyramidal absorber, (b) wedge absorber, and (c) convoluted absorber (pictures from Microwave Vision Group).

(a)   (b)

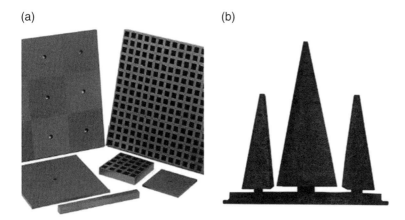

Figure 2.2 (a) Ferrite absorbers and (b) hybrid absorber (pictures from Microwave Vision Group).

dielectric absorbers (Figure 2.2b), and a hybrid absorber can have absorption from 25 dB at 30 MHz to 50 dB at 18 GHz (normal incidence). Absorption at lower frequencies is dominated by the ferrite absorber while absorption at higher frequencies is dominated by the dielectric absorber. It should be noted that the air gaps between the ferrite tiles degrade the absorbing performance and should be carefully controlled [32].

The working principle of the absorbing material can be understood from an ideal absorber. It is well known that the wave impedance of free space is

$$Z_0 = \frac{E}{H} = \sqrt{\frac{\mu_0}{\varepsilon_0}} \approx 377\,\Omega \tag{2.1}$$

where $\mu_0 = 4\pi \times 10^{-7}$ H m^{-1} and $\varepsilon_0 = 10^{-9}/(36\pi)$ F m^{-1} are the permeability and the permittivity of free space, respectively, and E and H are the magnitudes of the electric and magnetic fields. Perfect impedance matching can be realised if the electric permittivity and the magnetic permeability are equal when the reflection coefficient is zero:

$$\Gamma = \frac{\frac{Z_M}{Z_0} - 1}{\frac{Z_M}{Z_0} + 1} = 0 \tag{2.2}$$

where Z_M is the wave impedance of the absorber. Thus we have

$$\frac{Z_M}{Z_0} = \sqrt{\frac{\mu_r}{\varepsilon_r}} = 1 \tag{2.3}$$

where ε_r and μ_r are the complex relative permittivity and the permeability of the absorber:

$$\varepsilon_r = \frac{\varepsilon' - j\varepsilon''}{\varepsilon_0}, \ \mu_r = \frac{\mu' - j\mu''}{\mu_0} \tag{2.4}$$

where ε' and ε'' are the real part and imaginary part of the permittivity, and μ' and μ'' are the real part and imaginary part of the permeability, respectively. Eq. (2.3) means $\varepsilon_r = \mu_r$, and the wave decays exponentially in the absorber with distance, i.e. $e^{-\alpha d}$ where d is the propagation distance and α is the attenuation constant [4]:

$$\alpha = \omega\sqrt{\varepsilon_0\mu_0}\left(a^2 + b^2\right)^{1/4}\sin\left(\frac{1}{2}\tan^{-1}\left(\frac{a}{b}\right)\right) \tag{2.5}$$

where $a = \left(\varepsilon_r'\mu_r' - \varepsilon_r''\mu_r''\right)$ and $b = \left(\varepsilon_r'\mu_r'' + \varepsilon_r''\mu_r'\right)$, and ω is angular frequency. To get a large amount of attenuation in a small thickness, α must be large, which means that ε', ε'', μ', and μ'' must be large.

However, the real and imaginary parts are not independent of each other and are also functions of frequency. The Kramers–Kronig relations give that [33]

$$\varepsilon_r'(\omega) = 1 + \frac{2}{\pi}\int_0^\infty \frac{\omega'\varepsilon_r''(\omega')}{\omega'^2 - \omega^2}d\omega' \tag{2.6}$$

$$\varepsilon_r''(\omega) = \frac{2\omega}{\pi}\int_0^\infty \frac{1 - \varepsilon_r'(\omega')}{\omega'^2 - \omega^2}d\omega' \tag{2.7}$$

The ideal absorber is hard to realise in a very broadband frequency range because it is challenging to synthesis material with equal complex relative permittivity and permeability, but this provides a direction for the design of the RAM.

2.2.2 Absorbing Material Simulation

The analysis of the RAM performance can be considered as a micro level problem of AC design. If the shape, permittivity, conductivity, and permeability of the RAM are given, theoretically the reflection coefficient can be determined. In this book, we use complex permittivity, which takes the conductivity of the material into account:

$$\varepsilon(\omega) = \varepsilon' - j\varepsilon'' = \varepsilon' - j\frac{\sigma}{\omega} \tag{2.8}$$

where σ is the conductivity.

There are many methods to analyse the absorption performance of an absorber. We first introduce the mode matching method for the calculation of reflection coefficient of planar structure [33, 34], and then review the other numerical methods without giving detailed formulations.

Suppose a multilayer planar absorber has complex permittivity ε_1, ε_2, ..., ε_n, complex permeability μ_1, μ_2, ..., μ_n and thickness d_1, d_2, ..., d_n for each layer respectively, as shown in Figure 2.3. The incident angle is θ_0, and the plane of incidence is the plane which contains the surface normal and the propagation vector of the incident wave. Here the ZOX plane is the plane of incidence.

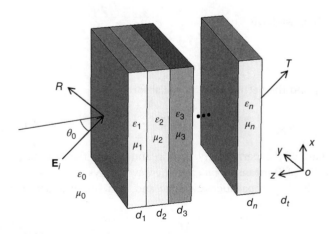

Figure 2.3 A multilayer planar absorber.

For the electric field (E-field) parallel to the plane of incidence (parallel polarisation), the magnetic field is in the y direction, and has the following expression:

$$\mathbf{H} = \hat{a}_y \exp[-\gamma_0(-z\cos\theta_0 + x\sin\theta_0)] \tag{2.9}$$

where $\hat{a}_y$ is the unit vector in the y direction and $\gamma_0 = \omega\sqrt{\mu_0\varepsilon_0}$ is the propagation constant in free space. The electric and magnetic fields in the ith layer have the forms [33, 34]:

$$H_{yi} = [X_i \exp(\gamma_{zi}z) + Y_i \exp(-\gamma_{zi}z)] \times \exp(-\gamma_0 x \sin\theta_0) \tag{2.10}$$

$$E_{xi} = \frac{\gamma_{zi}[-X_i \exp(\gamma_{zi}z) + Y_i \exp(-\gamma_{zi}z)]}{j\omega\varepsilon_i} \times \exp(-\gamma_0 x \sin\theta_0) \tag{2.11}$$

where $\gamma_{zi} = \gamma_i \cos\theta_i$. By applying the magnetic field continuity of $H_{yi} = H_{y(i+1)}$ at the interface between layers i and $i+1$, from (2.10) we have

$$X_i \exp\left(-\gamma_{zi}\sum_{m=1}^{i}d_m\right) + Y_i \exp\left(\gamma_{zi}\sum_{m=1}^{i}d_m\right)$$
$$= X_{i+1}\exp\left(-\gamma_{z(i+1)}\sum_{m=1}^{i}d_m\right) + Y_{i+1}\exp\left(\gamma_{z(i+1)}\sum_{m=1}^{i}d_m\right) \tag{2.12}$$

By applying the E-field continuity of $E_{xi} = E_{x(i+1)}$, from (2.11) we have

$$\varepsilon_{i+1}\gamma_{zi}\left[-X_i\exp\left(-\gamma_{zi}\sum_{m=1}^{i}d_m\right) + Y_i\exp\left(\gamma_{zi}\sum_{m=1}^{i}d_m\right)\right]$$
$$= \varepsilon_i\gamma_{z(i+1)}\left[-X_{i+1}\exp\left(-\gamma_{z(i+1)}\sum_{m=1}^{i}d_m\right) + Y_{i+1}\exp\left(\gamma_{z(i+1)}\sum_{m=1}^{i}d_m\right)\right] \tag{2.13}$$

From (2.12) and (2.13), we have

$$\begin{bmatrix} X_i\exp\left(-\gamma_{zi}\sum_{m=1}^{i}d_m\right) \\ \\ Y_i\exp\left(\gamma_{zi}\sum_{m=1}^{i}d_m\right) \end{bmatrix} = \mathbf{M}_{i(i+1)}\begin{bmatrix} X_{i+1}\exp\left(-\gamma_{z(i+1)}\sum_{m=1}^{i+1}d_m\right) \\ \\ Y_{i+1}\exp\left(\gamma_{z(i+1)}\sum_{m=1}^{i+1}d_m\right) \end{bmatrix} \tag{2.14}$$

where

$$\mathbf{M}_{i(i+1)} = \frac{1}{2}\left(1 + \frac{\varepsilon_i\gamma_{z(i+1)}}{\varepsilon_{i+1}\gamma_{zi}}\right)\begin{bmatrix} \exp\left(\gamma_{z(i+1)}d_{i+1}\right) & R_{i(i+1)}\exp\left(-\gamma_{z(i+1)}d_{i+1}\right) \\ \\ R_{i(i+1)}\exp\left(\gamma_{z(i+1)}d_{i+1}\right) & \exp\left(-\gamma_{z(i+1)}d_{i+1}\right) \end{bmatrix} \tag{2.15}$$

where $R_{i(i+1)}$ is the reflection coefficient at $z = -\Sigma_{m=1}^{i}d_m$:

$$R_{i(i+1)} = \frac{\varepsilon_{i+1}\gamma_{zi} - \varepsilon_i\gamma_{z(i+1)}}{\varepsilon_{i+1}\gamma_{zi} + \varepsilon_i\gamma_{z(i+1)}} \tag{2.16}$$

thus the transmission coefficient $T_\parallel$ and the reflection coefficient $\Gamma_\parallel$ can be obtained from the equation

$$\begin{bmatrix} \Gamma_\parallel \\ 1 \end{bmatrix} = \mathbf{M}_{01}\mathbf{M}_{12}\mathbf{M}_{23}, \ldots, \mathbf{M}_{nt}\begin{bmatrix} 0 \\ T_\parallel\exp\left(-\gamma_{zt}\sum_{m=1}^{t}d_m\right) \end{bmatrix} \tag{2.17}$$

The thickness of the last layer d_t can be an arbitrary value; it will be cancelled out with d_t in $\mathbf{M}_{nt}$. The transmission line approach can also be used to obtain the reflection and transmission coefficients, which have the same results [35].

For the E-field perpendicular to the plane of incidence (vertical polarisation), the E-field is in the y direction and the derivations are similar to the parallel polarisation case, we only need to replace E, H, μ, ε with $H, -E, \varepsilon, \mu$, and now $R_{i(i+1)}$ is

$$R_{i(i+1)} = \frac{\mu_{i+1}\gamma_{zi} - \mu_i\gamma_{z(i+1)}}{\mu_{i+1}\gamma_{zi} + \mu_i\gamma_{z(i+1)}} \tag{2.18}$$

When the last layer is metallic, the reflection coefficient of the last layer can be set as $R_{nt} = -1$.

Numerical methods such as the finite-difference time domain (FDTD) method [36], the frequency domain finite-difference (FDFD) method [37], the finite element method (FEM) [38], and the integral equation (IE) method [39, 40] are more general and can be applied to arbitrary shapes or more complex dielectric parameter distributions. When applying these methods, it is not necessary to simulate a large panel with many absorbers, the periodic boundary condition (PBC) can be applied and only one unit cell needs to be discretised and simulated. Figure 2.4 shows typical boundary settings in computer simulation technology (CST) software (www.cst.com) by using FEM, only one unit cell need to be simulated. The incident waves $E_{\parallel}$ and $E_{\perp}$ are shown in Figure 2.5. The incident waves can be obtained by setting all materials to vacuum and add an E-field monitor. After simulation, the reflection coefficients can be obtained from the defined Floquet port.

Other methods which simplify the structure under certain approximations can also be used, e.g. the transmission line method [41], the homogenisation method [9, 42, 43], and the rigorous coupled-wave analysis [44].

2.2.3 Absorbing Material Measurement

There are also many methods to measure the reflection coefficient of absorbing materials. For small pieces, the transmission line method can be used [45], for a large area of absorbers, the RCS method can be applied [46, 47]. If the oblique incidence is required, normally an arch system, as shown in Figure 2.6, is necessary [35, 48]. It should be noted that when using the arch system, the radius of the arch should be larger than

$$r = \frac{D^2}{\lambda} \tag{2.19}$$

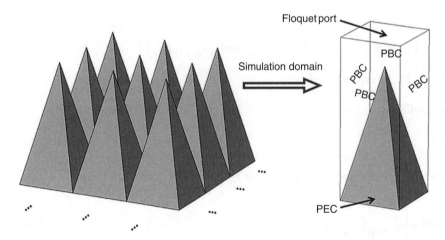

Figure 2.4 The simulation model of the pyramid absorber.

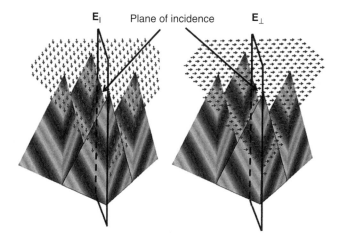

Figure 2.5 Two orthogonal incident waves/modes.

where D is the maximum aperture dimension of the antennas used and λ is the wavelength at the highest frequency of operation/interest [35]. The reason for the required distance is that the RAMs need to be placed in the far field region to have a uniform field illumination. For the RAM measurement, a distance between the near-field and the far-field is acceptable, thus D^2/λ can be used but not $2D^2/\lambda$ [35].

Although the setup in Figure 2.6 is easy to understand, in practice many factors need to be considered, such as the mutual coupling between antenna 1 and

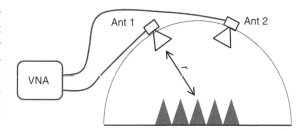

Figure 2.6 A typical arch system to measure the reflection coefficient of the absorbing material.

antenna 2, the edge diffraction of the finite size sample, the scattering from other objects, and the dynamic range of the equipment. For example, when the reflection coefficient is −50 dB, the size of the sample needs to be at least 20λ to reduce the edge effect [35], and the time-gating (TG) technique needs to be used to eliminate the mutual coupling and unexpected diffraction. An example is given below to demonstrate the working principle of the arch measurement system.

Suppose an arch system has a radius of $r = 100$ cm, the aperture D of the transverse electromagnetic (TEM) horn antennas antenna 1 and antenna 2 are 10 cm, the incident angle is set as 25°, a piece of absorbing material is placed in the middle of the arch system as shown in Figure 2.7a, the material properties are randomly chosen and are given in Figure 2.7b, and the thickness of the absorber is 5 cm. Antenna 1 is used as the transmitting (Tx) antenna and antenna 2 is used as the receiving (Rx) antenna. The finite integration time domain (FITD) method in CST software (www.cst.com) is used to simulate the field behaviour in this system, the frequency range is set as 2–10 GHz, and the excitation time domain signal is given in Figure 2.8a. The ground plane is set as a perfect electric conductor (PEC). After simulation, the received waveforms with and without absorber (PEC) are given in Figure 2.8b. The details of Figure 2.8b in different time spans are shown separately in Figure 2.9.

If we check the field plot in Figure 2.10 at different times, it can be found that the reflection from Figure 2.9a is from the mutual coupling of antennas, the waves have hardly interacted with the absorbing material, and there is little difference between with and without absorber. Figure 2.9b is the expected response: the wave reflected by the absorber is much smaller than that reflected by the PEC. After 12 ns, the signals are from the scattering of the environment (including antennas and the edge of the absorber). To obtain the reflection coefficient of the

(a)

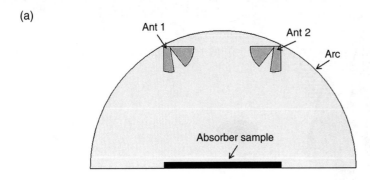

(b)

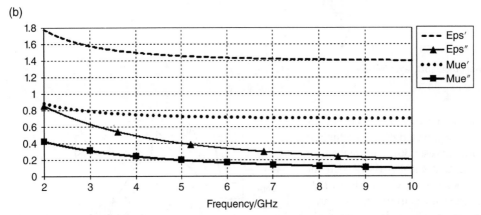

Figure 2.7 (a) An arch system with absorber sample and (b) the permittivity and permeability of the sample. ε' and $-\varepsilon''$ are the real and imaginary parts of the relative permittivity, $\varepsilon(\omega) = \varepsilon'(\omega) - j\varepsilon''(\omega)$, μ' and $-\mu''$ are the real and imaginary parts of the relative permeability, $\mu(\omega) = \mu'(\omega) - j\mu''(\omega)$.

absorber, we can apply the Fourier transform (FT) to the reflected signal and compare it to the reference (the PEC):

$$|\Gamma(\omega)|_{\text{NoTG}} = \left| \frac{\text{FT}(o2, 1_{\text{absorber}})}{\text{FT}(o2, 1_{\text{PEC}})} \right| \tag{2.20}$$

where $o2, 1_{\text{absorber}}$ is the time domain reflected signal when the absorber exists and $o2, 1_{\text{PEC}}$ is the time domain reflected signal when there is no absorber. Note in (2.20) we did not remove the interference from the mutual coupling of antennas and the other extraneous signals. The only part of the signal we need is in Figure 2.8b, thus if we introduce the TG operator to the original signals and set all other values to zero outside the time span of 6–12 ns, (2.20) becomes

$$|\Gamma(\omega)|_{\text{TG}} = \left| \frac{\text{FT}[\text{TG}(o2, 1_{\text{absorber}})]}{\text{FT}[\text{TG}(o2, 1_{\text{PEC}})]} \right| \tag{2.21}$$

Another method to simulate the reflection coefficient is to use the FEM with PBCs, as shown in Figure 2.4. If this method is used, it actually means that the size of the absorber is infinite and there is no edge diffraction, thus the reflection coefficient obtained from the FEM can be used as the reference result. Results from the FEM, (2.20) and (2.21) are all shown in Figure 2.11 to have a comparison. As can be seen, a very good agreement is obtained between the FEM and TG curves, while large fluctuations can be observed if no TG is applied because

(a)

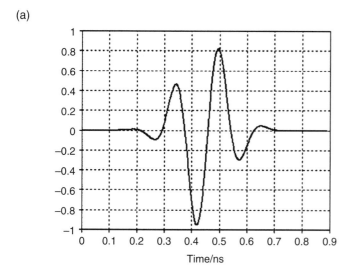

(b)

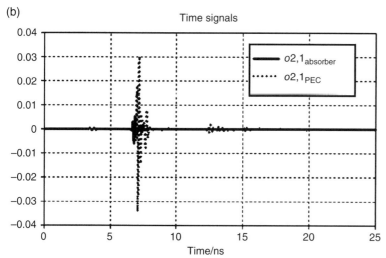

Figure 2.8 (a) Excitation waveform and (b) received signal with absorber and without absorber (PEC). *o*2,1 means the output signal from antenna 2 when antenna 1 is excited.

time domain interference from other sources is included. There is also a small deviation between the result from the TG technique and the FEM, which occurs because the received signal from the Rx antenna in the time span of 6–12 ns is also superimposed with extraneous signals. Although this demonstration is simulated in the time domain, in practice the measurement can be done in the frequency domain since the frequency domain measurement can be transformed into the time domain by using the inverse Fourier transform (IFT). The frequency domain measurement can also provide a much larger dynamic range than the time domain measurement. Thus (2.21) becomes:

$$|\Gamma(\omega)|_{\text{TG}} = \left| \frac{\text{FT}\left\{ \text{TG}\left[\text{IFT}\left(\tilde{S}2,1_{\text{absorber}} \right) \right] \right\}}{\text{FT}\left\{ \text{TG}\left[\text{IFT}\left(\tilde{S}2,1_{\text{PEC}} \right) \right] \right\}} \right| \tag{2.22}$$

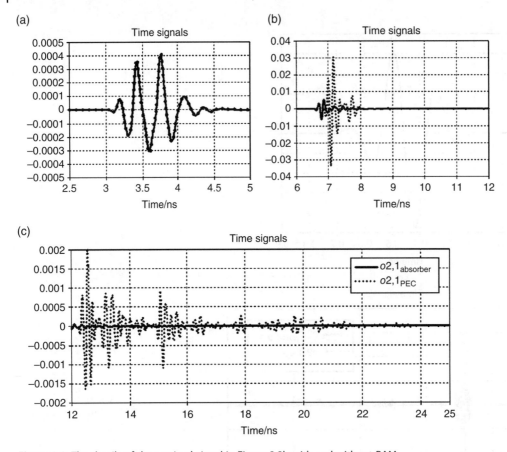

Figure 2.9 The details of the received signal in Figure 2.8b with and without RAM.

where $\widetilde{S}2,1_{\mathrm{absorber}}$ means the filtered S parameters between two antennas when the absorber exists and $\widetilde{S}2,1_{\mathrm{PEC}}$ means the filtered S parameters when there is no absorber. A band pass filter is used to select the frequency band of interest and also can be used to shape the excitation waveform in the time domain. A similar procedure can also be found in the chamber decay time measurement in the reverberation chamber (RC).

From this example, it can be concluded that the TG technique is based on a good understanding of time domain signal response, and the system needs to be carefully designed to separate the response in the time domain. If the scattered wave arrives at the Rx antenna at the same time as the reflected wave, the TG technique cannot filter it. In practice, the dynamic range of a vector network analyser (VNA) needs to be considered. The reflected wave could be too small to be recorded while a power amplifier is necessary to increase the dynamic range.

In the RC the TG technique can also be used to filter the unwanted response and can improve the field uniformity (FU) of the RC.

It is interesting to note that the average reflection coefficient over all incident angles and polarisations can be measured in an RC [49]. This measurement method is presented in Chapter 7 of this book.

For most of the time, the performance of the RAM is already known or commercially available. It is therefore not necessary to care about the micro level parameters such as permittivity and permeability since as long as the reflection coefficients are known, the designer can move on to the macro level design directly.

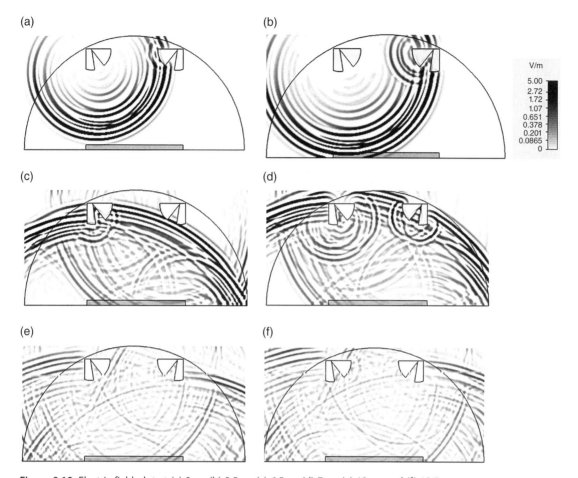

Figure 2.10 Electric field plot at (a) 3 ns, (b) 3.5 ns, (c) 6.5 ns, (d) 7 ns, (e) 12 ns, and (f) 12.5 ns.

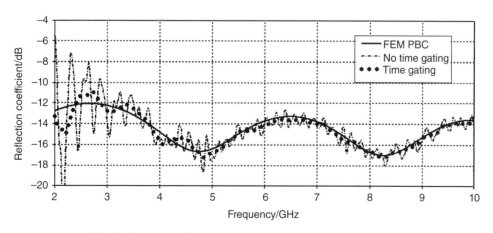

Figure 2.11 Simulated reflection coefficients from the FEM method, FITD with TG, and FITD without TG.

2.3 CEM Algorithms Overview

At the macro level, when the whole chamber performance is considered, full wave methods can also be applied. Because of the complexity and the large electrical size of the whole chamber, it requires a lot of memory and computation time thus simplifications have been made to increase the efficiency and reduce the time and memory requirement. For example, in [50], to avoid the calculation of Green's function in such a complex environment, conductive wire meshes have been used to imitate the RAM in the moment method (MoM), in [51], large cells were introduced to increase the time step and reduce the memory requirement in the FDTD method, and the homogenisation method can also be combined with the transmission line matrix (TLM) to boost efficiency [52]. Brute-force full-wave models without any simplifications have been applied as well, such as the TLM (https://www.cst.com/Applications/Article/Intelligent-Representation-Of-Anechoic-Chamber-Wall-Cuts-Electromagnetic-Simulation-Time-95), hybrid MoM/FEM [53], and FDTD [54]. Even with a high-performance computer, large electrical size problems with complex material scenarios are not easy to solve due to the large memory and time requirements.

High-frequency approximation methods such as GO have been proven to be a fast and efficient way to simulate the macro level problem [55–60]. However, RAM modelling at the system level is normally simplified by using a cosine approximation [56], an effective medium [61], a homogenisation model [9, 42] or a multi-layer model [57]. Thus, the RAM is not fully described, and to find an accurate equivalent analytical model over a wide frequency range and a wide incident angle is challenging.

The mesh type, complexity and electrical size for different computational electromagnetics methods (CEMs) are summarised in Figure 2.12, where *N* means the number of freedoms/unknowns. The electrical size and the physical size of the problem are shown in the table; the limits of the algorithms are mainly from the memory requirement and the effectiveness of the algorithm. High-frequency approximation methods can handle electrically large but not electrically small problems, thus dashed lines are used at the low frequency range. Typical mesh types such as triangle, tetrahedron, and hexahedra are also given, but we have to note that one algorithm may support many types of mesh.

It can be found that using only one method to simulate the wave propagation in an AC could be too complex (considering all material properties) or too simple (the RAM behaviour is simplified). To balance these two levels, at the macro level we apply the GO algorithm while at the micro level the full wave method FEM is applied. As well as being efficient and effective, this combination has commercial benefits: absorber

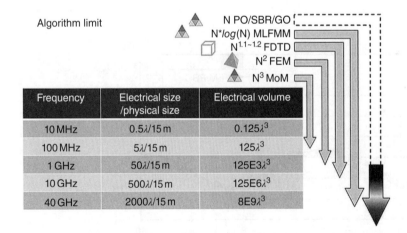

Figure 2.12 Comparison of different CEM methods. PO, physical optics; SBR, shooting bouncing ray; GO, geometric optics; MLFMM, multilevel fast multipole method algorithm; FDTD, finite-difference time domain method; FEM, finite element method; MoM, method of moment.

manufacturers do not need to disclose the material property distribution of the RAM and system designers do not need to care about the micro level issues. The connection between the GO and FEM is realised by using the reflection coefficient. In the next section GO theory is introduced, and software implementations are given in the next chapter.

2.4 GO Theory

GO theory has two versions: analytical and approximated. Both versions are given in this section, and it can be seen that the analytical version is not easy to implement numerically but the approximated one is more general and simpler.

2.4.1 GO from Maxwell Equations

The GO formula can be obtained from the high-frequency approximation of Maxwell's equations. The derivation can be found in [62] from J. A. Kong. The electric field $\mathbf{E}$ and the magnetic field $\mathbf{H}$ in free space have the forms

$$\mathbf{E}(\mathbf{r}) = \mathbf{E}e^{-jkL(\mathbf{r})} \tag{2.23}$$

$$\mathbf{H}(\mathbf{r}) = \mathbf{H}e^{-jkL(\mathbf{r})} \tag{2.24}$$

where $k = \omega/c$ is the free space wave number, ω is the angular frequency, c is the speed of light (actually the wave) in the media, $L(\mathbf{r})$ is the phase function, and the wavefronts can be described by equation $L(\mathbf{r})$ = constant. Substituting (2.23) and (2.24) in the source-free Maxwell's equations and making use of the vector identities

$$\nabla \times (\mathbf{A}\phi) = \phi(\nabla \times \mathbf{A}) - \mathbf{A} \times (\nabla\phi) \tag{2.25}$$

we have

$$\begin{aligned}\nabla \times \mathbf{E}(\mathbf{r}) &= \nabla \times \left[\mathbf{E}e^{-jkL(\mathbf{r})}\right] = e^{-jkL(\mathbf{r})}(\nabla \times \mathbf{E}) - \mathbf{E} \times \left(\nabla e^{-jkL(\mathbf{r})}\right) \\ &= e^{-jkL(\mathbf{r})}(\nabla \times \mathbf{E}) - \mathbf{E} \times \left[-jke^{-jkL(\mathbf{r})}\nabla L(\mathbf{r})\right] = e^{-jkL(\mathbf{r})}\left[\nabla \times \mathbf{E} + jk\mathbf{E} \times \nabla L(\mathbf{r})\right] \\ &= -j\omega\mu\mathbf{H}e^{-jkL(\mathbf{r})}\end{aligned} \tag{2.26}$$

thus $e^{-jkL(\mathbf{r})}$ can be eliminated on both sides of the equation and (2.26) becomes

$$\left[\nabla L(\mathbf{r})\right] \times \mathbf{E} - n\eta\mathbf{H} = \frac{-j}{k}\nabla \times \mathbf{E} \tag{2.27}$$

where $n = c_0\sqrt{\mu\varepsilon}$ is the refractive index, c_0 is the speed of light in free space, and $\eta = \sqrt{\mu/\varepsilon}$ is the characteristic impedance of the media. Similarly, from

$$\nabla \times \mathbf{H}(\mathbf{r}) = j\omega\varepsilon\mathbf{E}e^{-jkL(\mathbf{r})} \tag{2.28}$$

and repeating the similar derivation as in (2.26), we have

$$\left[\nabla L(\mathbf{r})\right] \times \mathbf{H} + \frac{n}{\eta}\mathbf{E} = \frac{-j}{k}\nabla \times \mathbf{H} \tag{2.29}$$

In the high-frequency limit, k is very large and the right-hand sides of (2.27) and (2.29) vanish

$$\left[\nabla L(\mathbf{r})\right] \times \mathbf{E} - n\eta\mathbf{H} = 0 \tag{2.30}$$

$$\left[\nabla L(\mathbf{r})\right] \times \mathbf{H} + \frac{n}{\eta}\mathbf{E} = 0 \tag{2.31}$$

We also have

$$\nabla \cdot (\mathbf{A}\phi) = \mathbf{A} \cdot (\nabla \phi) + \phi(\nabla \cdot \mathbf{A}) \qquad (2.32)$$

$$\nabla \cdot \mathbf{D}(\mathbf{r}) = \varepsilon \nabla \cdot \left(\mathbf{E} e^{-jkL(\mathbf{r})} \right) = \varepsilon \mathbf{E} \cdot \left[-jk e^{-jkL(\mathbf{r})} \nabla L(\mathbf{r}) \right] + \varepsilon e^{-jkL(\mathbf{r})} (\nabla \cdot \mathbf{E}) = 0 \qquad (2.33)$$

after simplifying (2.33) we have

$$\mathbf{E} \cdot [\nabla L(\mathbf{r})] = 0 \qquad (2.34)$$

Similarly, from $\nabla \cdot \mathbf{B}(\mathbf{r}) = 0$ we have

$$\mathbf{H} \cdot [\nabla L(\mathbf{r})] = 0 \qquad (2.35)$$

From (2.30) we have

$$\mathbf{H} = \frac{1}{n\eta} [\nabla L(\mathbf{r})] \times \mathbf{E} \qquad (2.36)$$

We substitute (2.36) into (2.34):

$$[\nabla L(\mathbf{r})] \times \mathbf{H} = \frac{1}{n\eta} [\nabla L(\mathbf{r})] \times \{[\nabla L(\mathbf{r})] \times \mathbf{E}\} = \frac{\nabla L(\mathbf{r})}{n\eta} [\nabla L(\mathbf{r})] \cdot \mathbf{E} - \frac{\mathbf{E}}{n\eta} [\nabla L(\mathbf{r})] \cdot [\nabla L(\mathbf{r})] \qquad (2.37)$$

Substitute (2.34) in (2.37):

$$-\frac{\mathbf{E}}{n\eta} [\nabla L(\mathbf{r})] \cdot [\nabla L(\mathbf{r})] = -\frac{n}{\eta} \mathbf{E} \qquad (2.38)$$

Finally, we have

$$|\nabla L(\mathbf{r})|^2 = n^2 \qquad (2.39)$$

which is the GO in differential form [62].

2.4.2 Analytical Expression of a Reflected Field from a Curved Surface

The analytical expressions for a reflected field from a curved surface can be obtained from the incident wave and the principle radii of curvature at the point of reflection [33]. For an incident wave $\mathbf{E}^i(Q_R)$ shown in Figure 2.13, θ_i is the incident angle and the reflected wave $\mathbf{E}^r(s)$ can be obtained from

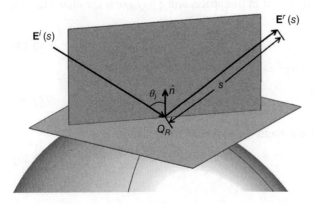

Figure 2.13 Incident and reflected waves for a curved surface.

$$\mathbf{E}^r(s) = \mathbf{E}^i(Q_R)\overline{\overline{\mathbf{R}}}\sqrt{\frac{\rho_1^r \rho_2^r}{(\rho_1^r + s)(\rho_2^r + s)}}e^{-j\beta s} \qquad (2.40)$$

$\overline{\overline{\mathbf{R}}}$ is the dyadic reflection coefficient, ρ_1^i and ρ_2^i are the principal radii of curvature of the incident wave front at the point of reflection [33], shown in Figure 2.14 and ρ_1^r and ρ_2^r are the principal radii of curvature of the reflected wave front at the point of reflection shown in Figure 2.15, and β is the propagation constant. The principal radii of curvature can be determined from

$$\rho_1^r = \left[\frac{1}{2}\left(\frac{1}{\rho_1^i} + \frac{1}{\rho_2^i}\right) + \frac{1}{f_1}\right]^{-1} \qquad (2.41)$$

$$\rho_2^r = \left[\frac{1}{2}\left(\frac{1}{\rho_1^i} + \frac{1}{\rho_2^i}\right) + \frac{1}{f_2}\right]^{-1} \qquad (2.42)$$

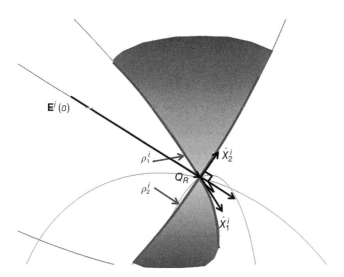

Figure 2.14 Incident wave front.

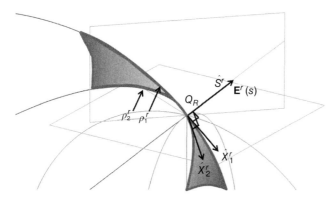

Figure 2.15 Reflected wave front.

where

$$\frac{1}{f_{1(2)}} = \frac{\cos\theta_i}{|\boldsymbol{\theta}|^2}\left(\frac{\theta_{22}^2 + \theta_{12}^2}{R_1} + \frac{\theta_{21}^2 + \theta_{11}^2}{R_2}\right)$$

$$\pm \frac{1}{2}\left\{\left(\frac{1}{\rho_1^i} - \frac{1}{\rho_2^i}\right)^2 + \left(\frac{1}{\rho_1^i} - \frac{1}{\rho_2^i}\right)\frac{4\cos\theta_i}{|\boldsymbol{\theta}|^2}\left(\frac{\theta_{22}^2 - \theta_{12}^2}{R_1} + \frac{\theta_{21}^2 - \theta_{11}^2}{R_2}\right)\right.$$

$$\left. + \frac{4\cos^2\theta_i}{|\boldsymbol{\theta}|^4}\left[\left(\frac{\theta_{22}^2 + \theta_{12}^2}{R_1} + \frac{\theta_{21}^2 + \theta_{11}^2}{R_2}\right)^2 - \frac{4|\boldsymbol{\theta}|^2}{R_1 R_2}\right]\right\}^{1/2} \quad (2.43)$$

where + is used for f_1 and − for f_2, and R_1 and R_2 are the radii of curvature of the reflecting surface at Q_R. In (2.43), $|\boldsymbol{\theta}|$ is the determinant of

$$\boldsymbol{\theta} = \begin{bmatrix} \hat{\mathbf{X}}_1^i \cdot \hat{\mathbf{u}}_1 & \hat{\mathbf{X}}_1^i \cdot \hat{\mathbf{u}}_2 \\ \hat{\mathbf{X}}_2^i \cdot \hat{\mathbf{u}}_1 & \hat{\mathbf{X}}_2^i \cdot \hat{\mathbf{u}}_2 \end{bmatrix} \quad (2.44)$$

or

$$|\boldsymbol{\theta}| = \left(\hat{\mathbf{X}}_1^i \cdot \hat{\mathbf{u}}_1\right)\left(\hat{\mathbf{X}}_2^i \cdot \hat{\mathbf{u}}_2\right) - \left(\hat{\mathbf{X}}_2^i \cdot \hat{\mathbf{u}}_1\right)\left(\hat{\mathbf{X}}_1^i \cdot \hat{\mathbf{u}}_2\right) \quad (2.45)$$

The vectors $\hat{\mathbf{X}}_1^i$ and $\hat{\mathbf{X}}_2^i$ represent the principal direction of the incident wave front at the reflection point Q_R with principal radii of curvature ρ_1^i and ρ_2^i. $\hat{\mathbf{u}}_1$ and $\hat{\mathbf{u}}_2$ are the unit vectors with principal radii of curvature R_1 and R_2, respectively. To define $\hat{\mathbf{X}}_1^r$ and $\hat{\mathbf{X}}_2^r$, we first introduce [33]

$$\mathbf{Q}^r = \begin{bmatrix} Q_{11}^r & Q_{12}^r \\ Q_{21}^r & Q_{22}^r \end{bmatrix} \quad (2.46)$$

where $\mathbf{Q}^r$ is defined as the curvature matrix for the reflected wave front whose entries are

$$Q_{11}^r = \frac{1}{\rho_1^i} + \frac{2\cos\theta_i}{|\boldsymbol{\theta}|^2}\left(\frac{\theta_{22}^2}{R_1} + \frac{\theta_{21}^2}{R_2}\right) \quad (2.47)$$

$$Q_{12}^r = Q_{21}^r = -\frac{2\cos\theta_i}{|\boldsymbol{\theta}|^2}\left(\frac{\theta_{22}\theta_{11}}{R_1} + \frac{\theta_{11}\theta_{21}}{R_2}\right) \quad (2.48)$$

$$Q_{22}^r = \frac{1}{\rho_2^i} + \frac{2\cos\theta_i}{|\boldsymbol{\theta}|^2}\left(\frac{\theta_{12}^2}{R_1} + \frac{\theta_{11}^2}{R_2}\right) \quad (2.49)$$

$\hat{\mathbf{X}}_1^r$ and $\hat{\mathbf{X}}_2^i$ can be written as

$$\hat{\mathbf{X}}_1^r = \frac{\left(Q_{22}^r - \frac{1}{\rho_1^r}\right)\hat{\mathbf{x}}_1^r - Q_{12}^r\,\hat{\mathbf{x}}_2^r}{\sqrt{\left(Q_{22}^r - \frac{1}{\rho_1^r}\right)^2 + \left(Q_{12}^r\right)^2}} \quad (2.50)$$

$$\hat{\mathbf{X}}_2^r = -\hat{\mathbf{s}}^r \times \hat{\mathbf{X}}_1^r \quad (2.51)$$

where $\hat{\mathbf{x}}_1^r$ and $\hat{\mathbf{x}}_2^r$ are unit vectors perpendicular to the reflected ray, and can be obtained using

$$\hat{\mathbf{x}}_1^r = \hat{\mathbf{X}}_1^i - 2\left(\hat{\mathbf{n}} \cdot \hat{\mathbf{X}}_1^i\right)\hat{\mathbf{n}} \quad (2.52)$$

$$\hat{\mathbf{x}}_2^r = \hat{\mathbf{X}}_2^i - 2\left(\hat{\mathbf{n}} \cdot \hat{\mathbf{X}}_2^i\right)\hat{\mathbf{n}} \quad (2.53)$$

$\hat{\mathbf{n}}$ is the unit vector normal to the surface at the reflection point. The dyadic reflection coefficient $\overline{\overline{\mathbf{R}}}$ is determined by the material property at the point Q_R, Figure 2.16 illustrates the decomposition of the incident and reflected waves:

Figure 2.16 Decomposition of incident and reflected waves.

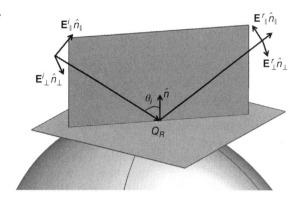

$$\begin{bmatrix} E_\parallel^r \\ E_\perp^r \end{bmatrix} = \bar{\bar{\mathbf{R}}} \begin{bmatrix} E_\parallel^i \\ E_\perp^i \end{bmatrix} = \begin{bmatrix} R_{\parallel\parallel} & R_{\parallel\perp} \\ R_{\perp\parallel} & R_{\perp\perp} \end{bmatrix} \begin{bmatrix} E_\parallel^i \\ E_\perp^i \end{bmatrix} \tag{2.54}$$

Generally, the principal radii of a surface $\mathbf{r} = \mathbf{r}(u, v)$ can be obtained as follows. The surface normal vector $\hat{\mathbf{n}}$ can be obtained by

$$\hat{\mathbf{n}} = \frac{\mathbf{r}_u \times \mathbf{r}_v}{|\mathbf{r}_u \times \mathbf{r}_v|} \tag{2.55}$$

where

$$\mathbf{r}_u = \frac{\partial \mathbf{r}(u, v)}{\partial u}, \mathbf{r}_v = \frac{\partial \mathbf{r}(u, v)}{\partial v} \tag{2.56}$$

The principal radii R_1 and R_2 are the solutions of

$$\left(EG - F^2\right)R^2 - (LG - 2MF + NE)R + \left(LN - M^2\right) = 0 \tag{2.57}$$

where

$$E = \mathbf{r}_u \cdot \mathbf{r}_u, F = \mathbf{r}_u \cdot \mathbf{r}_v, G = \mathbf{r}_v \cdot \mathbf{r}_v, L = \mathbf{r}_{uu} \cdot \hat{\mathbf{n}}, M = \mathbf{r}_{uv} \cdot \hat{\mathbf{n}}, N = \mathbf{r}_{vv} \cdot \hat{\mathbf{n}}$$

$$\mathbf{r}_{uu} = \frac{\partial^2 \mathbf{r}(u, v)}{\partial u^2}, \mathbf{r}_{uv} = \frac{\partial^2 \mathbf{r}(u, v)}{\partial u \partial v}, \mathbf{r}_{vv} = \frac{\partial^2 \mathbf{r}(u, v)}{\partial v^2} \tag{2.58}$$

The principal radii are illustrated in Figure 2.17. It should be noted that principal radii are also the maximum and minimum radius of curvature of the normal section curve, which is defined by the curve intercepted by the principal planes.

Specifically, when the material is a PEC at the point of reflection, by applying the boundary conditions of vanishing the tangential components of the E-field at the boundary $E_\perp^r + E_\perp^i = 0$, we have

$$\bar{\bar{\mathbf{R}}}_{PEC} = \begin{bmatrix} 1 & 0 \\ 0 & -1 \end{bmatrix} \tag{2.59}$$

The above equations are for general surfaces if the surface equation is known. This method can be used in the forward algorithm but not (easily) in the inverse algorithm (these

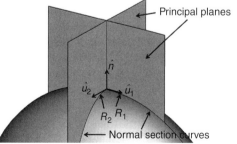

Figure 2.17 Definition of principal radii, principal planes, and normal section curves.

two algorithms are explained in the next chapter). In practice, the equations of the surfaces are normally unknown, and interpolation or surface reconstruction at reflection point is necessary when using this method.

2.4.3 Alternative GO Form

In the previous section, the analytical form of GO was given. To obtain the reflected field analytically, the surface equation must be known. In practice, the surface cannot be expressed analytically. Thus, an alternative form is needed to make the GO easy to implement numerically.

An alternative form of GO can be expressed as [63]

$$\mathbf{E} = \left\{ \prod_i \overline{\overline{\mathbf{R}}}_i \right\} \left\{ \prod_i \overline{\overline{\mathbf{T}}}_i \right\} \mathbf{E}_0 \left\{ \prod_i e^{-\gamma_i l_i} \right\} SF \qquad (2.60)$$

where the E-field is assumed to propagate like light, as shown in Figure 2.18, and A_0 and A are the cross-sectional areas of the ray tubes at the source point and the field point of interest, respectively, which will be used to calculate the spreading factor ($SF = \sqrt{A_0}/\sqrt{A}$). $\mathbf{E}_0$ is the E-field at the source point (reference point) and $\mathbf{E}$ is the E-field at the field point. $\prod_i \overline{\overline{\mathbf{R}}}_i$ and $\prod_i \overline{\overline{\mathbf{T}}}_i$ are the reflection and transmission coefficient dyads along the whole ray path, respectively, and $\prod_i e^{-\gamma_i l_i}$ is the total phase variations and losses along the whole path.

It can be seen from Figure 2.18 that once the ray path is obtained, it is easy to calculate the E-field at the field point. For the AC simulation, the waves are confined in a shielded cavity with very good shielding effectiveness, thus no transmission coefficient needs to be considered. This simplifies (2.60) to

$$\mathbf{E} = \left\{ \prod_i \overline{\overline{\mathbf{R}}}_i \right\} \mathbf{E}_0 \left\{ \prod_i e^{-\gamma_i l_i} \right\} SF \qquad (2.61)$$

And the definition of $\overline{\overline{\mathbf{R}}}$ for each reflection is given in (2.54). It can be found that if the transmission rays are ignored, the ray number can be reduced greatly. The ray number will not increase during the ray tracing procedure; this also simplifies the work in code programming.

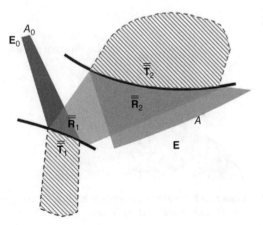

Figure 2.18 Radio wave propagation in GO. The wave front is reflected twice with reflection coefficient dyads $\overline{\overline{\mathbf{R}}}_1$ and $\overline{\overline{\mathbf{R}}}_2$. Transmitted waves are also shown with transmission coefficient dyads $\overline{\overline{\mathbf{T}}}_1$ and $\overline{\overline{\mathbf{T}}}_2$.

2.5 GO-FEM Hybrid Method

From the overview of CEM algorithms, it can be seen that using a single algorithm to solve electrically large problems with complex material distribution cannot be efficient. A hybrid method could combine the advantages of two or more different algorithms with the sacrifice of acceptable errors. FEM is famous for solving electrically small/medium problems with complex material configurations. The material in the solving domain can be anisotropic and inhomogeneous, and, by applying the PBCs, periodical structures can be solved efficiently [38, 64–68]. We can apply the FEM analysis to the RAM simulation. After the FEM analysis, the reflection coefficient dyad $\overline{\overline{R}}$ can be obtained for different types of RAMs. For each type of RAM, $\overline{\overline{R}}$ can be considered as a function of incident angle (θ, φ) and frequency (ω):

$$\overline{\overline{R}} = \begin{bmatrix} R_{\parallel\parallel}(\theta,\varphi,\omega) & R_{\parallel\perp}(\theta,\varphi,\omega) \\ R_{\perp\parallel}(\theta,\varphi,\omega) & R_{\perp\perp}(\theta,\varphi,\omega) \end{bmatrix} \tag{2.62}$$

Once $\overline{\overline{R}}$ is obtained, it can be saved into a database and used later in the GO algorithms. $\overline{\overline{R}}$ can also be obtained from measurement directly.

The advantages of this asynchronous hybrid method are:

1) There is no repeat calculation for the RAM reflection coefficient in the macro level simulation (GO algorithm); only interpolation is used when calculating the reflection coefficient of the RAM, which saves a lot of time for the calculation of each ray.
2) Each type of RAM corresponds to a specific $\overline{\overline{R}}$, and the data can be reused once it has been obtained from the manufacture, simulation, or measurement. A database can be built for ease of future use.
3) It encapsulates the information of the RAM, the material properties are not necessary. This protects the intellectual property (IP) of the RAM provider, which is commercially important.

The workflow of this geometric optics – finite element method (GO-FEM) method is illustrated in Figure 2.19. After the FEM analysis, one can move on the GO analysis with more information on chamber structure, RAM layout, and excitation source.

We finish this section with the details of the interpolation procedure of the reflection coefficient dyad $\overline{\overline{R}}$. In the measurement or simulation of $\overline{\overline{R}}$, finite samples of incident angles (θ, φ) and frequencies (ω) are used. However, in the GO algorithm, the incident angle and frequency of interest can be arbitrary, and interpolations are necessary to obtain $\overline{\overline{R}}$ for arbitrary incident angles and frequencies.

Suppose $\overline{\overline{R}}$ is known for incident angles (θ_0, φ_0) and (θ_1, φ_1), and frequency samples ω_0 and ω_1, that is, $\overline{\overline{R}}_{000}(\theta_0,\varphi_0,\omega_0)$, $\overline{\overline{R}}_{010}(\theta_0,\varphi_1,\omega_0)$, $\overline{\overline{R}}_{100}(\theta_1,\varphi_0,\omega_0)$, $\overline{\overline{R}}_{110}(\theta_1,\varphi_1,\omega_0)$, $\overline{\overline{R}}_{001}(\theta_0,\varphi_0,\omega_1)$, $\overline{\overline{R}}_{011}(\theta_0,\varphi_1,\omega_1)$, $\overline{\overline{R}}_{101}(\theta_1,\varphi_0,\omega_1)$, and $\overline{\overline{R}}_{111}(\theta_1,\varphi_1,\omega_1)$ are known. Suppose the wave incident angle is (θ, φ) and the frequency of interest is ω, where $0 \le \theta_0 \le \theta \le \theta_1 \le 90°$, $0 \le \varphi_0 \le \varphi \le \varphi_1 \le 360°$, and $\omega_0 \le \omega \le \omega_1$. θ is the polar angle and φ is the azimuth angle. Because the incident angle can only arrive from the upper half plane, the maximum

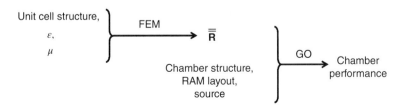

Figure 2.19 Workflow of the GO-FEM hybrid method.

value of θ is 90°. To obtain the interpolated value $\bar{\bar{\mathbf{R}}}(\theta,\varphi,\omega)$, we can first apply the bilinear interpolation to (θ, φ) to give [69]:

$$\bar{\bar{\mathbf{R}}}_{\theta\varphi0}(\theta,\varphi,\omega_0) = \frac{1}{(\theta_1-\theta_0)(\varphi_1-\varphi_0)}[\theta_1-\theta \;\; \theta-\theta_0] \begin{bmatrix} \bar{\bar{\mathbf{R}}}_{000} & \bar{\bar{\mathbf{R}}}_{010} \\ \bar{\bar{\mathbf{R}}}_{100} & \bar{\bar{\mathbf{R}}}_{110} \end{bmatrix} \begin{bmatrix} \varphi_1-\varphi \\ \varphi-\varphi_0 \end{bmatrix} \tag{2.63}$$

$$\bar{\bar{\mathbf{R}}}_{\theta\varphi1}(\theta,\varphi,\omega_1) = \frac{1}{(\theta_1-\theta_0)(\varphi_1-\varphi_0)}[\theta_1-\theta \;\; \theta-\theta_0] \begin{bmatrix} \bar{\bar{\mathbf{R}}}_{001} & \bar{\bar{\mathbf{R}}}_{011} \\ \bar{\bar{\mathbf{R}}}_{101} & \bar{\bar{\mathbf{R}}}_{111} \end{bmatrix} \begin{bmatrix} \varphi_1-\varphi \\ \varphi-\varphi_0 \end{bmatrix} \tag{2.64}$$

To simplify the expression we use $\bar{\bar{\mathbf{R}}}_{000}$ to represent $\bar{\bar{\mathbf{R}}}_{000}(\theta_0,\varphi_0,\omega_0)$ and similarly for other notations. Finally, the linear interpolation can be applied to ω to obtain

$$\bar{\bar{\mathbf{R}}}(\theta,\varphi,\omega) = \bar{\bar{\mathbf{R}}}_{\theta\varphi0} + (\omega-\omega_0)\frac{\bar{\bar{\mathbf{R}}}_{\theta\varphi1}-\bar{\bar{\mathbf{R}}}_{\theta\varphi0}}{\omega_1-\omega_0} \tag{2.65}$$

where $\bar{\bar{\mathbf{R}}}_{\theta\varphi0}$ and $\bar{\bar{\mathbf{R}}}_{\theta\varphi1}$ are $\bar{\bar{\mathbf{R}}}_{\theta\varphi0}(\theta,\varphi,\omega_0)$ and $\bar{\bar{\mathbf{R}}}_{\theta\varphi1}(\theta,\varphi,\omega_1)$ in (2.63) and (2.64).

2.6 Summary

In this chapter, we have considered the AC design problem from the theoretical point of view, no realisation issues have been considered. The working principle of RAM, the analysis methods, and measurements have been reviewed. The solutions for AC chamber design have been discussed and a number of CEM modelling methods have been compared for chamber simulation. Two forms of GO are given. To have an efficient solution, an appropriate CEM algorithm needs to be chosen; we have chosen the combination of the FEM and GO method as the proposed solution. The details on implementation and realisation of the proposed solution are given in the next chapter.

References

1 Hemming, L. H. (2002). *Electromagnetic Anechoic Chambers: A Fundamental Design and Specification Guide.* New York, NY: Wiley-IEEE Press.

2 Emerson, W. H. (1973). Electromagnetic wave absorbers and anechoic chambers through the years. *IEEE Transactions on Antennas and Propagations* 21 (4): 484–490.

3 Tatem, P.A., Marshall, P.D., and Williams, F.W. (1977). Modified Smoldering Test of Urethane Foams Used in Anechoic Chambers. NRL Report 8093.

4 Saville, P. (2005). *Review of Radar Absorbing Materials.* Defence R&D Canada – Atlantic Technical Memorandum.

5 Van der Plas, G., Barel, A., and Schweicher, E. (1989). A spectral iteration technique for analyzing scattering from circuit analog absorbers. *IEEE Transactions on Antennas and Propagation* 37 (10): 1327–1332.

6 Gaylor, K. (1989). *Radar Absorbing Materials – Mechanisms and Materials.* DSTO Materials Research Laboratory.

7 Chambers, B. (1999). A smart radar absorber. *Smart Materials and Structures* 8 (1): 64–72.

8 Zheng, Y., Wang, S., Feng, J. et al. (2006). Regulation mechanism of EM parameters in natural ferrite and its application in microwave absorbing materials. *Science in China: Series E Technological Sciences* 49 (1): 38–49.

9 Holloway, C. L. and Kuester, E. F. (1994). A low-frequency model for wedge or pyramid absorber arrays-II: computed and measured results. *IEEE Transactions on Electromagnetic Compatibility* 36 (4): 307–313.

10 Truong, V.-Van, Turner, B.D., Muscat, R. F., and Russo, M. S. (1997). Conducting-polymer-based radar absorbing materials. Proceedings of the SPIE (Smart Materials, Structures, and Integrated Systems Conference. Volume 3241.

11 Qiu, K., Feng, S., Wu, C. et al. (2016). Calculation of effective permittivity and optimization of absorption property of honeycomb cores with absorbing coatings. *Materials Science* 22 (3): 317–322.

12 Tao, H., Landy, N. I., Bingham, C. M. et al. (2008). A metamaterial absorber for the terahertz regime: design, fabrication and characterization. *Optical Express* 16 (10): 7191–7188.

13 Kuester, E.F. and Holloway, C.L. (1991). Electromagnetic Pyramidal Cone Absorber with Improved Low Frequency Design. US Patent, 5016185.

14 Chuang, B.-K. and Chuah, H.-T. (2003). Modeling of RF absorber for application in the design of anechoic chamber. *Progress in Electromagnetic Research* 43: 273–285.

15 Ponnekanti, S., Al-khaled, F. S. H., and Sali, S. (1995). An effective optimization method for the design of broadband microwave absorbers. *International Journal of Numerical Modelling: Electronic Networks, Devices and Fields* 8: 447–454.

16 Chambers, B., (1999). Characteristics of radar absorbers with tapered thickness, IEEE National Conference on Antennas and Propagation, York. pp. 27–30.

17 Chen, Z. N., Liu, D., Nakano, H. et al. (2016). *Handbook of Antenna Technologies*. Springer Reference.

18 Bucci, O. and Franceschetti, G. (1971). Scattering from wedge-tapered absorbers. *IEEE Transactions on Antennas and Propagation* 19 (1): 96–104.

19 Enayati, A. and Fallahi, A. (2014). *Full-Wave Modelling of Wedge Absorbers*. ATMS (Antenna Test & Measurement Society) India.

20 Nornikman, H., Soh, P. J., and Malek, F. et al. (2010). Microwave wedge absorber design using rice husk – an evaluation on placement variation. Asia-Pacific International Symposium on Electromagnetic Compatibility, Beijing. pp. 916–919.

21 Nornikman, H., Soh, P.J., Azremi, A.A.H., and Anuar, M.S. (2009). Performance simulation of pyramidal and wedge microwave absorbers. Third Asia International Conference on Modelling & Simulation, Bali. pp. 649–654.

22 Kim, D., Takahashi, M., Anzai, H., and Jun, S. Y. (1996). Electromagnetic wave absorber with wide-band frequency characteristics using exponentially tapered ferrite. *IEEE Transactions on Electromagnetic Compatibility* 38 (2): 173–177.

23 DeWitt, B. T. and Burnside, W. D. (1988). Electromagnetic scattering by pyramidal and wedge absorber. *IEEE Transactions on Antennas and Propagation* 36 (7): 971–984.

24 Lehto, A., Tuovinen, A., and Raisanen, A. (1991). Reflectivity of absorbers in 100–200 GHz range. *Electronics Letters* 27 (19): 1699–1700.

25 Lehto, A., Tuovinen, J., and Raisanen,A. (1991). Reflectivity of commercial absorbers at 100–200 GHz, Antennas and Propagation Society Symposium Digest, London, Ontario. Volume 2. pp. 1202–1205.

26 Amano, M. and Kotsuka, Y. (2003). A method of effective use of ferrite for microwave absorber. *IEEE Transactions on Microwave Theory and Techniques* 51 (1): 238–245.

27 Naito, Y. and Suetake, K. (1970). Application of ferrite to electromagnetic wave absorber and its characteristics, G-MTT 1970 International Microwave Symposium, Newport Beach, CA. pp. 273–278.

28 Shimada, K., Hayashi, T., and Tokuda, M. (2000). Fully compact anechoic chamber using the pyramidal ferrite absorber for immunity test. IEEE International Symposium on Electromagnetic Compatibility. Symposium Record, Washington, DC. Volume 1. pp. 225–230.

29 Shin, J. Y. and Oh, J. H. (1993). The microwave absorbing phenomena of ferrite microwave absorbers. *IEEE Transactions on Magnetics* 29 (6): 3437–3439.

30 Hatakeyama, K. and Inui, T. (1984). Electromagnetic wave absorber using ferrite absorbing material dispersed with short metal fibers. *IEEE Transactions on Magnetics* 20 (5): 1261–1263.

31 Feng, Z., Huang, A., and He, H. (2002).Wide-band electromagnetic wave absorber of rubber-ferrite. 3rd International Symposium on Electromagnetic Compatibility. pp. 420–423.

32 Liu, K. (1996). Analysis of the effect of ferrite tile gap on EMC chamber having ferrite absorber walls. Proceedings of Symposium on Electromagnetic Compatibility, Santa Clara, CA. pp. 156–161.

33 Balanis, C. A. (2012). *Advanced Engineering. Electromagnetics*, 2nde. Wiley.

34 Xia, X. and Deng, F. (2007). Reflection characteristic analysis of multilayer absorbing material coating. *Electro-Optical Technology Application* 22 (5): 19–23.

35 IEEE Standards Board. (2012). IEEE Recommended Practice for Radio-Frequency (RF) Absorber Evaluation in the Range of 30 MHz to 5 GHz, IEEE Standard 1128–1998.

36 Elmahgoub, K., Yang, F., Elsherbeni, A. Z. et al. (2010). FDTD analysis of periodic structures with arbitrary skewed grid. *IEEE Transactions on Antennas and Propagation* 58 (8): 2649–2657.

37 Sun, W., Liu, K., and Balanis, C. A. (1996). Analysis of singly and doubly periodic absorbers by frequency-domain finite-difference method. *IEEE Transactions on Antennas and Propagation* 44 (6): 798–805.

38 Jiang, Y. and Martin, A.Q. (1999). The design of microwave absorbers with high order hybrid finite element method. IEEE Antennas and Propagation Society International Symposium, Orlando, FL. Volume 4. pp. 2622–2625.

39 Janaswamy, R. (1992). Oblique scattering from lossy periodic surfaces with application to anechoic chamber absorbers. *IEEE Transactions on Antennas and Propagation* 40 (2): 162–169.

40 Marly, N., Baekelandt, B., Zutter, D. D., and Pues, H. F. (1995). Integral equation modeling of the scattering and absorption of multilayered doubly-periodic lossy structures. *IEEE Transactions on Antennas and Propagation* 43 (11): 1281–1287.

41 Kubytskyi, V., Sapoval, B., Dun, G., and Rosnarho, J. (2012). Fast optimization of microwave absorbers. *Microwave and Optical Technology Letters* 54 (11): 2472–2477.

42 Kuester, E. F. and Holloway, C. L. (1994). A low-frequency model for wedge or pyramid absorber arrays-I: theory. *IEEE Transactions on Electromagnetic Compatibility* 36 (4): 300–306.

43 Khajehpour, A. and Mirtaheri, S. A. (2008). Analysis of pyramid EM wave absorber by FDTD method and comparing with capacitance and homogenization methods. *Progress In Electromagnetics Research Letters* 3: 123–131.

44 Moharam, M. G., Pommet, D. A., Grann, E. B., and Gaylord, T. K. (1995). Stable implementation of the rigorous coupled-wave analysis for surface-relief gratings: enhanced transmittance matrix approach. *Journal of the Optical Society of America A* 12 (6): 1077–1086.

45 Zhou, T., Wel, D., Yang, S. et al. (2016). Measurement and characterization of flexible absorbing materials for applications in wireless communication. *Journal of Scientific and Industrial Metrology* 1 (2): 1–7.

46 Hassan, N., Adbullah, H., and Malek, M. et al. (2010). Measurement of pyramidal microwave absorbers using RCS methods. International Conference on Intelligent and Advanced Systems, Kuala Lumpur. pp. 1–5.

47 Fischer, B. E. and Lahaie, I. J. (2008). Recent microwave absorber wall-reflectivity measurement methods. *IEEE Antennas and Propagation Magazine* 50 (2): 140–147.

48 Feng, J., Zhang, Y., Wang, P., and Fan, H. (2016). Oblique incidence performance of radar absorbing honecombs. *Composites Part B Engineering* 99: 465–471.

49 Xu, Q., Huang, Y., Xing, L. et al. (2016). Average absorption coefficient measurement of arbitrarily shaped electrically large objects in a reverberation chamber. *IEEE Transactions on Electromagnetic Compatibility* 58 (6): 1776–1779.

50 Sasaki, T., Watanabe, Y., and Tokuda, M. (2006). NSA calculation of anechoic chamber using method of moment. Progress In Electromagnetics Research Symposium, Cambridge. pp. 200–205.

51 Kawabata, M., Shimada, Y., Shmada, K., and Kuwabara, N. (2008). FDTD method for site attenuation analysis of compact anechoic chamber using large-cell concept. *Electrical Engineering in Japan* 162 (4): 9–16.

52 Bellamine, F. (2006). Simulation of anechoic chamber using a combined TLM and homogenization method in the frequency range 30–200 MHz. First European Conference on Antennas and Propagation (EuCAP), Nice. pp. 1–6.

53 Campbell, D., Gampala, G., and Reddy, C. J. (2012) Modeling and analysis of anechoic chamber using CEM tools. Proceedings of AMTA Conference, Bellevue.

54 Kantartzis, N. V. and Tsiboukis, T. D. (2005). Wideband numerical modelling and performance optimisation of arbitrarily-shaped anechoic chambers via an unconditionally stable time-domain technique. *Electrical Engineering* 88 (1): 55–81.

55 Chung, B. K., The, C. H., and Chuah, H. T. (2004). Modeling of anechoic chamber using a beam-tracing technique. *Progress In Electromagnetics Research (PIER)* 49: 23–38.

56 Mishra, S. R. and Pavlasek, T. J. F. (1984). Design of absorber-lined chamber for EMC measurements using a geometrical optics approach. *IEEE Transactions on Electromagnetic Compatibility* EMC-26 (3): 111–119.

57 Mansour, M.K. and Jarem, J. (1990). Anechoic chamber design using ray tracing and theory of images. IEEE Southeastcon '90 Proceedings, New Orleans. pp. 689–695.

58 Holloway, C. L. and Kuester, E. F. (1996). Modeling semi-anechoic electromagnetic measurement chambers. *IEEE Transactions on Electromagnetic Compatibility* 38 (1): 79–94.

59 Lin, M., Ji, J., Hsu, C. G., and Hsieh, H.(2007). Simulation and analysis of EMC chambers by ray tracing method. IEEE International Symposium on Electromagnetic Compatibility, Honolulu. pp. 1–4.

60 Razavi, S. M. J., Khalaj-Amirhosseini, M., and Cheldavi, A. (2010). Minimum usage of ferrite tiles in anechoic chambers. *Progress In Electromagnetics Research B (PIER B)* 19: 367–383.

61 Chung, B. K. and Chuah, H. T. (2003). Modeling of RF absorber for application in the design of anechoic chamber. *Progress In Electromagnetics Research (PIER)* 43: 273–285.

62 Kong, J. A. (2008). *Electromagnetic Wave Theory*. Cambridge: EMW.

63 Yang, C., Wu, B., and Ko, C. (1998). A ray-tracing method for modeling indoor wave propagation and penetration. *IEEE Transactions on Antennas and Propagation* 46 (6): 907–919.

64 Lou, Z. and Jin, J.-M. (2004). Finite Element Modeling of Periodic Structures. WPS/Lecture Note Series.

65 Lou, Z. and Jin, J.-M. (2003). High-order finite element analysis of periodic absorbers. *Microwave and Optical Technology Letters* 37 (3).

66 Zhu, Y. and Cangellaris, A. C. (2006). *Multigrid Finite Element Methods for Electromagnetic Field Modeling*. IEEE Press.

67 Jin, J.-M. (2014). *The Finite Element Method in Electromagnetics*, 3rde. Wiley-IEEE Press.

68 McGrath, D. T. (1995). *Electromagnetic Analysis of Periodic Structure Reflection and Transmission by the Hybrid Finite Element Method*, PL-TR-95-1103. Phillips Laboratory.

69 Press, W. H., Teukolsky, S. A., Vetterling, W. T., and Flannery, B. P. (2007). *Numerical Recipes: The Art of Scientific Computing*, 3rde. Cambridge University Press.

3

Computer-aided Anechoic Chamber Design

3.1 Introduction

In this chapter, we focus on the software implementation of the anechoic chamber (AC) design. When an electromagnetic (EM) problem is modelled using a computer, the first thing is to define the problem and identify what information is needed. Similar to existing commercial computer-aided design (CAD) tools, to define an EM problem we need to digitise a 3D model, define the material properties, assign boundary conditions, define the excitation source, choose the right solver, and post process the results. These steps are detailed in this chapter: the framework of the solution is given first and then followed by the detailed techniques. It is shown in this chapter that the development of CAD tools is not just an EM problem but a multi-disciplinary project that involves electromagnetics, software engineering, and computer graphics.

3.2 Framework

The framework is shown in Figure 3.1, which includes four parts from top to bottom. We name the software based on this solution the Fast AC Evaluation Tool (FACET). The graphical user interface (GUI) was developed using Microsoft Visual Studio .NET (https://www.visualstudio.com) and the computational engine was developed using MATLAB (https://www.mathworks.com). They are connected using component object model (COM) technology. Like other computational electromagnetics (CEM) tools, it includes preprocessing, simulation, and post-processing parts. The two different algorithms share the same preprocessing but different post-processing parts. Each part will be explained in detail.

3.3 Software Implementation

3.3.1 3D Model Description

To model the AC using a computer, the first thing we need to do is to describe the 3D structure of the chamber. To make it compatible with other 3D modelling software, the STL (STereoLithography) file is chosen to describe the 3D model because it is widely used for computer-aided manufacturing and widely supported by many other software packages (https://en.wikipedia.org/wiki/STL_(file_format)). The STL file has two formats: the ASCII (American Standard Code for Information Interchange) format and the binary format. We use the ASCII format since it is very direct and also easy to create and modify. A typical ASCII STL format is shown in Figure 3.2.

Anechoic and Reverberation Chambers: Theory, Design, and Measurements, First Edition. Qian Xu and Yi Huang.
© 2019 John Wiley & Sons Ltd. Published 2019 by John Wiley & Sons Ltd.

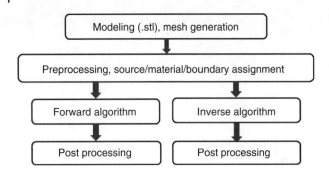

Figure 3.1 Framework of the CAD tool.

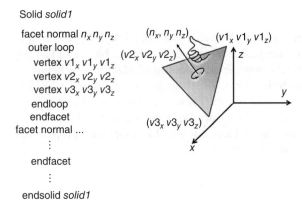

Figure 3.2 A typical ASCII STL file format. The normal vector and vertices are ordered by the right-hand rule.

As can be seen, the model is already discretised into triangular meshes when using the STL file to describe it. An ASCII STL file begins with the key word *solid*, and then the name of the model is given (*solid1* in Figure 3.2). The file continues with any number of triangles (shown in the right-hand side of Figure 3.2). *Facet normal* means the normal vector (n_x, n_y, n_z) of the triangle; the coordinates of the vertices of each triangle are given after *vertex* $((v1_x, v1_y, v1_z), (v2_x, v2_y, v2_z),$ and $(v3_x, v3_y, v3_z))$. The normal vector and vertices are ordered by the right-hand rule shown in Figure 3.2 (https://en.wikipedia.org/wiki/STL_(file_format)).

When importing this type of STL model, the program needs to scan the data line by line to import the normal vectors and coordinates of the vertices. A code snippet is given in Appendix A.1 for reference. A typical STL file after importing is shown in Figure 3.3. The triangle surface mesh is used to determine the intersection point between the ray and the model.

3.3.2 Algorithm Complexities

There are two algorithms to realise the geometric optics (GO) method: the forward and inverse algorithms. In the computer graphics community GO is also called ray tracing [1]. Since the preprocessing part for the forward and inverse algorithms is different, we review both the forward and inverse algorithms. An intuitive understanding is shown in Figure 3.4.

For the forward algorithm in Figure 3.4a, rays are launched from the source to all directions; the electric field (E-field) on a predefined monitor plane is recorded. If we consider the pseudo code for this forward algorithm it will look like this:

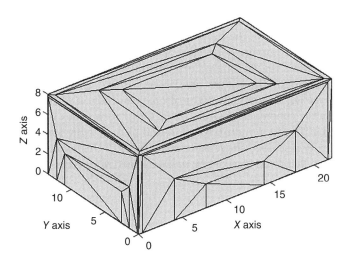

Figure 3.3 Discretised chamber model represented by triangles. Unit: metre.

Figure 3.4 Two different approaches in ray tracing/GO: (a) forward algorithm and (b) inverse algorithm.

(a)

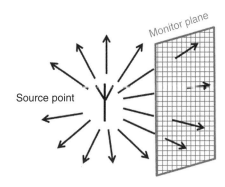

(b)

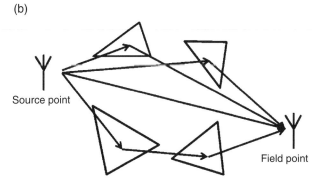

> ***For each*** *ray*
>   *Find the intersection point between the ray and the model/triangles,*
>   *Check if this ray is intercepted by the monitor plane,*
>   *Calculate the E-field and record it.*
> ***End***

This forward algorithm can deal with point-to-area (P2A) problems in 3D, while in 2D it is a point-to-line (P2L) problem, as the rays are launched in a 2D plane and the monitor is a line [2]. The bottleneck for the forward algorithm is the speed of finding the intersection point between the ray and the model. For a single ray, without using any acceleration technique, each triangle/patch of the model needs to be checked to see if it intersects the ray, thus the complexity is $O(NM)$, N is the number of patches describing the model, and M is the reflection order, which is the number of many times the rays are reflected by the model. By using the well-known octree method [3] (introduced in Section 3.3.6), the complexity can be reduced to $O(NM/2^H)$, where H is the depth of the octree. By using the kd-tree [4], the complexity can also be reduced to $O(M\log_2 N)$. It can be further reduced by taking advantage of the ray coherence theorem [5, 6]. Other techniques include parallelisation with a graphics processing unit (GPU) [7] and a multi-resolution grid to reduce the total ray number [8]. It is important to note that the complexity discussed above is for a single ray. If the ray number is T and the surface area of the chamber is S, only one reflection is considered; to have a good resolution at least 100 rays/λ^2 is necessary [9]. Approximately we have

$$\frac{180}{\Delta\theta} \times \frac{360}{\Delta\varphi} \geq 100 \times \frac{S}{\lambda^2} = T \tag{3.1}$$

Suppose the same angle steps are used for both the polar angle θ and the azimuth angle φ, we have

$$\Delta\theta \leq 648 \frac{\lambda^2}{S} \tag{3.2}$$

The number T is proportional to the square of the frequency f, $T \sim O(f^2)$. Finally, if a frequency sweep with F points is used, the complexity of the forward algorithm needs to be multiplied by a factor of $O(Ff^2)$. An example is used to evaluate the number of rays. Suppose an AC has dimensions of 10 m × 20 m × 10 m, so the surface area is 1000 square metres. At a frequency of 15 GHz, the wavelength is 0.02 m and the ray number will be at least

$$T = 100 \times \frac{1000}{0.02 \times 0.02} = 2.5 \times 10^8 \tag{3.3}$$

As can be seen, this is a huge number, and it is just for one frequency with a lower limit for one order reflection. When the order of reflections increases, the wave front also expands; it will require an even finer step size for the angles and even more rays.

For the inverse algorithm in Figure 3.4b, if the position of the receiving (Rx) antenna is already known, the inverse algorithm can be more efficient than the forward algorithm. The inverse algorithm is famous for solving point-to-point (P2P) problems, only the E-field of the predefined points is recorded [5]. In Figure 3.4b, both the source point and the field point are defined; triangles are tested to check if there are visible paths connecting these two points. Different from the forward algorithm, the inverse algorithm checks the triangles to find the possible reflected/image rays between the source point and the field point. The inverse algorithm is also called the path finding algorithm, and pseudo code for the inverse algorithm will look like this:

> **For each** triangle
>   *Find the reflected rays/path connect the source point and the field point*
>   *Calculate the E-field and record it.*
> **End**

The bottleneck for the inverse algorithm is finding the path that connects the source and field points with different orders. The maximum complexity is $O(N^M)$ where N is the number of patches/triangles describing the model and M is the reflection order [10]. As can be seen, when M is large it is also difficult to find all the rays/images connecting the predefined points. However, in the AC simulation we rarely need to trace high-order waves, as it is a highly attenuated environment. Another important characteristic is that the ray path between the source point and the field point can be recorded after path finding, which makes the complexity independent of frequency!

From the complexity review of the two algorithms, the designer can choose a suitable one for a specific problem. If the field distribution in a specific region is of interest, the forward algorithm is preferred; if only the field at some discrete points needs to be known, the inverse algorithm is more efficient. Different from other full-wave methods, the beauty of GO is that the fields contributed by different orders can be separated. The designer can identify where the unexpected field comes from by analysing the field with different orders.

Both algorithms are detailed with acceleration techniques. Before digging into the details of the two algorithms, preparations are necessary, including source definition, material definition, and boundary condition assignment.

3.3.3 Far-Field Data

To protect the intellectual property (IP) of the antenna supplier, the detailed structure of the antenna does not need to be known. Only the far-field pattern is useful, and the far-field pattern is used as the excitation source of the chamber. The E-field at 3 m distance is calculated as $\mathbf{E}_0$ in (2.53), and the E-field data can be obtained from either simulation or measurement of the antenna. To make it reusable, once the far-field pattern is obtained it will be saved into a library/database that can be reused for future simulations. Four matrices are used to save the complex electric far-field at each frequency. They are $\mathbf{E}_{\theta\text{mag}}$, $\mathbf{E}_{\theta\text{phase}}$, $\mathbf{E}_{\varphi\text{mag}}$, and $\mathbf{E}_{\varphi\text{phase}}$ with dimensions $N_R \times N_C$ where N_R is the number of points in the φ (azimuthal angle) direction and N_C is the number of points in the θ (polar angle) direction. $\mathbf{E}_0$ can be expressed using

$$\mathbf{E}_0 = \mathbf{E}_{\theta\text{mag}} \exp\left(j\mathbf{E}_{\theta\text{phase}}\right)\hat{\mathbf{a}}_\theta + \mathbf{E}_{\varphi\text{mag}} \exp\left(j\mathbf{E}_{\varphi\text{phase}}\right)\hat{\mathbf{a}}_\varphi \tag{3.4}$$

where $\mathbf{a}_\theta$ is the unit vector in the θ direction and $\mathring{\mathbf{a}}_\varphi$ is the unit vector in the φ direction.

When launching the rays from the centre of the radiation pattern, interpolation is also necessary. Like the radio absorbing material (RAM) measurement, we have finite samples for $\mathbf{E}_0$. For arbitrary (θ, φ) angles, similar to (2.55), we have

$$\mathbf{E}_{\theta\text{mag}}(\theta, \varphi) = \frac{1}{(\theta_1 - \theta_0)(\varphi_1 - \varphi_0)}[\theta_1 - \theta \ \ \theta - \theta_0] \begin{bmatrix} \mathbf{E}_{\theta\text{mag00}} & \mathbf{E}_{\theta\text{mag01}} \\ \mathbf{E}_{\theta\text{mag10}} & \mathbf{E}_{\theta\text{mag11}} \end{bmatrix} \begin{bmatrix} \varphi_1 - \varphi \\ \varphi - \varphi_0 \end{bmatrix} \tag{3.5}$$

where (θ_0, φ_0) and (θ_1, φ_1) are sampled angles. $\mathbf{E}_{\theta\text{mag00}}$ means the magnitude of the E-field in θ polarisation with sample angles (θ_0, φ_0) and similarly for other notations. Note that (3.5) only applies interpolation to $\mathbf{E}_{\theta\text{mag}}$, we also need to repeat this procedure for matrices $\mathbf{E}_{\theta\text{phase}}$, $\mathbf{E}_{\varphi\text{mag}}$ and $\mathbf{E}_{\varphi\text{phase}}$.

In the case of re-orientation of the transmitting (Tx) antenna, the rotated version of the radiation pattern can be obtained by coordinate transformation, as shown in Figure 3.5. Suppose the far-field is rotated around the z-axis with angle α, then around the y-axis with angle β and finally around the x-axis with angle γ. The vector in the transformed coordinate system can be expressed as

$$Q_3 Q_2 Q_1 \begin{bmatrix} A_x \\ A_y \\ A_z \end{bmatrix} = \begin{bmatrix} T_x \\ T_y \\ T_z \end{bmatrix} \tag{3.6}$$

where $\mathbf{A} = [A_x \ A_y \ A_z]^T$ is the vector in the original coordinate system, $\mathbf{T} = [T_x \ T_y \ T_z]^T$ is the vector in the transformed coordinate system, and

$$Q_1 = \begin{bmatrix} \cos\alpha & \sin\alpha & 0 \\ -\sin\alpha & \cos\alpha & 0 \\ 0 & 0 & 1 \end{bmatrix} \tag{3.7}$$

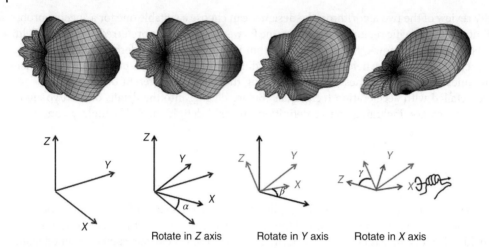

Rotate in Z axis   Rotate in Y axis   Rotate in X axis

Figure 3.5 Far-field rotations.

$$Q_2 = \begin{bmatrix} \cos\beta & 0 & \sin\beta \\ 0 & 1 & 0 \\ -\sin\beta & 0 & \cos\beta \end{bmatrix} \tag{3.8}$$

$$Q_3 = \begin{bmatrix} 1 & 0 & 0 \\ 0 & \cos\gamma & \sin\gamma \\ 0 & -\sin\gamma & \cos\gamma \end{bmatrix} \tag{3.9}$$

3.3.4 Boundary Conditions

In the full-wave simulation method, the boundary conditions normally have different types: Dirichlet (first type) boundary conditions specify the value of the function on the defined boundary, Neumann (second type) boundary conditions specify the normal derivative of the function on a surface, and Robin (third type) boundary conditions specify the linear combination of the function value and its normal derivative on the boundary. However, in the GO method, the boundary condition is very simple and only the reflection coefficient dyad needs to be specified. Once the reflection coefficient dyads are defined related to each type of the RAM, the boundary condition is specified by an index which marks the type of material of a triangle/patch.

Besides the general reflection coefficient dyad, there are three special types of boundary conditions, the perfect electric conductor (PEC), the perfect magnetic conductor (PMC), and the perfectly matched layer (PML). We have defined the reflection coefficient dyad of the PEC in (2.51), the PMC does not exist in practical ACs, and the PML can be set as a special type of material which has index of -1. All rays hit on the PML do not need to be traced further for the next order reflection, which can make it faster in the ray tracing.

In the program, we rarely assign the material types to triangles/patches one by one, a volume assignment can be performed by checking if triangles are inside a defined cube. Suppose a triangle has vertices with coordinates (P_{1x}, P_{1y}, P_{1z}), (P_{2x}, P_{2y}, P_{2z}), and (P_{3x}, P_{3y}, P_{3z}). A volume is selected to assign the material type. The pseudo code to assign the material type will have the following form:

> **For each** *triangle*
>   *Check if there is any vertex of the triangle inside the selected volume.*
>   *Set the material type index to the triangle.*
> **End**

Suppose the selected volume is in the range of $X_{min} \leq x \leq X_{max}$, $Y_{min} \leq y \leq Y_{max}$, and $Z_{min} \leq z \leq Z_{max}$ where X_{min}, Y_{min}, and Z_{min} represent the lower boundary of the selected volume, and X_{max}, Y_{max}, and Z_{max} represent the upper boundary of the selected volume. To test a triangle is inside the volume, we can evaluate the Boolean expression:

$$\left(X_{min} \leq P_{1x} \leq X_{max} \text{ and } Y_{min} \leq P_{1y} \leq Y_{max} \text{ and } Z_{min} \leq P_{1z} \leq Z_{max}\right) \text{ and}$$

$$\left(X_{min} \leq P_{2x} \leq X_{max} \text{ and } Y_{min} \leq P_{2y} \leq Y_{max} \text{ and } Z_{min} \leq P_{2z} \leq Z_{max}\right) \text{ and}$$

$$\left(X_{min} \leq P_{3x} \leq X_{max} \text{ and } Y_{min} \leq P_{3y} \leq Y_{max} \text{ and } Z_{min} \leq P_{3z} \leq Z_{max}\right)$$

3.3.5 RAM Description

A full numerical model is applied to describe the reflection coefficient of the RAM. The reflection coefficient dyad $\overline{\overline{\mathbf{R}}}$ includes four elements: $\mathbf{R}_{\parallel\parallel}$, $\mathbf{R}_{\parallel\perp}$, $\mathbf{R}_{\perp\parallel}$, and $\mathbf{R}_{\perp\perp}$, all of which are dependent on the incident angle and frequency, $\mathbf{R}_{ij} = \mathbf{R}_{ij}(\theta, \varphi, f)$ (i, j can be $\parallel$ or $\perp$). The 3D matrix shown in Figure 3.6 is used to save $\mathbf{R}_{ij}$ of each type of RAM, using three indices, θ, φ, and frequency.

Compared with the traditional simplified models [11–14], this approach needs more memory (10 MB/each RAM type) but it is much faster than the simplified models, which need more operations to calculate the reflection coefficient for each reflection (while this approach only needs interpolation). Also the numerical model encapsulates the detailed information of RAM which can be obtained from either simulation [11–19] or the arch method in measurements [20]. The information is saved in a library/database to make it reusable. Each type of RAM corresponds to four 3D matrices; once they are obtained, we do not need to know the micro level properties of the RAM (permittivity and shape). This is a general extraction procedure and can be used for any type of RAM.

A typical finite element method (FEM) analysis of a RAM unit can be performed with periodic boundary conditions (PBCs), shown in Figure 2.4. For each incident angle, two orthogonal incident waves/modes need to be analysed, as shown in Figure 2.5. A frequency sweep is adopted for each incident angle. When all the simulations/measurements are finished, $\overline{\overline{\mathbf{R}}}$ is ready. Since we can only simulate some discrete incident angles, for an arbitrary incident angle the reflection coefficient can be obtained using interpolation.

A typical value of $\overline{\overline{\mathbf{R}}}$ at 3 GHz is shown in Figure 3.7 (magnitude) and Figure 3.8 (phase) with different incident angles. A 5° step is used for both polar angle θ and azimuthal angle φ, and for other values 2D interpolation is used. Note that at the edge of the figure (large incident angle) the transition is not very smooth because we have used the nearest available value to interpolate it. As can be seen in Figure 3.7, the cross-polarisation $\mathbf{R}_{\parallel\perp}$, $\mathbf{R}_{\perp\parallel}$ are very small and $\mathbf{R}_{\parallel\parallel}$, $\mathbf{R}_{\perp\perp}$ are not sensitive to azimuthal angle φ since the pyramid is a relatively symmetric shape. For small angles of θ, $\mathbf{R}_{\parallel\parallel}$, and $\mathbf{R}_{\perp\perp}$ are very small (approximately −20 dB); when θ increases they decrease first and increase quickly with large θ angles. There is an optimised angle ($\theta \approx 60°$) which has relatively small reflection coefficient compared with other angles. Note this is not a general conclusion; this angle depends on the shape and material property of the RAM. Because of the reciprocity of the material properties, $\mathbf{R}_{\parallel\perp} = \mathbf{R}_{\perp\parallel}$.

The RAM layout definition process is the same as the boundary condition definition. The type index is used to mark the RAM type on each triangle. The rotation of RAM can be considered by applying an angle offset to $\mathbf{R}_{ij}$ with $\Delta\varphi : \mathbf{R}_{ij}(\theta, \varphi + \Delta\varphi, f)$.

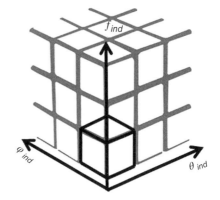

Figure 3.6 Data structure of the reflection coefficient.

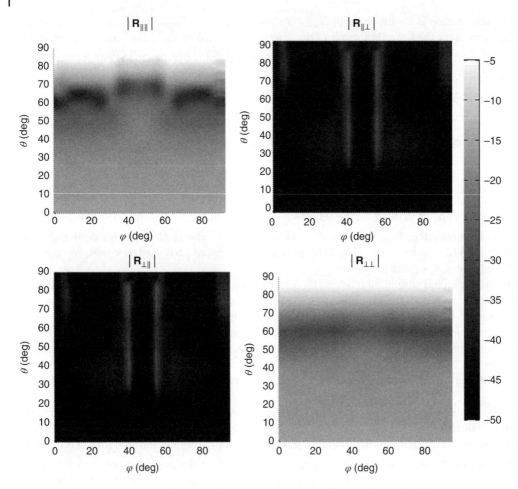

Figure 3.7 Typical magnitude values of each element of the reflection coefficient dyad. Unit: dB.

3.3.6 Forward Algorithm

In this section, the forward algorithm is detailed with acceleration techniques. First, the rays are launched from the source point, and then they are reflected and intercepted by the monitor plane, as shown in Figure 3.4a. Finally, results from different reflection orders are superimposed to obtain the total E-field.

To start the ray tracing procedure, the initial value of the E-field and the cross-sectional area need to be known. The sphere surrounding the source point is divided into triangular patches, as shown in Figure 3.9a. The vertices of the patch and the centre of the sphere form a tetrahedron, as shown in Figure 3.9b. The initial value $\mathbf{E}_0$ is determined by the E-field on the sphere using 2D interpolation of $\mathbf{E}_{\theta\text{mag}}$, $\mathbf{E}_{\theta\text{phase}}$, $\mathbf{E}_{\varphi\text{mag}}$, and $\mathbf{E}_{\varphi\text{phase}}$, and the radius of the sphere can be an arbitrary value, we use 3 m. The initial cross-sectional area is the triangular area A_0 which will be used to calculate the spreading factor (SF) later; different from the pyramid ray tube in [2], the tetrahedral ray tube is used. For a pyramid ray tube, the wave front is quadrilateral. There is a potential risk that the wave front will be distorted and self-intersected after reflecting by the model, but for the tetrahedral ray tube shown in Figure 3.10 the wave front is always kept in the shape of a triangle. After the ray and ray tube are traced, $\mathbf{E}$ at the field point can be determined using (2.53).

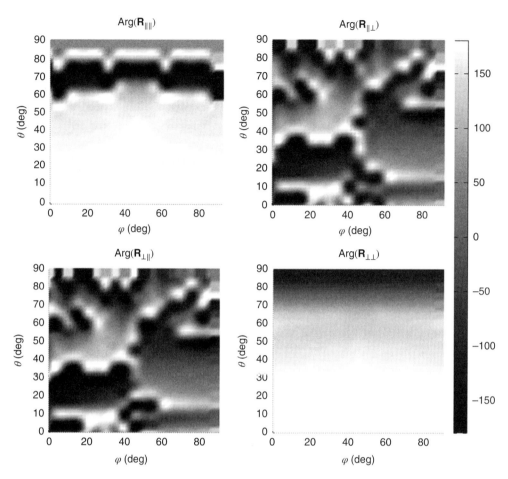

Figure 3.8 Typical phase values of each element of the reflection coefficient dyad. The phase reference plane is chosen at the bottom of the RAM. Unit: degree.

It is interesting to note that each tube shares the same vertex with its neighbour, but the SF can be different after the tube interacts with the model because the wave front area is different. This will make the E-field at the vertices ambiguous, as more than one wave front shares the same vertex. To eliminate this ambiguity, we only use the vertices to carry the SF information, the E-field is defined at the sample points inside each patch, as shown in Table 3.1, the tube triangle is divided into different orders, and the sample points are chosen to be the centre of each triangle. The coordinates of the sample points are given in Figures 3.11–3.13 for

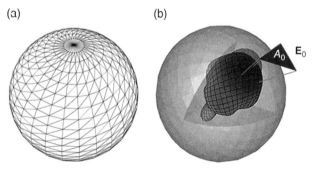

Figure 3.9 (a) Sphere division and (b) initial values.

the first three order divisions, where the vectors of the vertices of the triangle are $\mathbf{V}_1 = (x_1, y_1, z_1)$, $\mathbf{V}_2 = (x_2, y_2, z_2)$, and $\mathbf{V}_3 = (x_3, y_3, z_3)$.

To find the intersection point between the ray and the triangle meshes, an important date structure is the octree. Before the ray tracing is started, the octree needs to be built. We follow the same process given in [3], but

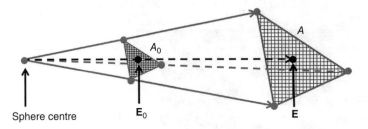

Sphere centre

Figure 3.10 E-field along the ray.

Table 3.1 Triangles with different division order.

Division order	Number of sample points	Triangle division
1	1	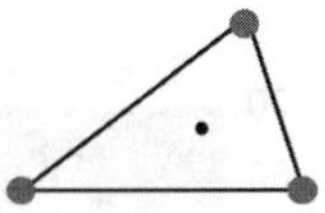
2	4	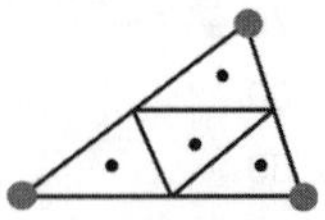
3	16	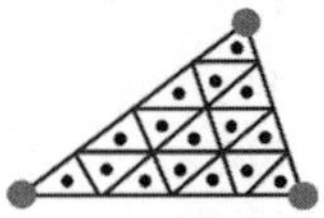
⋮	⋮	⋮
n	4^{n-1}	

$\vec{P}_1 =$	$\vec{V}_1$	$\vec{V}_2$	$\vec{V}_3$
	1/3	1/3	1/3

Figure 3.11 The coordinates of the centre of a triangle when the division order is 1.

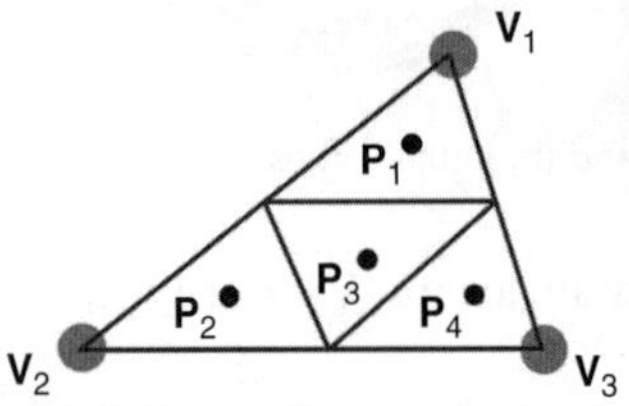

	$\vec{V}_1$	$\vec{V}_2$	$\vec{V}_3$
$\vec{P}_1 =$	2/3	1/6	1/6
$\vec{P}_2 =$	1/6	2/3	1/6
$\vec{P}_3 =$	1/3	1/3	1/3
$\vec{P}_4 =$	1/6	1/6	2/3

Figure 3.12 The coordinates of the centre of triangles when the division order is 2.

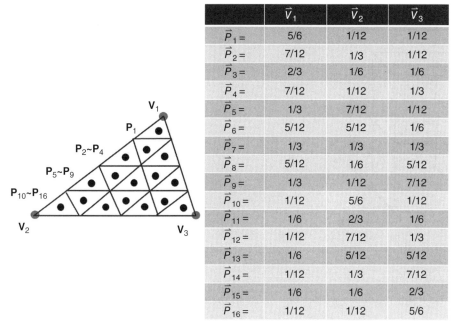

	$\vec{V}_1$	$\vec{V}_2$	$\vec{V}_3$
$\vec{P}_1 =$	5/6	1/12	1/12
$\vec{P}_2 =$	7/12	1/3	1/12
$\vec{P}_3 =$	2/3	1/6	1/6
$\vec{P}_4 =$	7/12	1/12	1/3
$\vec{P}_5 =$	1/3	7/12	1/12
$\vec{P}_6 =$	5/12	5/12	1/6
$\vec{P}_7 =$	1/3	1/3	1/3
$\vec{P}_8 =$	5/12	1/6	5/12
$\vec{P}_9 =$	1/3	1/12	7/12
$\vec{P}_{10} =$	1/12	5/6	1/12
$\vec{P}_{11} =$	1/6	2/3	1/6
$\vec{P}_{12} =$	1/12	7/12	1/3
$\vec{P}_{13} =$	1/6	5/12	5/12
$\vec{P}_{14} =$	1/12	1/3	7/12
$\vec{P}_{15} =$	1/6	1/6	2/3
$\vec{P}_{16} =$	1/12	1/12	5/6

Figure 3.13 The coordinates of the centre of triangles when the division order is 3.

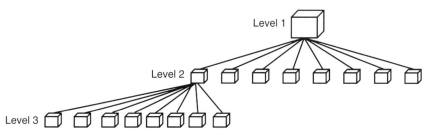

Figure 3.14 A schematic plot of an octree.

different from [3] we propose an adaptive octree depth instead of a fixed depth octree. An octree structure is shown in Figure 3.14, where each box can be divided into eight small boxes, like a tree. This can be understood from the geometric plot shown in Figure 3.15. Once we have divided the volume into boxes in layers, all the triangles constructed in the model can be allocated to the small boxes in the last level. A box can contain many triangles and one triangle can also be allocated to many boxes depending on the size of the boxes and the triangles. The pseudo code can have the following form:

> **For each** *triangle*
> *Check if it intersects the first level box of the octree.*
> *a: Divide the box into 8 sub-boxes*
> *Check if it intersects the sub-boxes*
> *If yes*
> **goto** *a.*
> **End If**

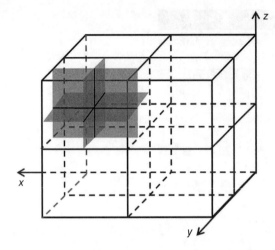

Figure 3.15 A geometric plot of an octree.

> **If** it is the last level
> > *Record the triangle index together with the box index in the last level.*
> **End If**
> **End**

The 'goto' statement in the pseudo code can be replaced by a 'while' loop as we rarely use 'goto' statements nowadays.

To test if a triangle can be allocated to a box, a few Boolean operations are necessary. Suppose a triangle and a box are as shown in Figure 3.16a, where the coordinates of the vertices of the triangle are (x_1, y_1, z_1), (x_2, y_2, z_2), and (x_3, y_3, z_3), and the volume of the box is $X_{\min} \leq x \leq X_{\max}$, $Y_{\min} \leq y \leq Y_{\max}$, and $Z_{\min} \leq z \leq Z_{\max}$. A boundary of the triangle can be found by finding the maximum and minimum values of the coordinates of the vertices (Figure 3.16b), which are $X'_{\min} \leq x \leq X'_{\max}$, $Y'_{\min} \leq y \leq Y'_{\max}$, and $Z'_{\min} \leq z \leq Z'_{\max}$, where $X'_{\min} = \min(x_1, x_2, x_3)$ and $X'_{\max} = \max(x_1, x_2, x_3)$ and we have similar expressions for $Y'_{\min}$, $Y'_{\max}$, $Z'_{\min}$, and $Z'_{\max}$. By checking if there is an overlapped region between these two boxes (Figure 3.16c), we can allocate the triangles into this hexahedron (box) in the octree. The Boolean expression is

$$\left(X_{\min} \leq X'_{\min} \leq X_{\max} \text{ or } X'_{\min} \leq X_{\min} \leq X'_{\max} \text{ or } X_{\min} \leq X'_{\max} \leq X_{\max} \text{ or } X'_{\min} \leq X_{\max} \leq X'_{\max}\right)$$

and

$$\left(Y_{\min} \leq Y'_{\min} \leq Y_{\max} \text{ or } Y'_{\min} \leq Y_{\min} \leq Y'_{\max} \text{ or } Y_{\min} \leq Y'_{\max} \leq Y_{\max} \text{ or } Y'_{\min} \leq Y_{\max} \leq Y'_{\max}\right)$$

and

$$\left(Z_{\min} \leq Z'_{\min} \leq Z_{\max} \text{ or } Z'_{\min} \leq Z_{\min} \leq Z'_{\max} \text{ or } Z_{\min} \leq Z'_{\max} \leq Z_{\max} \text{ or } Z'_{\min} \leq Z_{\min} \leq Z'_{\max}\right)$$

The Matlab code snippet is given in Appendix A.2. It can be seen that the condition in this intersection checking procedure is necessary but not sufficient, which means if the result returns *true* it is also possible that the triangle does not intersect the hexahedron, as shown in Figure 3.16b. However, this does not affect the result. In the ray tracing process, the intersection check between a ray and a hexahedron is performed first; only when the ray hits the last level of the octree is a precise intersection check between the ray and the triangle necessary.

We also need to consider the number of levels of the octree (the depth D). A good choice is when the smallest dimension of the hexahedron in the last level is comparable to the mean length of all the triangles, that is

(a)

(b)

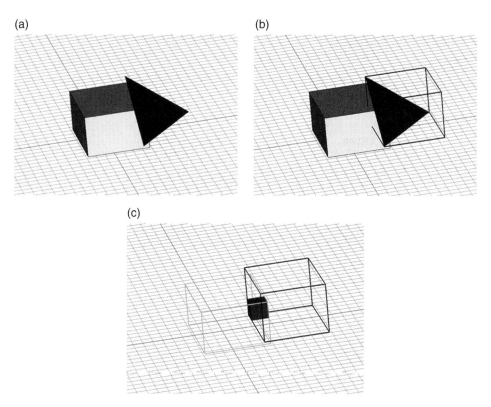

(c)

Figure 3.16 Intersection check between a triangle and a hexahedron.

$$D = 1 + \text{int} \left[\log_2 \frac{\max(W, H, L)}{\text{mean (triangle edge length)}} \right] \tag{3.10}$$

where W, H, and L represent the width, height, and length of the chamber, respectively, and 'int' means the integer part of a float number.

The complexity of the octree building process is $O(ND)$, where N is the number of triangles, because the triangles need to be checked one by one and the octree needs to be allocated level by level. For a specific 3D model, the octree only needs to be built once before the ray tracing. Although an acceleration strategy is not necessary in most cases, the octree building process can be easily parallelised. This parallelisation is shown in Figure 3.17, where different groups of the triangles can be built simultaneously on different threads, and in the last step all the octrees are combined and the final synthesised octree is obtained.

After the octree is built, rays can be launched and recorded on the monitor plane. To find the intersection point between the ray and the model, the well-known breadth-first search (BFS) algorithm [21] is used, as shown in Figure 3.18. The model is divided into boxes with hierarchy and the ray can be launched from anywhere in the model. The boxes at the top level are first checked. If the ray intersects with them, check the sub-level boxes. In the last level, the intersection between the ray and triangle is checked (the shaded area in Figure 3.18a) because the ray–box intersection check is much faster than the ray–triangle check. Figure 3.18b gives the searching direction and the sequence number; the shaded area means the box contains triangles that may potentially intersect with the ray.

To check if a ray intersects a box (hexahedron), we can first check if the ray intersects the six planes of the hexahedron, and then check if the intersection point is in the region of the box. Suppose a ray starts from the

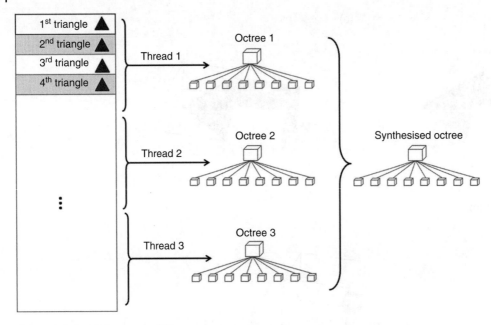

Figure 3.17 Parallel octree building.

(a)

(b)

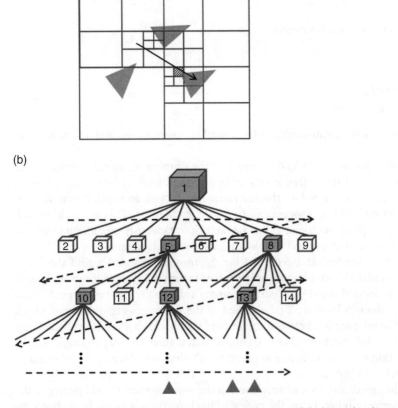

Figure 3.18 BFS searching algorithm: (a) geometrical view and (b) hierarchical view.

point $\mathbf{P}_0 = [p_{0x}\, p_{0y}\, p_{0z}]^T$ and has a direction vector $\mathbf{r} = [r_x\, r_y\, r_z]^T$, as shown in Figure 3.19. The hexahedron has a boundary of $X_{\min} \le x \le X_{\max}$, $Y_{\min} \le y \le Y_{\max}$, and $Z_{\min} \le z \le Z_{\max}$.

In solving the equation system,

$$\begin{cases} \mathbf{P}(t) = \mathbf{P}_0 + t\mathbf{r} \\ x = X_{\min} - 10^{-4} \text{ or } x = X_{\min} + 10^{-4} \text{ or} \\ y = Y_{\min} - 10^{-4} \text{ or } y = Y_{\min} + 10^{-4} \text{ or} \\ z = Z_{\min} - 10^{-4} \text{ or } z = Z_{\min} + 10^{-4} \end{cases} \qquad (3.11)$$

we can use 10^{-4} as error tolerance. To check if the ray intersects the hexahedron, the following Boolean can be evaluated, that is, when one of the six planes of the hexahedron intersects the ray and $t > 0$, the ray intersects the hexahedron:

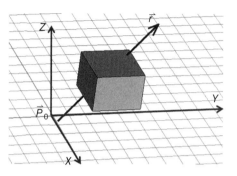

Figure 3.19 Ray–box intersection check.

$$\big(\text{when } z = Z_{\min} - 10^{-4} \text{ or } z = Z_{\max} + 10^{-4}, \text{check if } X_{\min} \le x \le X_{\max} \text{ and } Y_{\min} \le y \le Y_{\max}, \text{or}$$

$$\text{when } x = X_{\min} - 10^{-4} \text{ or } x = X_{\max} + 10^{-4}, \text{check if } Z_{\min} \le z \le Z_{\max} \text{ and } Y_{\min} \le y \le Y_{\max}, \text{or}$$

$$\text{when } y = Y_{\min} - 10^{-4} \text{ or } y = Y_{\max} + 10^{-4}, \text{check if } X_{\min} \le x \le X_{\max} \text{ and } Z_{\min} \le z \le Z_{\max}\big) \text{ and } t > 0.$$

The Matlab code snippet is given in Appendix A.3.

When the ray reaches the last level of the octree, we need to check if it intersects a triangle inside the hexahedron. Suppose a ray starts from the point $\mathbf{P}_0 = [p_{0x}\, p_{0y}\, p_{0z}]^T$, the direction is $\mathbf{r} = [r_x\, r_y\, r_z]^T$, the triangle vertices are $\mathbf{V}_0 = [v_{0x}\, v_{0y}\, v_{0z}]^T$, $\mathbf{V}_1 = [v_{1x}\, v_{1y}\, v_{1z}]^T$, and $\mathbf{V}_2 = [v_{2x}\, v_{2y}\, v_{2z}]^T$, as shown in Figure 3.20. The triangle equation can be expressed by

$$T(u,v) = \mathbf{V}_0 + u(\mathbf{V}_1 - \mathbf{V}_0) + v(\mathbf{V}_2 - \mathbf{V}_0), \quad u \ge 0,\ v \ge 0,\ u + v \le 1 \qquad (3.12)$$

by using (u, v) local coordinate system. To check if the ray intersects the triangle, the following equation system can be solved [22]:

$$\begin{bmatrix} t \\ u \\ v \end{bmatrix} = \begin{bmatrix} p_{0x} - v_{0x} \\ p_{0y} - v_{0y} \\ p_{0z} - v_{0z} \end{bmatrix} \begin{bmatrix} -r_x & v_{1x} - v_{0x} & v_{2x} - v_{0x} \\ -r_y & v_{1y} - v_{0y} & v_{2y} - v_{0y} \\ -r_z & v_{1z} - v_{0z} & v_{2z} - v_{0z} \end{bmatrix}^{-1}$$

$$(3.13)$$

If the solutions satisfy $t > 0$, $u \ge 0$, $v \ge 0$, $u + v \le 1$, the ray intersects the triangle. The Matlab code snippet is given in Appendix A.4.

It is important to note that if the model is over-divided by the octree (the octree level is too deep), the searching speed will deteriorate. The ray–triangle and the ray–box checking numbers with different octree depths are given in Figure 3.21. The model contains 768 triangle meshes and 1000 random rays are launched to perform the benchmark. When the octree depth is 1, the model is not divided into an octree; the ray–box checking number is zero. As can be seen in Figure 3.21, although a deeper octree reduces the ray–triangle checking number, it increases

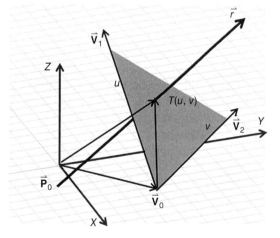

Figure 3.20 Ray–triangle intersection check.

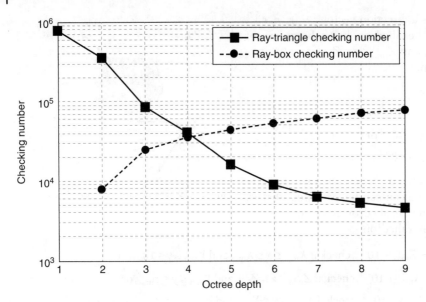

Figure 3.21 Ray–triangle and ray–box checking numbers with different octree depths.

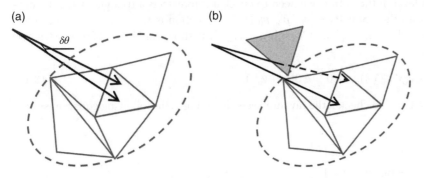

Figure 3.22 Check the triangles near the last intersection triangles.

the ray–box checking number thus more time is wasted on the ray–box intersection check. This is the reason why the adaptive octree has been proposed.

Another technique used to accelerate the ray–triangle intersection is to check the triangles near the last intersection first before using the octree, as shown in Figure 3.22a. This is because when the angle between the launched rays is very small ($\delta\theta$ in the figure), there is a high possibility that the next ray will intersect the same triangles as the previous one [23]. However, if the ray is blocked by another triangle before it hits the triangles, this method could give wrong result. When there is no extra scatterer in the free space of the AC this method can be used, otherwise the scenario in Figure 3.22b needs to be considered.

Once the intersection point between the ray and the triangle is found, Eq. (2.54) is used to calculate the reflected E-field and update the initial value and the wave front area A_0 for the next trace. Because the incident angle can be of an arbitrary value and the frequency of interest may not be exactly the same as the sample frequency in the full numerical model of RAM (where a set of frequencies is used), 1D and 2D interpolations are used to obtain the reflection coefficient value. First, each matrix in the reflection coefficient dyad $\overline{\overline{\mathbf{R}}}$ in (2.54) is interpolated with the frequency of interest, since each element in the matrix needs to be interpolated and it is time-consuming to do this in the ray tracing loop. This procedure can be moved out of the ray tracing loop and

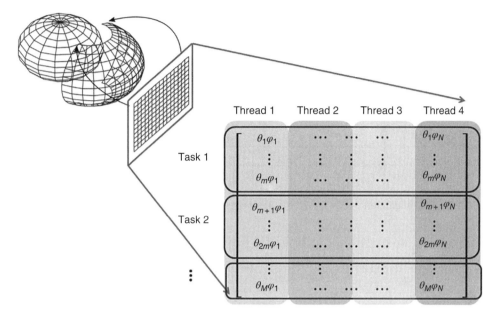

Figure 3.23 Parallelisation of ray tracing.

it only needs to be calculated once for each frequency of interest. Then, the 2D angle interpolation is applied to each matrix in $\overline{\overline{\mathbf{R}}}$. Considering the magnitude and phase, this makes only eight 2D interpolations for each ray–triangle intersection, which is much faster than the traditional RAM model.

One of the advantages of GO is that it is easy to parallelise. We have already used the octree algorithm to accelerate the intersection checking process. Further acceleration techniques are also considered, and both the distributed computing and multithreading techniques can be employed to divide the sphere in Figure 3.9a into sub regions, shown in Figure 3.23; rays in different sub regions are traced simultaneously in a different computing engine, and finally the results are combined.

The total ray number can also be reduced to accelerate the GO simulation. Two methods are used to reduce the total number of rays: one method is to set a threshold for the E-field value (e.g. –30 dB of the peak value). The rays below the threshold will be skipped. Another method is to limit the ray launching region; we rarely need to consider the rays close to the polar points in Figure 3.9a. By combing these two methods the speed can be improved significantly.

After the monitor plane is defined, the plane needs to be discretised into meshes. The mesh size is normally chosen to be smaller than $\lambda/10$ (λ is the wavelength of the frequency of interest). Figure 3.24a gives the procedure for the value assignment in each tube: the values on each grid are first initialised as zero, after the ray tube is intercepted by the monitor plane, the grid points in the tubes are checked, the distance between the grid points and the sample points are calculated, and the E-field of the nearest sample point is chosen to be the value of the grid point. Although this assignment for four grid points in the monitor plane is illustrated in Figure 3.24a, this procedure is repeated for all the grid points in the illuminated area. Figure 3.24b shows the superposition of the field values between two tubes, and the field values on the grid points shared by different tubes are superimposed.

Finally, the E-field from the same order rays with different threads is combined, and then the total E-field can to be obtained from the superposition of different orders. Figure 3.25 shows the 0-order E-field combination from different threads, the 1st-order E-field combination is shown in Figure 3.26, and the superposition of 0 order and 1st order is illustrated in Figure 3.27. The order of the ray means how many times the ray has been

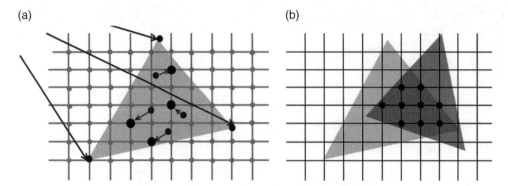

Figure 3.24 Superposition of the E-field in different tubes: (a) value assignment in each tube and (b) value update for the grid points shared by different tubes.

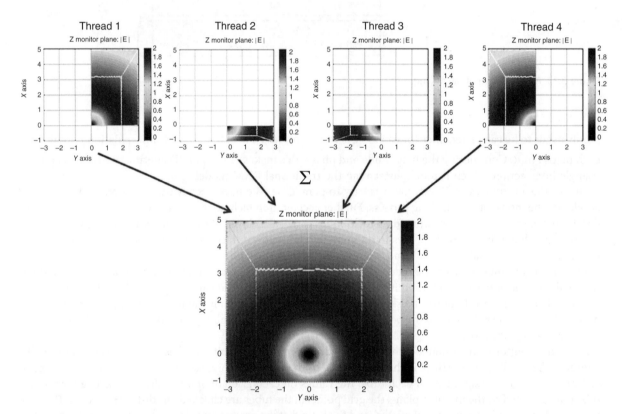

Figure 3.25 Superposition of the 0-order E-field from different threads, only the magnitude part is plotted.

reflected, 0 order means a direct hit and 1st order means the ray has been reflected once by the model. It can be seen from the figures that there are white gaps in the figures; this is because when the ray tubes hit the sharp corners of the cavity, the wave front is drastically deformed. A filter algorithm is used to filter these abnormal ray tubes, and it can be realised by checking the ratio between the wave front area and the square of the total ray length. If it is larger or smaller than a threshold value, it means the wave front area increases or decreases abnormally, thus the corresponding ray tube can be filtered.

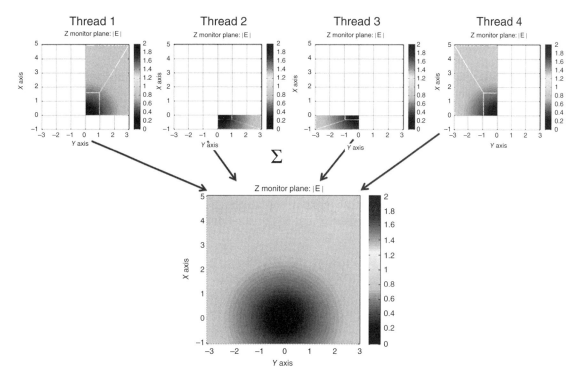

Figure 3.26 Superposition of the 1st-order E-field from different threads, only the magnitude part is plotted.

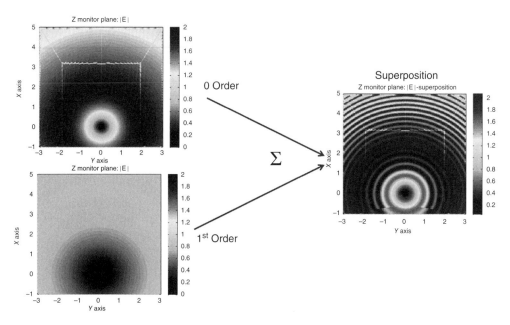

Figure 3.27 Superposition of the E-field from different orders, only the magnitude part is plotted.

3.3.7 Inverse Algorithm

Different from launching the rays in all directions in the forward algorithm, the inverse algorithm finds the path that connects the source point and the field point, which is called path finding. After the paths are found and saved, the initial value and the reflection coefficient can be obtained in the same way as in the forward algorithm, and the final E-field with different orders can also be easily superimposed.

Suppose each triangle in the model has a unique index number; the possible paths to be checked are shown in Figure 3.28. For the rays reflected once from source point (S) to field point (F), the possible paths are S-1-F, S-2-F, S-3-F and S-4-F. As can be seen in Figure 3.28a, only two paths are practical (visible), other images are invisible because the intersection points are outside the triangle. Similar procedures happen to the 2nd-order rays in Figure 3.28b, and the number of paths to be checked becomes 12. Generally, for the model with N triangles, the number of Mth-order paths to be checked is $N(N-1)^{M-1}$ [10, 23], which is the complexity of the inverse algorithm. It is important to note that without any acceleration strategy, it could be very time-consuming to check the images one by one when N and M are large numbers. The pseudo code can be:

(a)

(b)

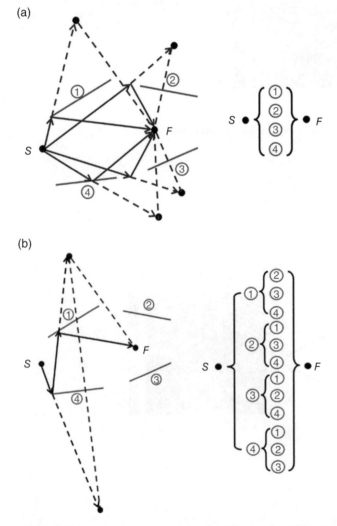

Figure 3.28 Path finding in the inverse algorithm: (a) 1st-order paths and (b) 2nd-order paths.

For each *triangle*
  For each *reflection order*
    Generate the image point from all possible paths with given reflection order.
    Check if there is a ray which connects the source point and the field point with the reflection order (if the image is visible from the source point).
    If yes, save the ray path.
  End
End

The ray–triangle intersection check is the same as that in the forward algorithm. The only difference is that in the forward algorithm the rays are launched to all directions, while in the inverse algorithm the direction is determined by the source point and the image of the field point. Another issue is that in the inverse algorithm no ray tube is used to calculate SF; the free-space attenuation is used for each ray (path). This could lead to inaccurate results when dealing with a curved surface, but in ACs we rarely have structures with very large curvatures.

Parallelisation can be used as shown in Figure 3.29. The engine can be a distributed computer or a thread in a single computer. The whole path tree is split into different parts which are checked by different engines simultaneously.

An additional acceleration strategy which takes advantage of the chamber shape is convex acceleration. In topology, the 3D model can be divided into two categories: the concave and the convex, as shown in Figure 3.30. For a convex shape, all line segments connecting any pair of points are inside the shape. For a concave shape, there is a possibility that the line segments will be intercepted by the model itself. This offers an opportunity to accelerate the path finding process. Generally speaking, each path needs to be checked to make sure it is not intercepted by the other triangles of the model, but for a convex model this checking procedure is not necessary since most of the chambers are convex. A benchmark has been determined to validate convex acceleration: for a model with 116 triangles, one thread is used and only the 1st-order reflection is considered. We found that the speed with the convex acceleration is 28 times faster than the general algorithm.

3.3.8 Post Processing

In post processing, the raw data of the E-field are converted to figures of merit to make them understandable. Typical figures of merit are site attenuation (SA) in a fully AC, or normalised site attenuation (NSA) in a semi-AC, site voltage-standing-wave ratio (SVSWR), and field uniformity (FU). All of these values are extracted from the E-field distribution in the post processing. If the transmitting (Tx) antenna is well matched and the input power is normalised to 1 W, the NSA value can be calculated by using [24]

$$NSA\,(dB) = 46.76 + G - 20\log_{10}f - 20\log_{10}E_{max} \qquad (3.14)$$

where G is the gain of the Tx antenna in free space in dBi, f is the frequency of interest in MHz, and E_{max} is the maximum E-field value in V m^{-1} measured by the receiving (Rx) antenna, when the height is scanning from 1 to 4 m.

The site attenuation is defined as the ratio of the power input to a matched and balanced lossless tuned dipole radiator to that at the output of a similarly balanced matched lossless tuned dipole

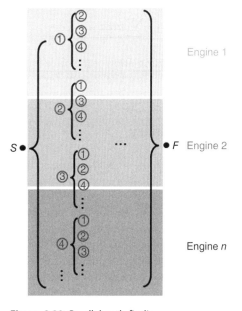

Figure 3.29 Parallel path finding.

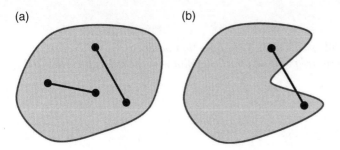

(a)   (b)

Figure 3.30 Topology of the model: (a) convex shape and (b) concave shape.

Table 3.2 Acceptability criterion.

Test performed	Acceptability criterion	Reference
NSA	±4 dB	[25]
SVSWR	≤6 dB	[26]
FU	−0 dB, +6 dB for 75%	[26]

receiving antenna. *Normalise* means the site attenuation is normalised to the transmitting and receiving antenna factors (Chapter 5 in [24]). The definitions of SVSWR and FU can be found in [25, 26].

For a semi-AC, the NSA is required to be in the range of ±4 dB of the values given in the standard [25], the SVSWR is required to be ≤6 dB, and the FU is required to be in the range of 0 dB to +6 dB for 75% of the sample points for each frequency (shown in Table 3.2).

The cost of RAM used in the chamber can be obtained by

$$cost_{\text{RAM}} = \sum_i area_i \times price_i \tag{3.15}$$

which is the summation of the price for all the RAMs used in the chamber, where i means the type index of the RAM. The cost is a vital parameter for commercial consideration, especially when the chamber is large.

3.4 Summary

In this chapter, the software implementation of the CAD tool has been detailed. Two different algorithms have been introduced: a forward algorithm and an inverse algorithm. They share the same pre-processing part, but different post-processing parts. The complexity and parallelisation have also been discussed. The forward algorithm is more efficient for P2A problems while the inverse algorithm is more suitable for P2P problems. The corresponding acceleration strategies have been explained in detail. The adaptive octree has been proposed for the forward algorithm, and a new acceleration strategy, named convex acceleration, has also been proposed for the inverse algorithm. Both have improved the efficiency significantly.

We have also borrowed the philosophy from software engineering: consider the RAM and the antenna as objects and the properties (reflection coefficient, position, radiation pattern, etc.) are encapsulated, like object-oriented programming (OOP). This approach does not require the detailed information of the antenna and the RAM. It provides a seamless connection between the micro level and the macro level designs, which makes the design procedure systematic. Compared with the traditional analytical model the proposed RAM model requires more memory to save it (around 10 MB for each type, which is still very small for computers nowadays), which is a trade-off of the proposed approach.

References

1 Glassner, A. S. (1989). *An Introduction to Ray Tracing*. Morgan Kaufmann.

2 Yang, C.-F., Wu, B.-C., and Ko, C.-J. (1998). A ray-tracing method for modeling indoor wave propagation and penetration. *IEEE Transactions on Antennas and Propagation* 46 (6): 907–919.

3 Jin, K., Suh, T., Suk, S. et al. (2006). Fast ray tracing using a space-division algorithm for RCS prediction. *Journal of Electromagnetic Waves and Applications* 20 (1): 119–126.

4 Tao, Y., Lin, H., and Bao, H. (2008). Kd-tree based fast ray tracing for RCS prediction. *Progress In Electromagnetics Research (PIER)* 81: 329–341.

5 Saeidi, C., Fard, A., and Hodjatkashani, F. (2012). Full three-dimensional radio wave propagation prediction model. *IEEE Transactions on Antennas and Propagation* 60 (5): 2462–2471.

6 Ohta, M. and Maekawa, M. (1987). Ray coherence theorem and constant time ray tracing algorithm. *Computer Graphics* 303–314.

7 Meng, H. (2011). Acceleration of Asymptotic Computational Electromagnetics Physical Optics – Shooting and Bouncing Ray (PO-SBR) Method using CUDA. MS thesis, Department of Electrical and Computer Engineering, University of Illinois at Urbana-Champaign., Illinois.

8 Bang, J. and Kim, B. (2007). Time consumption reduction of ray tracing for RCS prediction using efficient grid division and space division algorithms. *Journal of Electromagnetic Waves and Applications* 21 (6): 829–840.

9 Ling, H., Chou, R. C., and Lee, S. W. (1989). Shooting and bouncing rays: calculating the RCS of an arbitrarily shaped cavity. *IEEE Transactions on Antennas and Propagation* 37 (2): 194–205.

10 McKown, J. W. and Hamilton, R. L. (1991). Ray tracing as a design tool for radio networks. *IEEE Network* 5 (6): 27–30.

11 Kubytskyi, V., Sapoval, B., Dun, G., and Rosnarho, J. (2012). Fast optimization of microwave absorbers. *Microwave and Optical Technology Letters* 54 (11): 2472–2477.

12 Kuester, E. F. and Holloway, C. L. (1994). A low-frequency model for wedge or pyramid absorber arrays-I: theory. *IEEE Transactions on Electromagnetic Compatibility* 36 (4): 300–306.

13 Holloway, C. L. and Kuester, E. F. (1994). A low-frequency model for wedge or pyramid absorber arrays-II: computed and measured results. *IEEE Transactions on Electromagnetic Compatibility* 36 (4): 307–313.

14 Khajehpour, A. and Mirtaheri, S. A. (2008). Analysis of pyramid EM wave absorber by FDTD method and comparing with capacitance and homogenization methods. *Progress in Electromagnetics Research Letters* 3: 123–131.

15 Elmahgoub, K., Yang, F., Elsherbeni, A. Z. et al. (2010). FDTD analysis of periodic structures with arbitrary skewed grid. *IEEE Transactions on Antennas and Propagation* 58 (8): 2649–2657.

16 Sun, W., Liu, K., and Balanis, C. A. (1996). Analysis of singly and doubly periodic absorbers by frequency-domain finite-difference method. *IEEE Transactions on Antennas and Propagation* 44 (6): 798–805.

17 Jiang, Y. and Martin, A.Q. (1999). The design of microwave absorbers with high order hybrid finite element method. IEEE Antennas and Propagation Society International Symposium, Orlando.Volume 4. pp. 2622–2625.

18 Janaswamy, R. (1992). Oblique scattering from lossy periodic surfaces with application to anechoic chamber absorbers. *IEEE Transactions on Antennas and Propagation* 40 (2): 162–169.

19 Marly, N., Baekelandt, B., Zutter, D. D., and Pues, H. F. (1995). Integral equation modeling of the scattering and absorption of multilayered doubly-periodic lossy structures. *IEEE Transactions on Antennas and Propagation* 43 (11): 1281–1287.

20 IEEE Std 1128-1998 (1998). *IEEE Recommended Practice for Radio-Frequency (RF) Absorber Evaluation in the Range of 30 MHz to 5 GHz*. IEEE Standard.

21 Cormen, T. H., Leiserson, C. E., Rivest, R. L., and Stein, C. (2009). *Introduction to Algorithms*, 3e. The MIT Press.

22 Moller, T. and Trumbore, B. (1997). Fast, minimum storage ray/triangle intersection. *Journal of Graphics Tools* 2 (1): 21–28.

23 Catedra, M. F., Perez, J., Saez de Adana, F., and Gutierrez, O. (1998). Efficient ray-tracing techniques for three-dimensional analyses of propagation in mobile communications: application to picocell and microcell scenarios. *IEEE Antennas and Propagation Magazine* 40 (2): 15–28.

24 Kodali, V. P. (2001). *Engineering Electromagnetic Compatibility: Principles, Measurements, Technologies, and Computer Models*, 2e. New York, NY: Wiley-IEEE Press.

25 CISPR 16-1-4 (2012). *Specification for radio disturbance and immunity measuring apparatus and methods - Part 1–4: Radio disturbance and immunity measuring apparatus - Antennas and test sites for radiated disturbance measurements*, 3.1e. IEC Standard.

26 IEC 61000-4-3 (2008). *Electromagnetic Compatibility (EMC) – Part 4-3: Testing and Measurement Techniques – Radiated, Radio-Frequency, Electromagnetic Field Immunity Test*, 3.1e. IEC Standard.

4

Anechoic Chamber Design Examples and Verifications

4.1 Introduction

Anechoic chambers (ACs) may have various forms in applications, such as tapered chambers [1], compact chambers with reflectors [2], double horn chambers [3, 4], etc. Normally full ACs are used for antennas and radar cross section (RCS) measurements while semi-ACs are employed for electromagnetic compatibility (EMC) measurements. There are many ways and parameters to characterise the chamber performance: the site attenuation (SA, or normalised SA), site voltage standing wave ratio (SVSWR) and field uniformity (FU) are the three key parameters which are normally employed in the chamber-related standards [5, 6] to evaluate the chamber performance.

In this chapter, chamber measurement methods are introduced first, including normalised site attenuation (NSA), SVSWR and FU measurement, then practical design examples are given. The measurement and simulation results obtained from the developed computer aided design (CAD) tool (Fast AC Evaluation Tool (FACET) in Chapter 3) are compared. Discussions and conclusions are also provided. Since the fields in EMC semi-ACs are less uniform than in full ACs, the focus of our example is on EMC chambers.

4.2 Normalised Site Attenuation

4.2.1 NSA Definition

NSA is used to quantify how close a test site is to an ideal half plane with a perfect electrical conductor (PEC) ground. The measurement is performed in a semi-anechoic chamber (the ground plane is metallic, walls, and ceilings are lined with radio absorbing materials (RAMs)) and a schematic plot without the walls and ceiling is shown in Figure 4.1. The standard measurement distance can be 3, 10, or 30 m, depending on the relevant standards, the chamber size, and the size of the object under test.

In the measurement, the Tx antenna is connected to a signal generator and the Rx antenna is connected to a spectrum analyser (a receiver, or a voltage meter). All measurements are performed in a 50 Ω system. The Rx antenna is scanned with the height from 1 to 4 m (can be different for different types of antenna [5]) and the maximum signal V_{site} is measured. Then the reading of V_{direct} is taken with two coaxial cables disconnected from the two antennas and connected to each other via an adapter. The NSA value in dB can be obtained using [5]:

$$NSA\,(\text{dB}) = V_{\text{direct}} - V_{\text{site}} - AF_{\text{T}} - AF_{\text{R}} - \Delta AF \tag{4.1}$$

where AF_{T} is the antenna factor of the transmitting antenna (dB m^{-1}), AF_{R} is the antenna factor of the receiving antenna (dB m^{-1}), ΔAF is the mutual impedance correction factor (dB), and the voltages are in dBV. It should

Anechoic and Reverberation Chambers: Theory, Design, and Measurements, First Edition. Qian Xu and Yi Huang.
© 2019 John Wiley & Sons Ltd. Published 2019 by John Wiley & Sons Ltd.

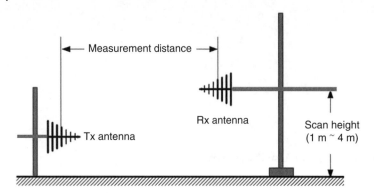

Figure 4.1 NSA measurement setup.

be noted that the mutual impedance correction factor is for a tunable dipole antenna; for a broadband antenna no correction factor is needed. In the measurement, if the tunable dipole antenna is used, the length of the antenna needs to be tuned to each frequency in the measurement. A broadband antenna such as a log-periodic dipole array (LPDA) or a biconical antenna does not have this issue and the continuous frequency sweep can be used in the whole frequency band. When broadband antennas are used, the swept frequency method can be applied to measure the NSA [5]. In this method, the Rx antenna height and frequency are scanned simultaneously; the frequency sweep will be much greater than the antenna height scan rate, which means the frequency sweep can be finished for an almost constant antenna height. During the height scanning process, the peak hold ('max hold') function is enabled, which records the maximum value for all the frequencies in the sweeping range after the height scan is finished.

For a good semi-anechoic chamber, the NSA value is required to be in the range of ±4 dB of the values given in the standard [5]. The NSA value can also be obtained from the electric field (E-field). If the transmitting antenna is well-matched and the input power is normalised to 1 W, the NSA value can be calculated using [7].

$$NSA\,(\text{dB}) = 46.76 + G - 20\log_{10}f - 20\log_{10}E_{\max} \tag{4.2}$$

where G is the gain of the Tx antenna in free space in dBi, f is the frequency of interest in MHz, $E_{\max}$ is the maximum E-field value in V m^{-1} measured by the receiving antenna when the height is scanning from 1 to 4 m. (4.1) and (4.2) are equivalent, in practical measurement. (4.1) is easy to use since the loss of cables can be corrected by measuring V_{direct}, but in simulation (4.2) is more convenient since it is easy to force/normalise the radiated power to 1 W for all frequencies.

4.2.2 NSA Simulation and Measurement

We first simulate an ideal semi-anechoic chamber (the RAMs absorb waves perfectly) and then move on to a real scenario. A semi-anechoic chamber as shown in Figure 4.2 is selected as an example for evaluation [8]. The size of the chamber is 22 m × 13.5 m × 8 m ($L \times W \times H$). The Rx antenna is positioned at the left side in the test region, the Tx antenna is 3 m away from the Rx antenna, the height is 2 m, and the coordinate is (12.62, 6.745, 2). Both Rx and Tx antennas are half-wave dipoles. All the boundary conditions are set as perfectly matched layers (PMLs) except the ground plane (set as a PEC). A monitor plane is set at $x = 15.45\ m$ across the centre of the test region (a circle) to record the E-field.

Since this is a perfect half-space problem, and the Green's function for the half space is well-known, only the Tx antenna needs to be discretised and the Moment Method or Method of Moment (MoM) is used to simulate it. The far-field data is exported to FACET to predict the chamber performance; the results obtained from FACET and MoM method are compared to verify the accuracy of the proposed solution. The magnitude of the total E-field on the monitor plane at 1 GHz is shown in Figure 4.3. As can be seen, they are in good agreement.

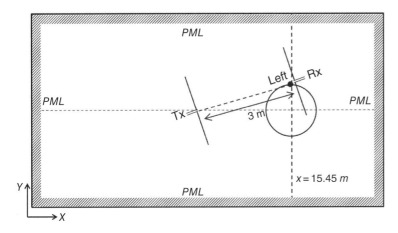

Figure 4.2 A cross-section (horizontal) view of an ideal semi-anechoic chamber as an example. The ground plane is set as PEC where other planes are set as PML.

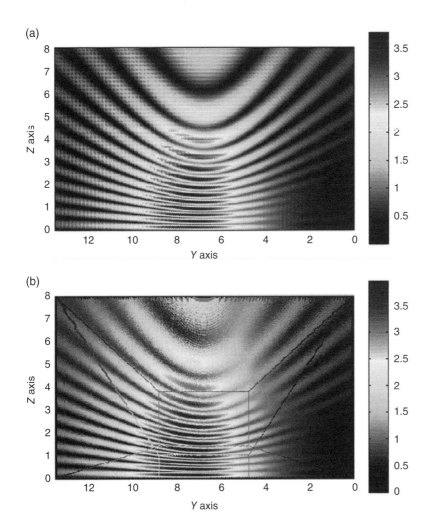

Figure 4.3 Total E-field in the semi-anechoic chamber on the monitor plane at 1 GHz: (a) obtained by using method of moment (MoM) and (b) obtained by using the forward algorithm. Unit: $V\,m^{-1}$.

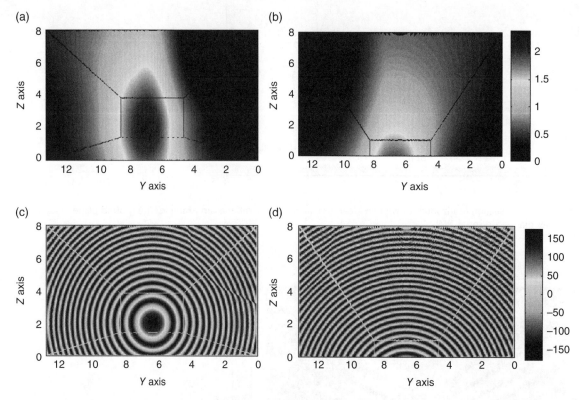

Figure 4.4 *y*-polarised E-field with different reflective orders: (a) 0-order magnitude, (b) 1st-order magnitude (unit V m^{-1}), (c) 0-order phase, and (d) 1st-order phase (unit: degree).

The small difference between Figure 4.3a and Figure 4.3b is due to the mutual coupling of the image antenna (because of the ground plane). The mutual coupling between the Tx antenna and its image was considered in MoM, but not in FACET. A unique feature of geometric optics (GO) is that the fields with different orders of reflections can be viewed separately, as shown in Figure 4.4. The superposition of the 0-order and 1st-order E-field makes the total E-field in Figure 4.3b. It is important to note that when the ray tubes hit the corners of the chamber, the wave front distorts drastically, which may produce unreasonable values for the further rays, thus these ray tubes are filtered and may result in small errors.

By using the inverse algorithm, the rays with 0- and 1st-order are shown in Figure 4.5; higher order rays do not exist for the ideal semi-anechoic chamber. Actually only the 0- and 1st-order rays contribute to the field superposition, higher order rays are absorbed by the PML boundary. The E-field values are extracted with the height scanning from 0 to 4 m at the centre of the test region. Both the forward and inverse algorithms are used; results are compared and shown in Figure 4.6. It can be seen that all the results are in good agreement except the height is close to the ceiling. This is because for the forward algorithm, the ray tubes hit the corners are filtered and the values become inaccurate when close to the corner (the height is 8 m).

After the E-field values are extracted, (4.2) is used to obtain the NSA values. The same procedure is repeated for each frequency of interest. Figure 4.7 gives the results obtained from the International Special Committee on Radio Interference (CISPR) standard [5], the MoM method and FACET. It is important to note that when we use the half-wave dipole antenna, mutual coupling correction factors should be used. As can be seen in Figure 4.7, all the curves agree well with each other. The difference (mainly at lower frequencies) between the MoM and the other two is mainly due to the non-typical balun of the Tx antenna, as already stated in [5].

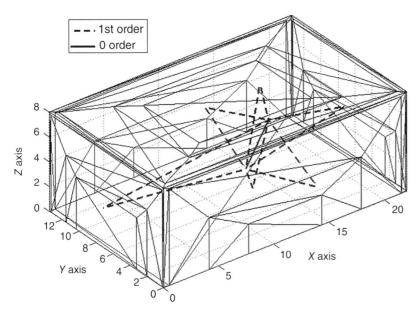

Figure 4.5 Paths connect the source point and the field point with different orders: 1 ray with 0 order and 6 rays with 1st order.

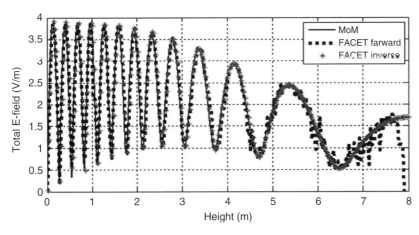

Figure 4.6 Total E-field comparison using different methods.

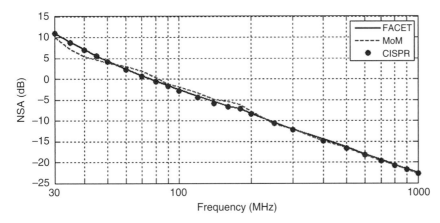

Figure 4.7 NSA comparison of the reference semi-anechoic chamber.

(a)

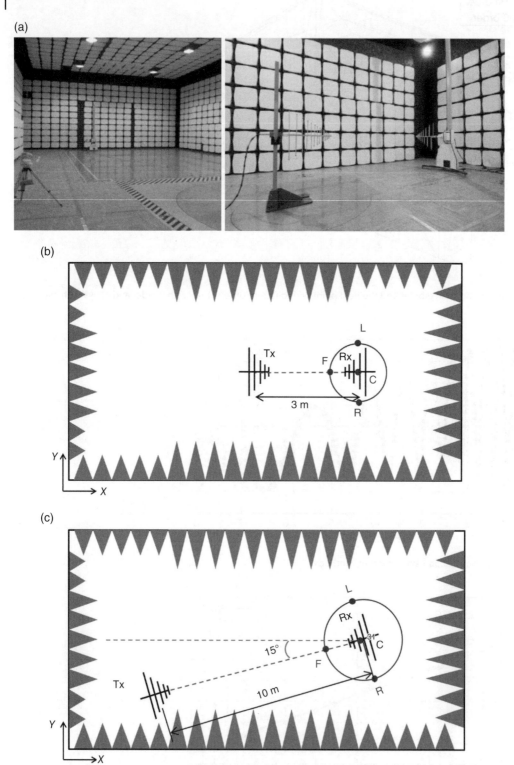

(b)

(c)

Figure 4.8 NSA measurement scenario: (a) a typical EMC chamber (pictures from Rainford EMC Systems Ltd), (b) 3 m NSA measurement, and (c) 10 m NSA measurement.

Also GO is a high frequency method: at lower frequencies, the accuracy of the proposed GO-based method is reduced. The NSA reference values for 3, 10, 30 m in [5] are plotted in Appendix B.

For this scenario, the overall simulation time using FACET for each polarisation is less than 1 minute with a normal personal computer (PC). The results validate the proposed method with confidence. It should be noted that for a practical chamber it is not realistic to simulate it using the MoM. It requires a huge amount of memory, and the complex material definition which is sometimes not known. Next we give a practical AC design and compare the simulation and measurement results.

A typical semi-anechoic chamber design is shown in Figure 4.8. The size is 22 m × 13.5 m × 8 m ($L \times W \times H$). The layout of the RAM is given in Figure 4.9 and the shading represents the RAM types. Each RAM was characterised and the complex reflection coefficients for different angles were obtained. The magnitude of the normal incident reflection coefficient is shown in Figure 4.10. Both 3 and 10 m NSA values are measured following the standard steps in [5]. Four different locations of the Rx antenna in the turntable region are tested: left (L), right (R), front (F), and centre (C). For each location, there are two height values and two polarisations for the Tx antenna. These make 2 × 2 × 4 = 16 cases for each distance, as shown in Table 4.1. A biconical antenna is

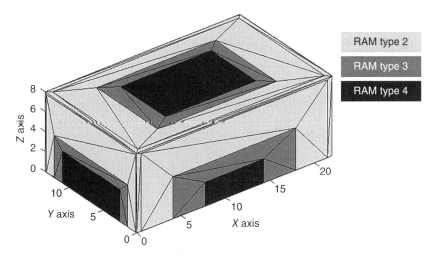

Figure 4.9 Discretised chamber model with different RAMs (represented by the shaded colours in different regions).

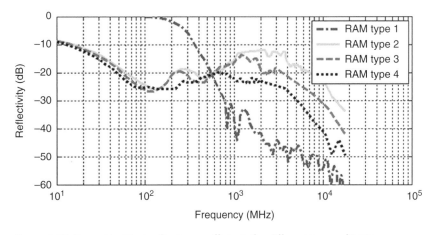

Figure 4.10 Normal incident reflection coefficients for different types of RAM.

Table 4.1 NSA test scenarios.

Polarisation	Tx height	Rx location
Horizontal (H)	Lower (L)	Left (L)
		Right (R)
Vertical (V)	Upper (U)	Front (F)
		Centre (C)

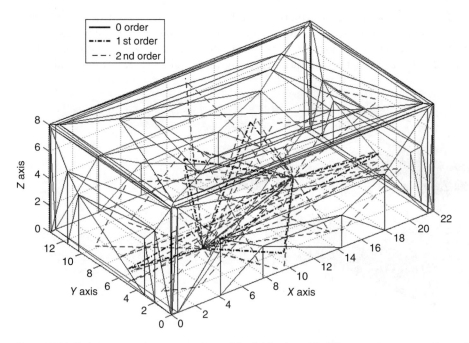

Figure 4.11 Paths connect the source point and the field point with different orders: 1 ray with 0 order, 6 rays with 1st order, and 18 rays with 2nd order.

Table 4.2 Resource comparison.

	CPU	Mesh no.	Memory requirement	Simulation time
GO inverse algorithm	2.33 GHz Two threads	116 Triangles	~200 MB	2.5 min
FDTD	3.0 GHz Four threads	~90 million Hexahedra	~10 GB	~8 h

used in the frequency range of 30 – 200 MHz and an LPDA is used in the frequency range of 200 MHz – 1 GHz. For the 10 m case, the Tx antenna is chosen to be 15° off the axis to test a more general scenario.

After the GO simulation, the paths with different orders for one of the receiving points are shown in Figure 4.11. A hundred sample points are used for the 1–4 m height scanning. The rays up to the 2nd order are considered, and the convex acceleration is used. For each scenario, the resource consumption using the inverse algorithm is shown in Table 4.2. The resource consumption using the traditional finite-difference

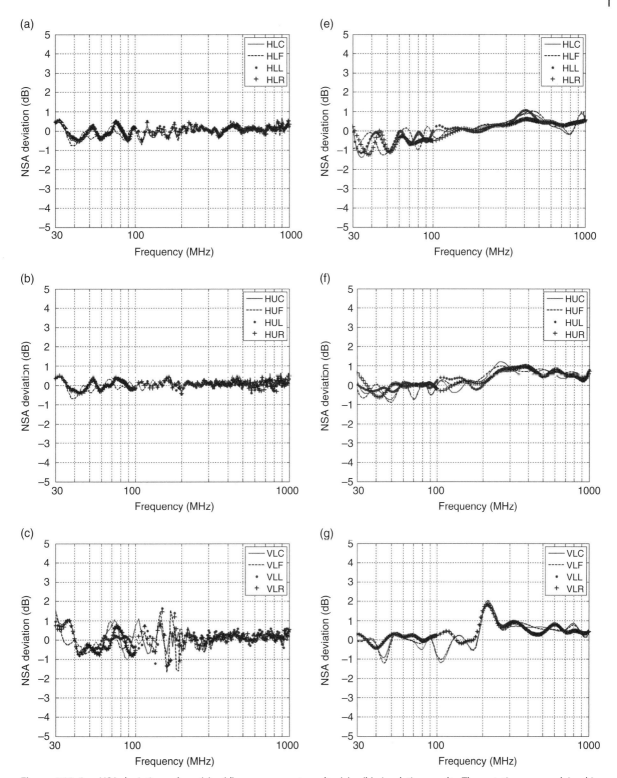

Figure 4.12 3 m NSA deviation values: (a) – (d) measurement results, (e) – (h) simulation results. The notations are explained in Table 4.1, e.g. HLC stands for horizontal polarisation (H), the Tx antenna at the lower (L) height and the Rx antenna is located at the centre (C) of the test region (measurement results from Rainford EMC Systems Ltd).

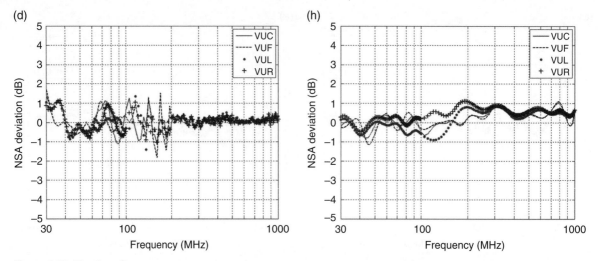

Figure 4.12 (Continued)

time domain (FDTD) has also been estimated and given in the table. It can be seen that the full wave method takes 8 hours and 10 GB memory to complete the simulation and is not a good choice for a chamber designer. For the frequency higher than 1 GHz or a larger chamber the problem will become even worse.

The measurement results for the 3 m NSA values deviated from the CISRP standard [5] are shown in Figure 4.12a–d and the simulation results using FACET are shown in Figure 4.12e–h. The measured and simulated results agree well with each other and the spikes in the measurement results may be due to the unexpected scatters from the complex environment (cables, masts, imperfect ground). Figure 4.13 gives the measured and simulated 10 m NSA deviation. It can be seen that the simulated and measured results are comparable; the peak value may have a slight shift in frequency which is due to the phase error at lower frequencies (mutual coupling, near-field effect). The differences between the simulation and measurement values are within ±2 dB.

4.3 Site Voltage Standing Wave Ratio

4.3.1 SVSWR Definition

The SVSWR is another important parameter to evaluate the performance of a chamber. In the ideal case, SVSWR is 1 (0 dB). The measurement scenario is shown in Figure 4.14. The E-field at the sample points (dots in Figure 4.14b) are recorded in the frequency range of 1 – 18 GHz. The Tx probe/antenna is swept along the measurement axis.

Take measurement positions F1 – F6 in Figure 4.14 as an example, F5 and F1 are measured relative to F6 as follows, moving away from the Tx antenna [5]:

1) F5 = F6 + 2 cm away from the Tx antenna
2) F4 = F6 + 10 cm away from the Tx antenna
3) F3 = F6 + 18 cm away from the Tx antenna
4) F2 = F6 + 30 cm away from the Tx antenna
5) F1 = F6 + 40 cm away from the Tx antenna.

It should be noted that the distances between them are not uniform. This is because we need to sample the E-field in a very wide frequency range; if they are sampled uniformly, the recorded E-field could be always at the

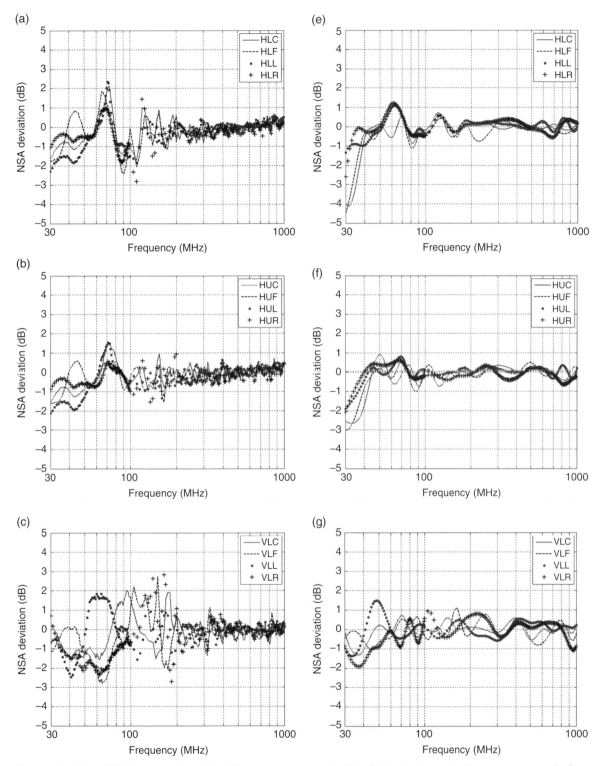

Figure 4.13 10 m NSA deviation values: (a) – (d) measurement results, (e) – (h) simulation results (measurement results from Rainford EMC Systems Ltd).

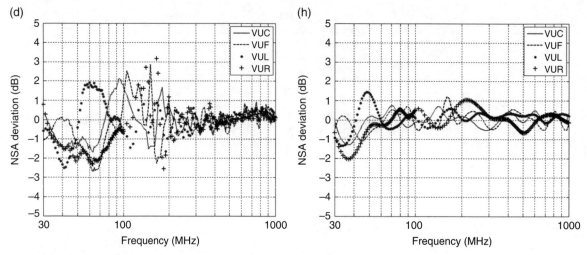

Figure 4.13 (Continued)

peak or trough points of the standing wave at some frequencies. Suppose E_{max} and E_{min} are the measured maximum and minimum E-fields, respectively, along the axis. The SVSWR can be obtained from

$$S_{VSWR} = \frac{E'_{max}}{E'_{min}} = \frac{V'_{max}}{V'_{min}} \tag{4.3}$$

where E'_{max} (V'_{max}) and E'_{min} (V'_{min}) are the normalised measured E-field (voltage)

$$E'_* = E_* \frac{d_{TxRx}}{d_{ref}} \tag{4.4}$$

where d_{TxRx} is the distance between the Tx antenna and the Rx antenna, and $*$ can be max or min. Because the free-space attenuation ($E \propto d^{-1}$) will not be accounted for the calculation of S_{VSWR}, d_{ref} can be chosen as the distance between the Tx antenna and the first receiving point.

The physical meaning of SVSWR is clear: suppose the amplitude of the direct field is E', the amplitude of the extraneous field is E_{extra}. If the direct field and the extraneous field are superimposed coherently, the magnitude of the total field is

$$E'_{max} = E' + E_{extra} \tag{4.5}$$

If the direct field and the extraneous field are superimposed incoherently (out of phase), the magnitude of the total field is

$$E'_{min} = E' - E_{extra} \tag{4.6}$$

Thus the peak-to-peak ripple can be observed as the SVSWR [9]:

$$S_{VSWR}(dB) = 20\log_{10}\frac{E'_{max}}{E'_{min}} = 20\log_{10}\frac{E' + E_{extra}}{E' - E_{extra}} \tag{4.7}$$

As an example, suppose a peak-to-peak ripple of 0.3 dB was observed, which means

$$S_{VSWR}(dB) = 20\log_{10}\frac{1 + E_{extra}/E'}{1 - E_{extra}/E'} = 0.3\,dB \tag{4.8}$$

(a)

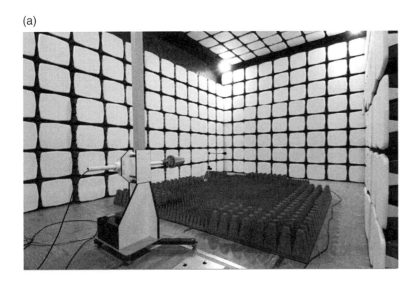

(b)

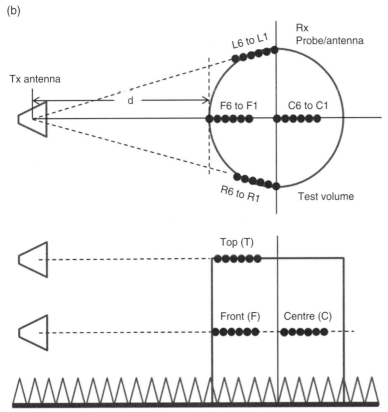

Figure 4.14 SVSWR measurement setup. (a) picture from Rainford EMC Systems Ltd and (b) typical SVSWR measurement scenarios (top view and side view).

thus the extraneous field level can be obtained:

$$20\log_{10}\frac{E_{\text{extra}}}{E'} = 20\log_{10}\frac{10^{0.3/20}-1}{10^{0.3/20}+1} = -35.3\,\text{dB} \tag{4.9}$$

which means the extraneous field level is –35.3 dB lower than the direct field. A plot of peak-to-peak ripple with extraneous field level is shown in Figure 4.15.

4.3.2 SVSWR Simulation and Measurement

A design example is shown in Figure 4.16, where the size of the chamber is 9 m × 6 m × 6 m ($L \times W \times H$). The reflection coefficients of different types of RAMs are given in Figure 4.10. To reduce reflections from the ground plane, part of the ground is covered by RAMs. After applying the inverse method, the rays connecting the Tx and Rx antenna can be found, as shown in Figure 4.17. By repeating the procedure for all the scenarios listed in Table 4.3, comparisons between the measurement and simulation results can be found and are shown in Figure 4.18.

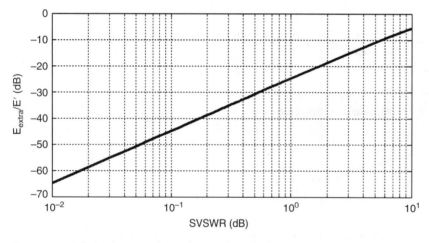

Figure 4.15 Relation between the peak-to-peak ripple (SVSWR) and the extraneous field level.

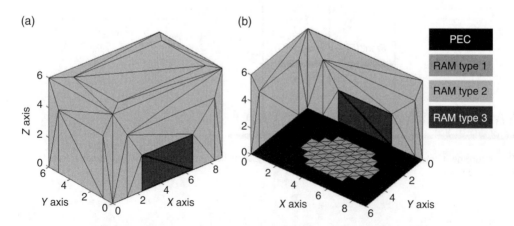

Figure 4.16 Simulation model, different colours represent different types of RAM: (a) top view and (b) bottom view.

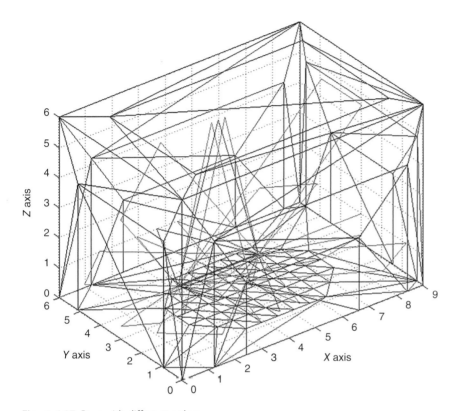

Figure 4.17 Rays with different orders.

Table 4.3 SVSWR test scenarios.

Polarisation	Rx position
Horizontal (H)	Front (F)
	Centre (C)
Vertical (V)	Right (R)
	Left (L)
	Top (T)

It can be seen that the simulated SVSWR is much smaller than the measured value at higher frequencies. This is because, at higher frequencies, the radiation pattern may not be exactly the same as that in the simulation, the tip scattering of RAM could be significant and hard to predict, the reflection coefficient is more sensitive to material properties, and also the supporting objects used in the measurement could become potential scattering sources. It can be seen that, when frequency is smaller than 8 GHz, the differences between the measured values and simulated values are smaller than 2 dB.

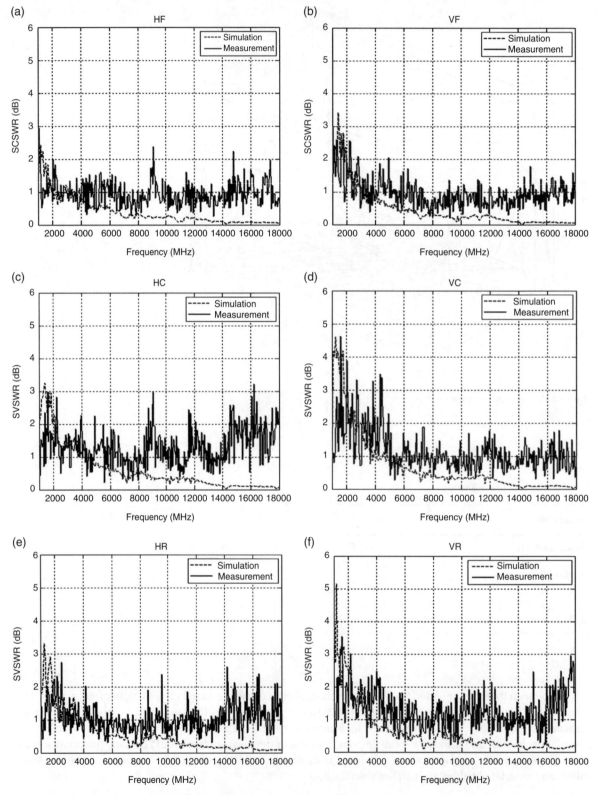

Figure 4.18 Simulated and measured SVSWR. HF means the Tx antenna is horizontal polarisation and the Rx is chosen at the front position in Table 4.3 (measurement results from Rainford EMC Systems Ltd).

(g)

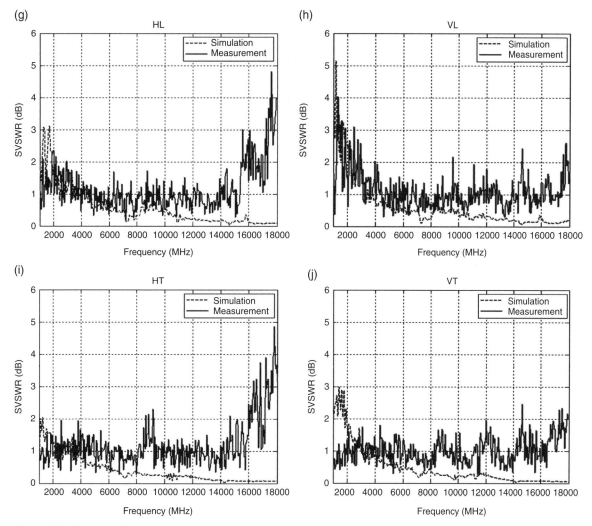

(h)

(i)

(j)

Figure 4.18 (Continued)

4.4 Field Uniformity

4.4.1 FU Definition

The FU measurement is performed by measuring the E-field magnitude at 16 points in the uniform field area (UFA) at each frequency [6], as shown in Figure 4.19. The transmitting antenna is placed at a distance of 3 m from the UFA plane with a height of 1.55 m. The E-field magnitude for at least 12 (75%, can be 100% if required) of the 16 points of the UFA are required to be within the tolerance of $^{-0}_{+6}$ dB [6]. The following steps can be used to extract the FU with 75%. The Matlab code is listed in Appendix A.5.

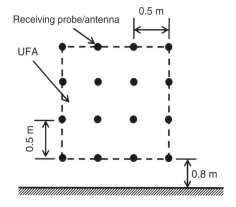

Figure 4.19 FU sample points of the UFA.

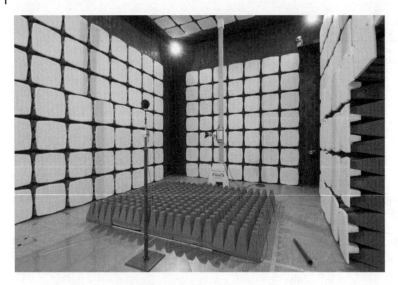

Figure 4.20 FU measurement scenario of an EMC chamber (picture from Rainford EMC Systems Ltd).

1) Sort the 16 field strength readings into an ascending order.
2) Select the first field strength as the reference and calculate the deviation from this for the 12th reading. Record this deviation as Δ_1.
3) Select the second field strength as the reference and calculate the deviation from this for the $2 + 11 = 13$th reading. Record this deviation as Δ_2.
4) Repeat the steps for Δ_i, which is the deviation of the $i + 11$th reading from the ith field strength. The maximum i is 5 when 12 of the 16 points are used.
5) Choose the minimum value of all Δ ($\min(\Delta_i, i = 1..5)$), which is the FU with 75%.

The measurement scenario is shown in Figure 4.20, which is a semi-anechoic chamber for EMC. To improve the FU, part of the ground is covered with RAMs (according to the standard [6]) to reduce the reflection from the ground.

4.4.2 FU Simulation and Measurement

The simulation model is shown in Figure 4.21, where, as can be seen, part of the ground is covered by RAMs. The normal incident reflection coefficients of different RAMs are shown in Figure 4.10. The rays connecting the Tx and Rx antennas are shown in Figure 4.22. The measurement and simulation are repeated for horizontal and vertical polarisations. Comparisons are given in Figure 4.23 with 75% sample points and 100% sample points in the frequency range of 80 MHz–18 GHz. Note that 75% of sample points fall in the range of $^{-0}_{+6}$dB, but not 100% of sample points. This is acceptable according to the standard, since in practice the size of the equipment under test (EUT) may not be the same as the UFA and not all the surface needs to be illuminated (only electronic devices). However, nowadays 100% sample points are also required, which normally means a stricter standard defined by the customer.

The results show that for frequencies higher than 200 MHz and lower than 8 GHz, the difference is smaller than 2 dB for both polarisations. For the frequencies lower than 200 MHz, the error could be due to the high frequency approximation of the GO method. For frequencies higher than 8 GHz, the error could be due to the difference between the simulation and measurement: the radiation pattern of the antenna, the material

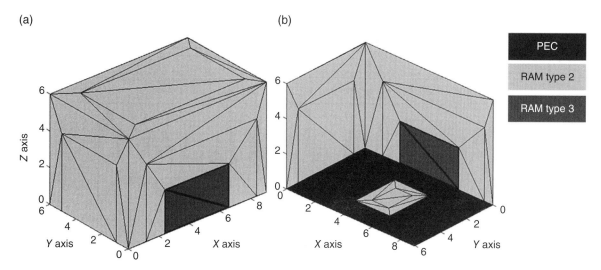

Figure 4.21 Simulation model, different colours represent different types of RAM: (a) top view and (b) bottom view.

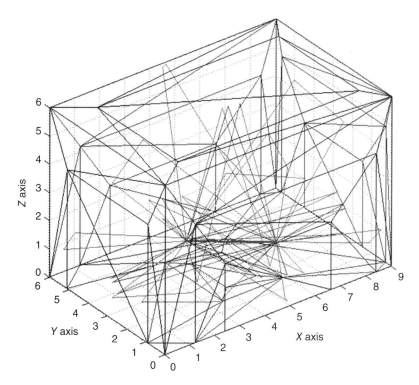

Figure 4.22 Rays with different orders.

(a)

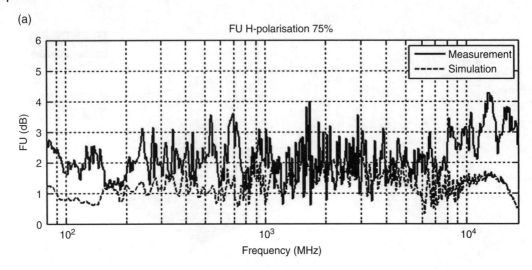

(b)

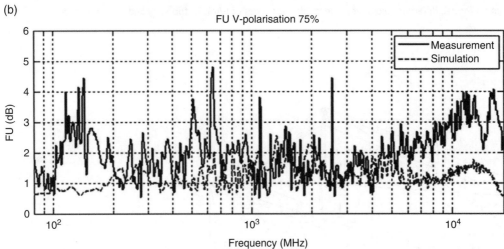

(c)

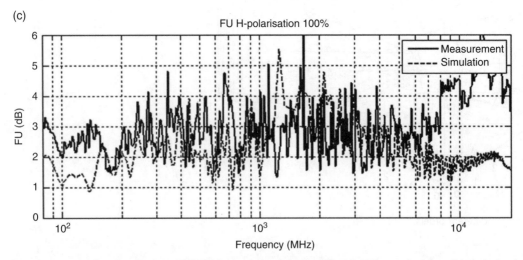

Figure 4.23 Simulated and measured FU: (a) horizontal polarisation FU with 75% sample points, (b) vertical polarisation FU with 75% sample points, (c) horizontal polarisation FU with 100% sample points, and (d) vertical polarisation FU with 100% sample points (measurement results from Rainford EMC Systems Ltd).

(d)

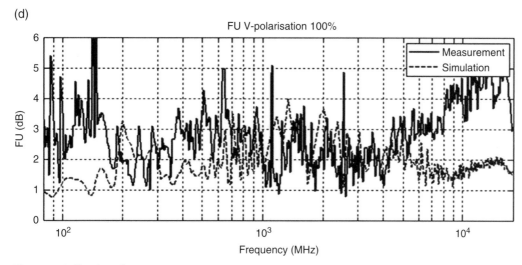

Figure 4.23 (Continued)

property of the RAM, or scattering/reflections from masts and cables. The tip scattering of the RAM is also not considered in simulation.

4.5 Design Margin

An important problem in ACs is the design margin. To quantify the design margin, we need to consider the relationship between the reflection coefficient and the figures of merit. Obviously, if all absorbers on the wall are perfect (no reflection), the site is an ideal site. When the absorbers are not perfect, how the figure of merit deteriorates with the increase in the reflection coefficient is the problem we need to answer, especially at low frequencies when the GO algorithm becomes less accurate.

We have used the GO method to compare the simulation results and measurement results. In this section, we use the finite integration time domain (FITD) method in the CST Microwave Studio® to investigate how the NSA deteriorates with the increase in the reflection coefficient. Suppose the model we investigated in this section is shown in Figure 4.24, where the dimensions are 9 m × 6 m × 6 m ($L \times W \times H$), which is the same as in Figure 4.17. The centre of the test area is 2.85 m from the wall and has a radius of 1 m, dipole antennas are used to perform the simulation, the height of the Tx antenna is 2 m and the Rx antenna is scanning from 1 to 4 m. For the ideal case, all the boundary conditions are set as open. We first simulate the idea case and then use this as the reference to investigate the NSA as a function of the reflection coefficient. In the simulation, the Rx antenna

(a)

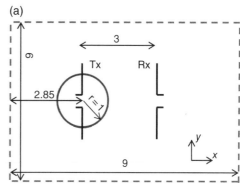

(b)

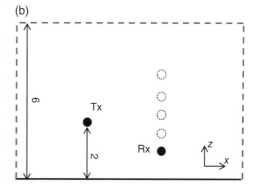

Figure 4.24 AC model with size 9 m × 6 m × 6 m. The open boundaries are represented by dashed lines, the ground plane is set as a PEC, antennas are placed with horizontal polarisation: (a) plan view and (b) longitudinal section view.

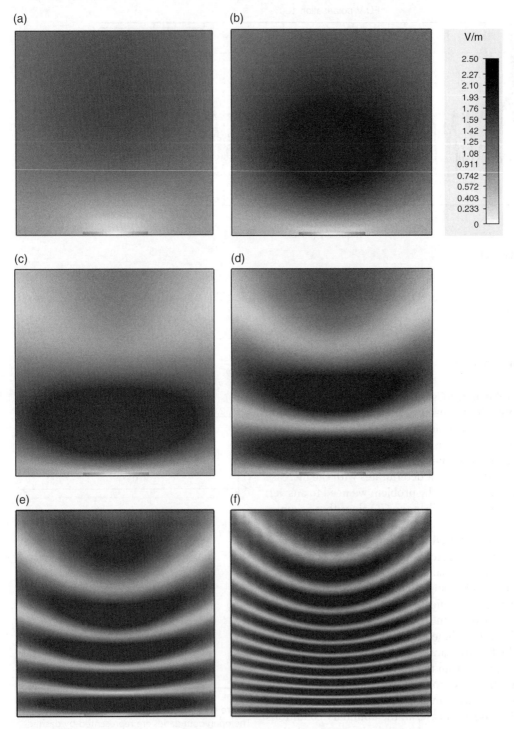

Figure 4.25 The magnitude of the E-field in the monitor plane. The monitor plane is set at 3 m from the Tx antenna and in the yoz plane, horizontal polarisation, Tx antenna is at the centre of the test area: (a) 30 MHz, (b) 60 MHz, (c) 100 MHz, (d) 200 MHz, (e) 400 MHz, and (f) 1000 MHz.

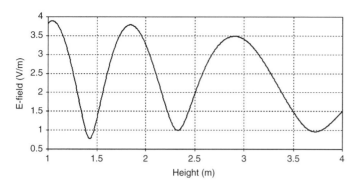

Figure 4.26 The magnitude of the E-field from 1 to 4 m height at 400 MHz.

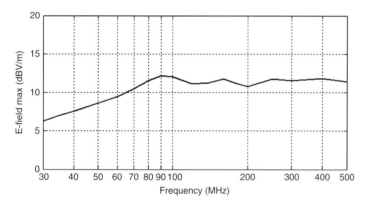

Figure 4.27 The maximum E-field in the height of 1–4 m in the frequency range of 30 – 500 MHz. The Tx antenna is horizontal polarised.

Figure 4.28 Normal reflection coefficient simulation of a planar absorber.

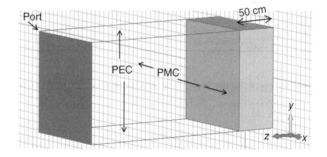

Figure 4.29 The whole chamber with planar absorber, the Tx antenna is horizontal polarised.

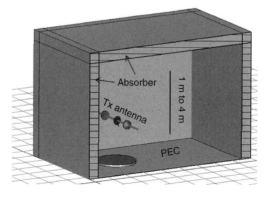

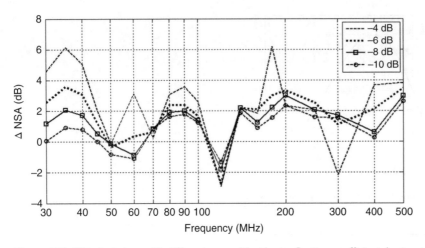

Figure 4.30 NSA deviations with different normal incident reflection coefficient, horizontal polarisation.

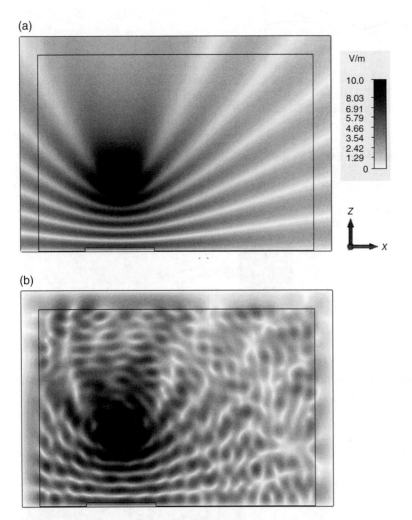

Figure 4.31 E-field magnitude distribution at 500 MHz, horizontal polarisation, (a) perfect absorber and (b) absorber with −10 dB normal incident reflection coefficient.

is not necessary; a field probe/monitor is used to record all the E-field values in the yoz plane of the Rx antenna. The length of the antenna is tuned to make sure the antenna is well-matched at each frequency. Although the full wave simulation is resource consuming, it can still handle this size of problem at 1000 MHz; about 7 GB memory is required to simulate this problem.

After simulation, the E-field in the monitor plane from 1 to 4 m can be obtained; the maximum value can be used to calculate the NSA value. As we are interested in the deviation of the NSA values when the absorbers are not perfect, it is not necessary to convert the maximum E-field to the NSA value, the maximum E-field can be compared directly. The magnitude of the E-field in the monitor plane is shown in Figure 4.25 at frequencies 30, 60, 100, 200, 400, and 1000 MHz. Because low frequency behaviour is concerned, the range of 30 − 500 MHz is chosen to study the NSA. It can also be found from Figures 4.12 and 4.13 that when the frequency is higher than 500 MHz, the NSA deviation is small. This is because the dielectric (foam) absorber can have a very good performance at high frequencies, while at low frequencies the ferrite tiles dominate the absorption.

When the E-field on the monitor plane is obtained, the E-field from height 1–4 m can be extracted. Figure 4.26 shows the E-field at 400 MHz from 1 to 4 m height. The maximum value can be obtained, which is 3.89 V m^{-1}. By extracting the maximum E-field for all the other frequencies specified in [5] in the range of 30 − 500 MHz (30, 35, 40, 45, 50, 60, 70, 80, 90, 100, 120, 140, 160, 180, 200, 250, 300, 400, 500 MHz), all the maximum E-field can be obtained, which is illustrated in Figure 4.27. Figure 4.27 can be used as the reference value to study the effect of the reflection coefficient.

The reflection coefficient of the imperfect absorbers can be realised by tuning the material parameters of the absorber. The thickness of the absorber is fixed at 50 cm and the reflection coefficient can be obtained by using the transverse electromagnetic (TEM) waveguide method. The boundaries at y max and y min are set as PEC while the boundaries at x max and x min are set as perfect magnetic conductors (PMCs), the excitation port is set as z max and z min is set as a PEC. Figure 4.28 illustrates the settings. We only consider the reflection coefficient at normal incident angle as it is mostly the parameter provided by the RAM supplier. After tuning the reflection coefficient to a specific value, the absorbers are attached to the inner surface of the chamber (shown in Figure 4.29), then the whole model is simulated.

The maximum E-field with different normal incident reflection coefficient can be obtained and compared to the reference value in Figure 4.27. We denote the E-field in Figure 4.27 as E_{maxref} and the E-field with a different normal incident reflection coefficient as $E_{\text{max}-*\text{dB}}$, where '*' means the magnitude of the reflection coefficient. By tuning the normal incident reflection coefficient, the deviation of the maximum E-field can be obtained as

$$\Delta NSA_{-*\text{dB}} = E_{\text{max}-*\text{dB}} - E_{\text{maxref}} \tag{4.10}$$

which is also the deviation of NSA values. ΔNSA_{-*dB} with different normal incident reflection coefficient is shown in Figure 4.30. From Figure 4.30, we can conclude that, for horizontal polarisation, if the NSA deviation is required to be smaller than ±4 dB, the normal incident reflection coefficient should be at least smaller than −6 dB. Because the normal incident reflection coefficient is lower than the oblique incident reflection, when the reflection coefficient is −10 dB, the maximum deviation is about 2.6 dB. The E-field magnitude distribution is also illustrated in Figures 4.31. As expected, compared to the ideal absorber (open boundary), the distribution is blurred because of the superposition from reflected waves.

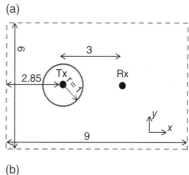

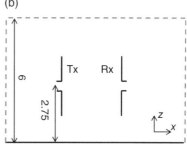

Figure 4.32 AC model with size 9 m × 6 m × 6 m. The open boundaries are represented by dashed lines, the ground plane is set as PEC, antennas are placed with vertical polarisation, (a) plan view and (b) longitudinal section view.

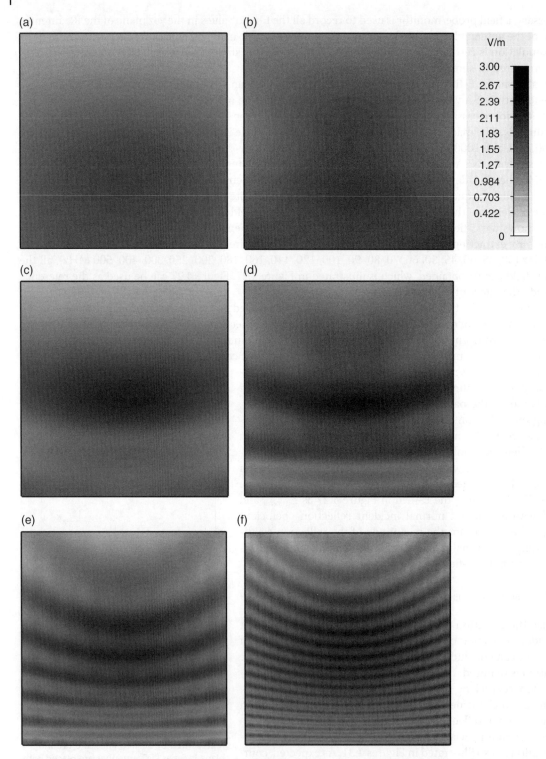

Figure 4.33 The magnitude of the E-field in the monitor plane. The monitor plane is set at 3 m from the Tx antenna and in the yoz plane, vertical polarisation, Tx antenna is at the centre of the test area: (a) 30 MHz, (b) 60 MHz, (c) 100 MHz, (d) 200 MHz, (e) 400 MHz, and (f) 1000 MHz.

For vertical polarisation scenario in Figure 4.32, the centre of the Tx antenna is fixed at a height of 2.75 m, the Rx antenna is swept from 1 to 4 m, but at low frequencies the minimum height is higher than 1 m to make sure the Rx is a certain distance from the ground [5]. The magnitude of the E-field in the monitor plane is shown in Figure 4.33, and the maximum field from 1 to 4 m is given in Figure 4.34. By repeating a similar procedure in horizontal polarisation, the deviation of NSA values with a different normal reflection coefficient is illustrated in Figure 4.35. It can be seen that when the normal reflection coefficient is −14 dB, we still have a maximum NSA deviation of 3 dB. Thus we have a 1 dB safety margin for −14 dB reflection coefficient. If a larger safety margin is required, the reflection coefficient should be even lower. For this scenario, the NSA of vertical polarisation is more sensitive to the normal reflection coefficient than the horizontal polarisation; the requirement for the vertical polarisation overwrites the requirement for the horizontal polarisation. The E-field magnitude distribution for vertical polarisation is illustrated in Figure 4.36.

In this section, we have investigated the design margin of the NSA value for a given size of chamber using the full wave method, and only one pair of Tx and Rx positions is simulated. It would be necessary to check more Tx and Rx positions in practice, and this could lead to a stricter requirement for the reflection coefficient. The safety margin could also be different for different sizes of chamber. We assume the RAMs are uniform in the

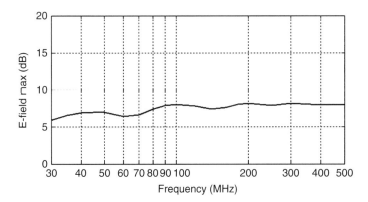

Figure 4.34 The maximum E-field in the height of 1–4 m in the frequency range of 30 – 500 MHz. The Tx antenna is vertical polarised.

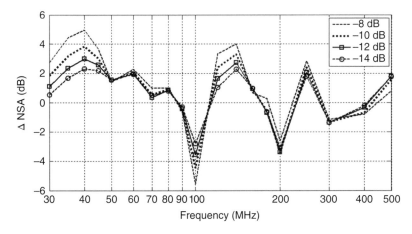

Figure 4.35 NSA deviations with different normal incident reflection coefficient, vertical polarisation.

(a)

(b)

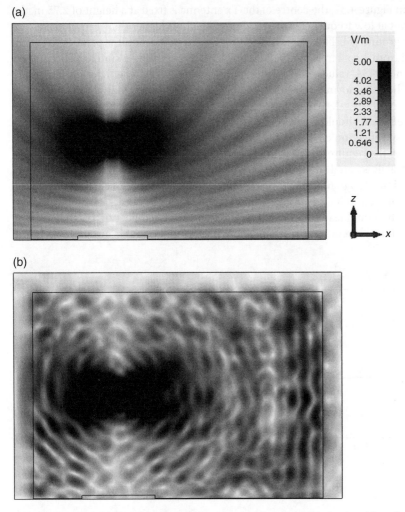

Figure 4.36 E-field magnitude distribution at 500 MHz, vertical polarisation: (a) perfect absorber and (b) absorber with −14 dB normal incident reflection coefficient.

simulation, and all the RAMs have the same thickness and the same normal reflection coefficient. In practice, nonuniform layout of RAMs could also be possible and improve the performance of the chamber.

4.6 Summary

In this chapter, the definitions of NSA, SVSWR, and FU have been given, typical chamber designs have been presented, measurement results and simulation results obtained from the developed CAD tool have been compared. For the NSA values, the difference between the simulated values and measured values can be smaller than ±2 dB in the whole frequency range (30 MHz – 1 GHz) in the selected example. For the SVSWR values, a good agreement has been obtained when the frequency is between 1 and 14 GHz, and the difference is smaller than 2 dB when frequency is lower than 8 GHz. For the FU values, the difference is smaller than 2 dB in the frequency range of 300 MHz – 8 GHz. A typical template of a test report is included in Appendix C.

It is important to note that the GO is a high frequency approximation method. In the measurement part we have compared the simulation and measurement results which are in good agreement. A potential problem is that at lower frequencies the near-field mutual coupling and diffraction may affect the results (the prediction accuracy is expected to be reduced using the GO method). Another issue is that the tip scattering of RAM at higher frequencies becomes significant and it is not considered in this model. When the tip scattering becomes the major contribution for the unexpected field (in millimetre waves), it will limit the boundary of the high frequency of this method. This could be in the region of statistical electromagnetics but not deterministic electromagnetics. There are also detailed structures which could not be included in the simulation because of the complexity of the measurement scenario (masts, cables, etc.).

The design margin for a given size of chamber has been quantified and discussed; a full wave method has been used to simulate the NSA deviations. For the given size of the chamber, in order to have a less than 3 dB deviation of NSA values, the normal reflection coefficient should be at least −14 dB. Although the margin is limited for a specific scenario – the size of the chamber, the Rx, and Tx antenna positions, it still provides a guideline to analyse the design margin at low frequencies (≤500 MHz). In practice, because of other scatterers (masts, cables, etc.) and imperfect ground (discontinuity at the turntable circumference), a certain safety margin is necessary which makes a stricter requirement for the RAM reflection coefficient. The safety margin of SVSWR and FU at high frequencies can also be investigated by using the GO method; the procedures are similar and not repeated. Although we have not selected a full AC for this investigation, the approach presented here is applicable and better results are expected.

References

1 King, H. E., Shimabukuro, F. I., and Wong, J. L. (1967). Characteristics of a tapered anechoic chamber. *IEEE Transactions on Antennas and Propagation* 15 (3): 488–490.

2 Galindo-Israel, V., Rengarajan, S. R., Imbriale, W. A., and Mittra, R. (1991). Offset dual-shaped reflectors for dual chamber compact ranges. *IEEE Transactions on Antennas and Propagation* 39 (7): 1007–1013.

3 Hemming, L.H. (1985). Anechoic chamber. US Patent 4 507 660, March 26 1985.

4 Sanchez, G.A. (1997). Geometrically optimized anechoic chamber. US Patent 5 631 661, May 20, 1997.

5 CISPR 16-1-4. (2012). Specification for radio disturbance and immunity measuring apparatus and methods – Part 1–4: Radio disturbance and immunity measuring apparatus – Antennas and test sites for radiated disturbance measurements. 3.1e. IEC Standard.

6 IEC 61000-4-3. (2008). Electromagnetic compatibility (EMC) – Part 4–3: Testing and measurement techniques – Radiated, radio-frequency, electromagnetic field immunity test. 3.1e. IEC Standard.

7 Kodali, V. P. (2001). *Engineering Electromagnetic Compatibility: Principles, Measurements, Technologies, and Computer Models*, 2e. New York: Wiley-IEEE Press.

8 Xu, Q., Huang, Y., Zhu, X. et al. (2016). Building a better anechoic chamber: a geometric optics-based systematic solution, simulated and verified. *IEEE Antennas and Propagation Magazine* 58 (2): 94–119.

9 Hemming, L. H. (2002). *Electromagnetic Anechoic Chambers: A Fundamental Design and Specification Guide*. New York: Wiley-IEEE Press.

5

Fundamentals of the Reverberation Chamber

5.1 Introduction

Unlike an anechoic chamber (AC), a reverberation chamber (RC) has a completely different boundary and works in a very different way although it is also an indoor facility for electromagnetic measurements. In this chapter, fundamental theories for a RC are reviewed and summarised. In the AC, deterministic theory is used, but in the RC, statistical theories are normally applied. There are reasons why statistical theory is pursued: (i) the problem could be too complex to solve by using deterministic methods, (ii) the problem is too sensitive to some parameters which make the deterministic solution meaningless, and (iii) in practice, not all parameters are well-known. Especially in electrically large electromagnetic system, these three factors are obvious.

In order to have a step-by-step understanding, the deterministic theory is introduced first, and then followed by the statistical theory. Frequently used physical quantities are introduced, and it can be found that most physical quantities in the RC have statistical meaning. This is very different from the quantities in the AC. Frequently used figures of merit in RC are also introduced and explained.

5.2 Resonant Cavity Model

An RC can be considered as an electrically large resonant cavity. For some regular shape cavities such as rectangular, cylindrical, and spherical, the analytical solutions can be obtained by using the well-known method of separation of variables (Fourier method) [1–5]. Generally, in an arbitrarily shaped cavity shown in Figure 5.1, the electromagnetic wave satisfies the wave equation in the frequency domain

$$\left(\nabla^2 + k^2\right)\mathbf{E} = 0 \tag{5.1}$$

where k is the wave number $k = 2\pi/\lambda = \omega\sqrt{\varepsilon\mu}$, ε and μ are the material permittivity and permeability in the cavity, respectively, and $\mathbf{E}$ is the electric field (E-field) in the frequency domain $\mathbf{E}(j\omega)$. If we consider a source free problem in the cavity, k is also called eigenvalues. In the volume we have

$$\nabla\bullet\mathbf{E}\big|_{\text{in V}} = 0 \tag{5.2}$$

If the boundary is a perfect electric conductor (PEC), the tangential component of the E-field on the boundary is zero:

$$\hat{\mathbf{n}} \times \mathbf{E}\big|_{\text{on S}} = 0 \tag{5.3}$$

Specifically, for a rectangular shape with dimensions of $a \times b \times d$ shown in Figure 5.2, (5.1)–(5.3) have transverse electric (TE) and transverse magnetic (TM) solutions.

Anechoic and Reverberation Chambers: Theory, Design, and Measurements, First Edition. Qian Xu and Yi Huang.
© 2019 John Wiley & Sons Ltd. Published 2019 by John Wiley & Sons Ltd.

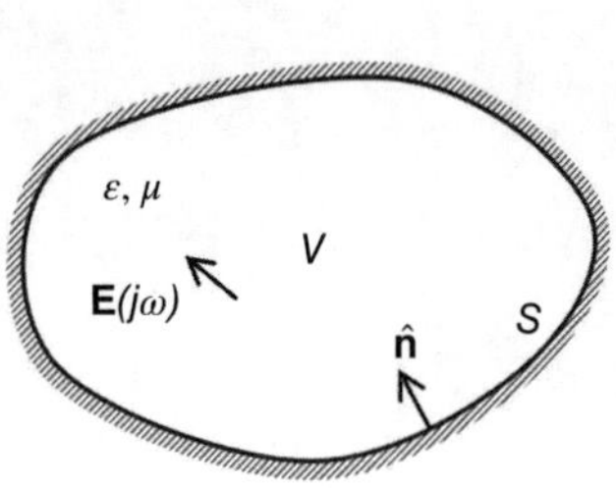

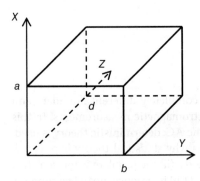

Figure 5.1 Arbitrarily shaped cavity.

Figure 5.2 Rectangular cavity.

For the TE mode

$$E_{xmnp}^{TE} = -\frac{j\omega_{mnp}\mu k_y H_0}{k_{mnp}^2 - k_z^2} \cos\left(\frac{m\pi}{a}x\right) \sin\left(\frac{n\pi}{b}y\right) \sin\left(\frac{p\pi}{d}z\right)$$

$$E_{ymnp}^{TE} = \frac{j\omega_{mnp}\mu k_x H_0}{k_{mnp}^2 - k_z^2} \sin\left(\frac{m\pi}{a}x\right) \cos\left(\frac{n\pi}{b}y\right) \sin\left(\frac{p\pi}{d}z\right)$$

$$E_{zmnp}^{TE} = 0$$

$$H_{xmnp}^{TE} = -\frac{k_x k_y H_0}{k_{mnp}^2 - k_z^2} \sin\left(\frac{m\pi}{a}x\right) \cos\left(\frac{n\pi}{b}y\right) \cos\left(\frac{p\pi}{d}z\right)$$

$$H_{ymnp}^{TE} = \frac{k_y k_z H_0}{k_{mnp}^2 - k_z^2} \cos\left(\frac{m\pi}{a}x\right) \sin\left(\frac{n\pi}{b}y\right) \sin\left(\frac{p\pi}{d}z\right)$$

$$H_{zmnp}^{TE} = H_0 \cos\left(\frac{m\pi}{a}x\right) \cos\left(\frac{n\pi}{b}y\right) \sin\left(\frac{p\pi}{d}z\right) \qquad (5.4)$$

The allowable values of the mode numbers are $m = 0, 1, 2, ..., n = 0, 1, 2, ...,$ and $p = 1, 2, 3, ...$ with the exception that $m = n = 0$ is not allowed.

For the TM mode

$$E_{xmnp}^{TM} = -\frac{k_x k_z E_0}{k_{mnp}^2 - k_z^2} \cos\left(\frac{m\pi}{a}x\right) \sin\left(\frac{n\pi}{b}y\right) \sin\left(\frac{p\pi}{d}z\right)$$

$$E_{ymnp}^{TM} = \frac{k_y k_z E_0}{k_{mnp}^2 - k_z^2} \sin\left(\frac{m\pi}{a}x\right) \cos\left(\frac{n\pi}{b}y\right) \sin\left(\frac{p\pi}{d}z\right)$$

$$E_{zmnp}^{TM} = E_0 \sin\left(\frac{m\pi}{a}x\right) \sin\left(\frac{n\pi}{b}y\right) \cos\left(\frac{p\pi}{d}z\right)$$

$$H_{xmnp}^{TM} = -\frac{j\omega_{mnp}\varepsilon k_y E_0}{k_{mnp}^2 - k_z^2} \sin\left(\frac{m\pi}{a}x\right) \cos\left(\frac{n\pi}{b}y\right) \cos\left(\frac{p\pi}{d}z\right)$$

$$H_{ymnp}^{TM} = \frac{j\omega_{mnp}\varepsilon k_x E_0}{k_{mnp}^2 - k_z^2} \cos\left(\frac{m\pi}{a}x\right) \sin\left(\frac{n\pi}{b}y\right) \cos\left(\frac{p\pi}{d}z\right)$$

$$H_{zmnp}^{TM} = 0 \tag{5.5}$$

The allowable values of the mode numbers are $m = 1, 2, 3, \ldots, n = 1, 2, 3, \ldots,$ and $p = 0, 1, 2, \ldots.$ $m, n,$ and p satisfy the equation

$$k_{mnp}^2 = \left(\frac{m\pi}{a}\right)^2 + \left(\frac{n\pi}{b}\right)^2 + \left(\frac{p\pi}{d}\right)^2 \tag{5.6}$$

The resonant frequencies f_{mnp} can be obtained as

$$f_{mnp} = \frac{k_{mnp}}{2\pi\sqrt{\varepsilon\mu}} = \frac{1}{2\sqrt{\varepsilon\mu}}\sqrt{\left(\frac{m}{a}\right)^2 + \left(\frac{n}{b}\right)^2 + \left(\frac{p}{d}\right)^2} \tag{5.7}$$

When $m, n,$ and p are all nonzero, the TE mode and TM mode have the same resonant frequency, which means the two modes are degenerate. If $a < b < d$, the lowest resonant frequency occurs for the TE_{011} mode ($m = 0$, $n = 1$, $p = 1$).

From the resonant mode theory it can be understood that the field in the cavity can be decomposed into the superposition of modes with different weight (amplitude)

$$\mathbf{E}(x,y,z) = \sum_{m,n,p} A_{mnp}\mathbf{E}_{mnp}^{TE} + B_{mnp}\mathbf{E}_{mnp}^{TM} \tag{5.8}$$

where the amplitude of each mode can be found by using the mode orthogonality

$$\iiint_V \mathbf{E}_{mnp}^{TE}\cdot\mathbf{E}_{m'n'p'}^{TE*}dV = 0, \iiint_V \mathbf{E}_{mnp}^{TE}\cdot\mathbf{E}_{m'n'p'}^{TM*}dV = 0, \iiint_V \mathbf{E}_{mnp}^{TM}\cdot\mathbf{E}_{m'n'p'}^{TM*}dV = 0 \tag{5.9}$$

when $m \neq m'$ or $n \neq n'$ or $p \neq p'$. '$*$' means the complex conjugation. By applying this orthogonality to both sides of (5.8), A_{mnp} and B_{mnp} can be expressed as

$$A_{mnp} = \iiint_V \mathbf{E}(x,y,z)\cdot\mathbf{E}_{mnp}^{TE*}dV, B_{mnp} = \iiint_V \mathbf{E}(x,y,z)\cdot\mathbf{E}_{mnp}^{TM*}dV \tag{5.10}$$

More detailed information about the electromagnetic fields generated by a given source/antenna inside an RC can be found from references such as [5–7]. It is important to understand that the total fields are the combination of all the cavity modes. When a stirrer is used to stir the field inside an RC, the weighting coefficients for many modes are changed.

In practice, we are normally interested in the mode number rather than the detailed E-field distribution of a single mode. The total number of modes N with eigenvalues k_{mnp} less than or equal to k can be counted by using a computer, and the smooth approximation has been obtained in [8]:

$$N_s(k) = \frac{abd}{3\pi^2}k^3 - \frac{a+b+d}{2\pi}k + \frac{1}{2} \tag{5.11}$$

where the subscript 's' means smooth. It can be written as a function of frequency:

$$N_s(f) = \frac{8\pi abd}{3}\frac{f^3}{c^3} - (a+b+d)\frac{f}{c} + \frac{1}{2} \tag{5.12}$$

where c is the speed of light in the cavity. The smoothed mode density can be obtained by differentiating (5.12) with respect to f:

$$D_s(f) = \lim_{\Delta f \to 0}\frac{\Delta N_s(f)}{\Delta f} = 8\pi abd\frac{f^2}{c^3} - \frac{a+b+d}{c} \tag{5.13}$$

The mode density tells us how many modes can be excited in a given bandwidth. If the mode density is low, the excited mode number could be low, which means the independent sample number is low and this will lead a statistical nonuniformity.

Generally, for cavities of general shape, according to Weyl's law [9, 10], the smoothed mode number depends on the volume of the cavity $V = abd$, but not the detailed dimensions:

$$N_W(f) = \frac{8\pi V}{3}\frac{f^3}{c^3} \tag{5.14}$$

and the smoothed mode density is

$$D_W(f) = 8\pi V\frac{f^2}{c^3} \tag{5.15}$$

As expected, when the cavity is electrically large, (5.14) keeps the high order term and agrees well with (5.12). Results for the cylindrical and spherical shapes can be found in [3] and they also agree with Weyl's law. The number of degenerated modes should be considered as we expect the modes to be uniformly distributed over different frequencies. If there are too many degenerated modes the uniformity could not be guaranteed for different frequencies. Thus the spherical cavities are rarely used as an RC. The smoothed mode density of (5.15) is illustrated in Figure 5.3 with different volumes of 1, 10, 100, and 1000 m^3. As can be seen, when the volume is 1 m^3, if the mode density is required to be greater than 1 mode/MHz, the frequency needs to be higher than 1 GHz. If the frequency is lowered by a factor of x, the volume of the RC needs to be increased by a factor of x^2 to maintain the mode density.

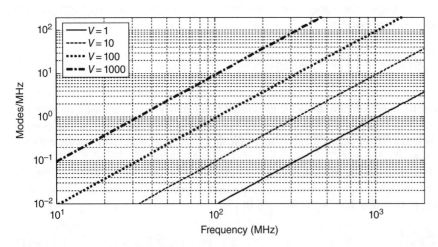

Figure 5.3 Mode density vs frequency of different RC volume.

Figure 5.4 Mode number counting when m, n and p are large.

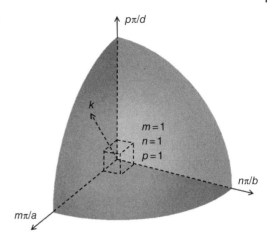

An interesting method to count the mode number in (5.14) is to use the graph shown in Figure 5.4 [8]. The volume of a quarter of a sphere with radius k is $\pi k^3/3$, each small cube has a volume of $\pi^3/(abd)$, and the vertices of the small cube represent each possible combination of m, n and p. Since each vertex is shared by eight neighbour cubes and each cube has eight vertices, the total number of vertices in the sphere can be obtained as

$$N = \frac{\pi k^3}{3} \bigg/ \frac{\pi^3}{abd} \times 8/8 = \frac{k^3 abd}{3\pi^2} \tag{5.16}$$

which is the same as (5.14) if written in the function of frequency.

The quality factor (Q factor) is defined as the ratio between the stored energy and the dissipated energy in one cycle (a factor of 2π is included):

$$Q = 2\pi \frac{\bar{W}}{TP_d} = \omega \frac{\bar{W}}{P_d} \tag{5.17}$$

where $\bar{W}$ is the time-averaged total stored energy in the cavity, T is the period, ω is the angular frequency, and P_d is the dissipated power. For rectangular cavities, the Q factor can be derived analytically for TE and TM modes [1–3]:

$$Q_{mnp}^{TE} = \omega \frac{\mu \iiint\limits_{V} \mathbf{H}_{mnp}^{TE} \cdot \mathbf{H}_{mnp}^{TE*} dV}{R_s \oiint\limits_{S} \mathbf{H}_{mnp}^{TE} \cdot \mathbf{H}_{mnp}^{TE*} dS} \tag{5.18}$$

where μ is the material dielectric permeability in the cavity, R_s is the surface resistance

$$R_s \cong \mathrm{Re}(\eta) \cong \sqrt{\frac{\omega \mu_0}{2\sigma}} \tag{5.19}$$

$\mu_0 = 4\pi \times 10^{-7}$ H/m is the permeability of free space, and σ is the conductivity of the wall of the cavity. By substituting (5.4) into (5.18), the Q factor of the TE mode can be obtained as [1–3]

$$Q_{mnp}^{TE} = \omega \frac{\eta\, abd k_{xy}^2\, k_{mnp}^3}{4R_s \left[bd \left(k_{xy}^4 + k_y^2 k_z^2 \right) + ad \left(k_{xy}^4 + k_z^2 \right) + ab k_{xy}^2 k_z^2 \right]} \tag{5.20}$$

where $\eta = \sqrt{\mu/\varepsilon}$ and $k_{xy}^2 = k_x^2 + k_y^2$. Similarly, for TM mode

$$Q_{mnp}^{TM} = \omega \frac{\mu \iiint_V \mathbf{H}_{mnp}^{TM} \cdot \mathbf{H}_{mnp}^{TM*} dV}{R_s \oiint_S \mathbf{H}_{mnp}^{TM} \cdot \mathbf{H}_{mnp}^{TM*} dS} = \frac{\eta abd k_{xy}^2 k_{mnp}}{4R_s \left[k_x^2 b(a+d) + k_y^2 a(b+d) \right]} \tag{5.21}$$

In the RC, when the cavity is electrically large, we are more interested in the composite Q factor when many modes are excited. This can be found by averaging the Q factors over a small range of k [8]:

$$\widetilde{Q} = \frac{1}{\langle 1/Q \rangle} = \frac{3\eta kabd}{4R_s A} \frac{1}{1 + \frac{3\pi}{8k} \left(\frac{1}{a} + \frac{1}{b} + \frac{1}{d} \right)} \tag{5.22}$$

where $A = 2(ab + bd + ad)$ is the surface area.

Generally, for electrically large cavities of general shape, the composite Q factor has been obtained as [3, 11]

$$\widetilde{Q} = \frac{3V}{2\mu_r \delta A} \tag{5.23}$$

where $\delta = \sqrt{2/\omega\mu\sigma}$ is the skin depth, $\mu_r = \mu/\mu_0$ is the relative permeability, and A is the inner surface area of the cavity.

A quantity closely related to Q factor is the decay constant (or time constant) τ:

$$\tau = \frac{Q}{\omega} \tag{5.24}$$

This can be derived from the differential equation of

$$d\bar{W} = -P_d dt \tag{5.25}$$

which means the reduced energy is the dissipated power times the time duration. By applying (5.17) we have

$$\frac{d\bar{W}}{dt} = -\frac{\omega}{Q} \bar{W} \tag{5.26}$$

With a given initial condition $\bar{W}|_{t=0} = \bar{W}_0$, (5.26) has the solution of

$$\bar{W} = \bar{W}_0 \exp(-t/\tau), \text{where } \tau = \frac{Q}{\omega} \tag{5.27}$$

As can be seen, the energy decays exponentially, and the decay speed is determined by the decay constant or Q factor.

From the Q factor, the average mode bandwidth is defined as the bandwidth over which the excited power in a particular cavity mode with resonant frequency f_0 is larger than half the excited power at f_0 [3, 12, 13]:

$$\frac{\Delta f}{f} = \frac{1}{Q} \tag{5.28}$$

The average mode bandwidth can be used to estimate how many modes are excited in an RC, and the excited mode number $N_{\Delta f}$ can be obtained from (5.15) and (5.28) [14]:

$$N_{in \, \Delta f} = D_w \Delta f = \frac{8\pi V f^3}{Qc^3} \tag{5.29}$$

The cavity model has also been used in the full wave simulation of the RC (the RC simulation is discussed in Chapter 6) and the use of the cavity Green's function has proved the effectiveness of the source stirring technique [6, 7].

5.3 Ray Model

The RC can be modelled by using the ray theory or image theory, and the effectiveness of this method has been proven in [15–21]. The wave propagation in the RC can also be simulated by using the ray tracing method in Section 2.4.3.

The E-field at the Rx antenna can be obtained from the summation of the E-field on all the rays, as shown in Figure 5.5. The ray number can be huge, and only a few rays are illustrated. From the simulation of the ray model, the time domain behaviour can be fitted by tuning the reflection coefficient R, and the time constant can be controlled. Suppose the RC has dimensions $a \times b \times d$ and d is the smallest dimension, the time between two reflections is at least d/c, and the time constant can be related to the reflection coefficient as [17]

$$\tau \approx -\frac{d}{c\ln R} \tag{5.30}$$

where c is the speed of light, or [3, 22]

$$\tau \approx -\frac{l_c}{c\ln R} \tag{5.31}$$

where l_c is the mean-free path between wall reflections. For a rectangular room [23]

$$l_c = \frac{4V}{A} \tag{5.32}$$

where A is the inner surface area of the room. From (5.30)–(5.32), Q can also be controlled by using $Q = \omega\tau$. This is very useful, as all the loss in the RC can be equivalent to the loss on the walls and this equivalence does not affect the statistical behaviour of the RC.

The mean-free path length can be related to the characteristic room time t_c as $t_c = 2l_c/c$, and it has been shown that the reverberation condition occurs after about $5t_c$ [24–27].

Figure 5.5 Ray model of the RC.

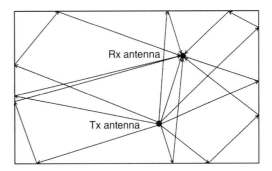

5.4 Statistical Electromagnetics

5.4.1 Plane-Wave Spectrum Model

In the ray model, the superposition of finite plane waves is considered. If the plane wave number is generalised to infinite, we have the well-developed plane-wave spectrum model from D.A. Hill [3].

The E-field in the RC can be represented by the superposition of plane waves coming from random angles θ, φ with random amplitude, phase, and polarisation $\mathbf{F}(\Omega)$, as shown in Figure 5.6.

$$\mathbf{E}(\mathbf{r}) = \int_0^{2\pi} \int_0^{\pi} \mathbf{F}(\theta, \varphi) \exp(-j\mathbf{k} \cdot \mathbf{r}) \sin\theta d\theta d\varphi \tag{5.33}$$

Equation (5.33) is also written in the short form

$$\mathbf{E}(\mathbf{r}) = \iint_{4\pi} \mathbf{F}(\Omega) \exp(-j\mathbf{k} \cdot \mathbf{r}) d\Omega \tag{5.34}$$

where Ω is the solid angle, the vector wave number $\mathbf{k}$ is

$$\mathbf{k} = -k(\hat{\mathbf{x}} \sin\theta \cos\varphi + \hat{\mathbf{y}} \sin\theta \sin\varphi + \hat{\mathbf{z}} \cos\theta) \tag{5.35}$$

$\mathbf{F}(\Omega)$ is the angular spectrum and can be written in two polarisations

$$\mathbf{F}(\Omega) = \hat{\boldsymbol{\theta}} F_\theta(\Omega) + \hat{\boldsymbol{\varphi}} F_\varphi(\Omega) \tag{5.36}$$

$\hat{\boldsymbol{\theta}}$ and $\hat{\boldsymbol{\varphi}}$ are orthogonal to each other and to $\mathbf{k}$. Both $F_\theta(\Omega)$ and $F_\varphi(\Omega)$ can be further written in terms of their real and imaginary parts

$$F_\theta(\Omega) = F_{\theta r}(\Omega) + jF_{\theta i}(\Omega) \text{ and } F_\varphi(\Omega) = F_{\varphi r}(\Omega) + jF_{\varphi i}(\Omega) \tag{5.37}$$

In the RC, $\mathbf{F}(\Omega)$ can be considered as a random variable. For each stirrer position we have a different $\mathbf{F}(\Omega)$, that is, for each stirrer position we have a sample from a random variable with a certain probability distribution.

If the RC is well-stirred (ideally random), we have [3]

$$\langle F_\theta(\Omega) \rangle = \langle F_\varphi(\Omega) \rangle = 0 \tag{5.38}$$

$$\langle F_{\theta r}(\Omega_1) F_{\theta i}(\Omega_2) \rangle = 0, \langle F_{\varphi r}(\Omega_1) F_{\varphi i}(\Omega_2) \rangle = 0, \left\langle \left\{ \begin{matrix} F_{\theta r}(\Omega_1) \\ F_{\theta i}(\Omega_1) \end{matrix} \right\} \times \left\{ \begin{matrix} F_{\varphi r}(\Omega_2) \\ F_{\varphi i}(\Omega_2) \end{matrix} \right\} \right\rangle = 0 \tag{5.39}$$

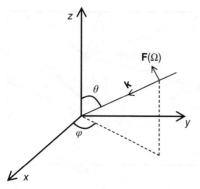

Figure 5.6 Plane-wave spectrum model.

$$\langle F_{\theta r}(\Omega_1)F_{\theta r}(\Omega_2)\rangle = \langle F_{\theta i}(\Omega_1)F_{\theta i}(\Omega_2)\rangle =$$
$$\langle F_{\varphi r}(\Omega_1)F_{\varphi r}(\Omega_2)\rangle = \langle F_{\varphi i}(\Omega_1)F_{\varphi i}(\Omega_2)\rangle = C_E\delta(\Omega_1 - \Omega_2) \tag{5.40}$$

'< >' means the ensemble average of the random variable which can be understood as the average value over different stirrer positions in the RC; '{ }' means only one element in the bracket is chosen. (5.38) means the angular spectrum polarisation component can be positive or negative, thus there is no positive or negative bias and the mean value is zero. (5.39) means the real part and imaginary part of the two polarisations are uncorrelated. (5.40) means the waves from two different arriving angles have different scattering paths and are uncorrelated. C_E is a constant, δ is the Dirac delta function. From (5.39) and (5.40) we also have [3]

$$\left\langle F_\theta(\Omega_1)F_\varphi^*(\Omega_2)\right\rangle = 0 \tag{5.41}$$

$$\left\langle F_\theta(\Omega_1)F_\theta^*(\Omega_2)\right\rangle = \left\langle F_\varphi(\Omega_1)F_\varphi^*(\Omega_2)\right\rangle = 2C_E\delta(\Omega_1 - \Omega_2) \tag{5.42}$$

From (5.34) and (5.38), the mean value of the total E-field in the RC can be obtained:

$$\langle \mathbf{E}(\mathbf{r})\rangle = \iint_{4\pi} \mathbf{F}(\Omega) \exp(-j\mathbf{k}\cdot\mathbf{r})d\Omega = 0 \tag{5.43}$$

The square of the absolute of the total E-field can be expressed as

$$|\mathbf{E}(\mathbf{r})|^2 = \iint_{4\pi}\iint_{4\pi} \mathbf{F}(\Omega_1)\cdot\mathbf{F}^*(\Omega_2)\exp[-j(\mathbf{k}_1 - \mathbf{k}_1)\cdot\mathbf{r}]d\Omega_1 d\Omega_2 \tag{5.44}$$

By applying (5.41) and (5.42), we have [3]

$$\left\langle |\mathbf{E}(\mathbf{r})|^2\right\rangle = 16\pi C_E \equiv E_0^2 \tag{5.45}$$

which is the spatial uniformity of an ideal RC: the mean-square value of the E-field is independent of position $\mathbf{r}$. The value is denoted by E_0^2, and E_0^2 is used as the reference value for other quantities.

Similarly, derivations have been given in [3] to obtain the rectangular components of the E-field ($|\mathbf{E}(\mathbf{r})|^2 = |E_x|^2 + |E_y|^2 + |E_z|^2$):

$$\langle E_x\rangle = \langle E_y\rangle = \langle E_z\rangle = 0 \tag{5.46}$$

$$\left\langle |E_x|^2\right\rangle = \left\langle |E_y|^2\right\rangle = \left\langle |E_z|^2\right\rangle = \frac{E_0^2}{3} \tag{5.47}$$

The rectangular components do not change with the direction of the axis, which is the isotropy property of an ideal RC. Measurement results have also confirmed this [28]. The magnetic field (H-field) has the same statistical spatial uniformity and isotropy [3]:

$$\langle \mathbf{H}(\mathbf{r})\rangle = 0 \tag{5.48}$$

$$\left\langle |\mathbf{H}(\mathbf{r})|^2\right\rangle = \frac{E_0^2}{\eta^2} \tag{5.49}$$

$$\langle H_x\rangle = \langle H_y\rangle = \langle H_z\rangle = 0 \tag{5.50}$$

$$\left\langle |H_x|^2\right\rangle = \left\langle |H_y|^2\right\rangle = \left\langle |H_z|^2\right\rangle = \frac{E_0^2}{3\eta^2} \tag{5.51}$$

where η is the characteristic impedance of the medium. In free space, it is 377 Ohms. The mean value of the energy density W has also been derived [3]:

$$W = \frac{1}{2}\left[\varepsilon|\mathbf{E}(\mathbf{r})|^2 + \mu|\mathbf{H}(\mathbf{r})|^2\right] \tag{5.52}$$

$$\langle W \rangle = \varepsilon E_0^2 \tag{5.53}$$

The mean value of power density or Poynting vector $\mathbf{S} = \mathbf{E} \times \mathbf{H}$ is [3]

$$\langle \mathbf{S}(\mathbf{r}) \rangle = 0 \tag{5.54}$$

Similar to the plane wave in free space, a scalar power density can be defined as

$$S = c\langle W \rangle = \frac{E_0^2}{\eta} \tag{5.55}$$

The probability distribution functions (PDFs) of the field quantities can be obtained from the central limit theorem (CLT). The rectangular components can be written in the form of

$$E_x = E_{xr} + jE_{xi}, E_y = E_{yr} + jE_{yi}, E_z = E_{zr} + jE_{zi} \tag{5.56}$$

where E_{xr}, E_{xi} are the real and imaginary parts of E_x. Because the waves superimpose randomly, from the CLT, the normal (Gaussian) distribution can be obtained and the PDF is [3]

$$PDF(E_{xr}) = \frac{1}{\sigma\sqrt{2\pi}}\exp\left(-\frac{x^2}{2\sigma^2}\right), x = E_{xr} \tag{5.57}$$

The same PDF also applies to other components in (5.56). The mean value is

$$\langle E_{xr} \rangle = \langle E_{yr} \rangle = \langle E_{zr} \rangle = \langle E_{xi} \rangle = \langle E_{yi} \rangle = \langle E_{zi} \rangle = 0 \tag{5.58}$$

The variances σ^2 are

$$\sigma^2 = \langle |E_{xr} - \langle E_{xr} \rangle|^2 \rangle = \langle E_{xr}^2 \rangle = \langle E_{yr}^2 \rangle = \langle E_{zr}^2 \rangle = \langle E_{xi}^2 \rangle = \langle E_{yi}^2 \rangle = \langle E_{zi}^2 \rangle = \frac{E_0^2}{6} \tag{5.59}$$

Because E_{xr} and E_{xi} are independent and follow the same PDF in (5.57), the magnitude of E_x has a chi distribution with two degrees of freedom (Rayleigh distribution):

$$PDF(|E_x|) = \frac{x}{\sigma^2}\exp\left(-\frac{x^2}{2\sigma^2}\right), x = |E_x| \tag{5.60}$$

The mean value is $\sigma\sqrt{\pi/2}$ and the standard deviation is $\sigma\sqrt{2-\pi/2}$. The square magnitude of E_x has a chi-square distribution with two degrees of freedom (exponential distribution):

$$PDF(|E_x|^2) = \frac{1}{2\sigma^2}\exp\left(-\frac{x}{2\sigma^2}\right), x = |E_x|^2 \tag{5.61}$$

The mean value and the standard deviation are the same and equal $2\sigma^2$. Measurements have been performed to verify this and it has been shown that the power received by any type of antenna in a well-stirred (ideal) RC follows the exponential distribution [3, 29, 30]. The total E-field magnitude is chi distributed with six degrees of freedom:

$$|\mathbf{E}| = \sqrt{|E_x|^2 + |E_y|^2 + |E_z|^2} = \sqrt{E_{xr}^2 + E_{xi}^2 + E_{yr}^2 + E_{yi}^2 + E_{zr}^2 + E_{zi}^2} \tag{5.62}$$

$$PDF(|\mathbf{E}|) = \frac{x^5}{8\sigma^6}\exp\left(-\frac{x^2}{2\sigma^2}\right), x = |\mathbf{E}| \tag{5.63}$$

The mean value is $15\sigma\sqrt{2\pi}/16$ ($\approx 2.35\sigma$), and the standard deviation is $\sigma\sqrt{6-225\pi/128}$ ($\approx 0.69\sigma$). The squared magnitude of the total E-field has a chi-square distribution with six degrees of freedom:

$$PDF\left(|\mathbf{E}|^2\right) = \frac{x^2}{16\sigma^6}\exp\left(-\frac{x}{2\sigma^2}\right), x = |\mathbf{E}|^2 \tag{5.64}$$

The mean value is $6\sigma^2$ and the standard deviation is $\sigma^2/\sqrt{12}$.

5.4.2 Field Correlations

The spatial correlation functions between different quantities have been derived in [3]. The correlation function ρ of the total E-field with distance r is given in [31–33] and has a similar expression in acoustic RCs [34]:

$$\rho(r) = \frac{\langle \mathbf{E}_1 \bullet \mathbf{E}_2^* \rangle}{\sqrt{\langle |\mathbf{E}_1|^2 \rangle \langle |\mathbf{E}_2|^2 \rangle}} = \frac{\sin(kr)}{kr} \tag{5.65}$$

$\rho(r)$ is plotted in Figure 5.7 with the x-axis in wavelengths. As can be seen, the first zero occurs at $r = 0.5\lambda$, which is defined as the correlation length.

The angular correlation of the E-field with angle γ is [3]

$$\rho(\hat{\mathbf{s}}_1, \hat{\mathbf{s}}_2) = \frac{\langle E_1 E_2^* \rangle}{\sqrt{\langle |E_1|^2 \rangle \langle |E_2|^2 \rangle}} = \cos(\gamma) \tag{5.66}$$

where $E_1 = \hat{\mathbf{s}}_1 \bullet \mathbf{E}$, $E_2 = \hat{\mathbf{s}}_2 \bullet \mathbf{E}$, and $\hat{\mathbf{s}}_1$, and $\hat{\mathbf{s}}_2$ are unit vectors with an angular separation γ. The angular correlation is illustrated in Figure 5.8. As expected, when $\gamma = 90°$ E_1 and E_2 are uncorrelated.

The spatial correlation for the longitudinal E-field is [3]

$$\rho_l(r) = \frac{\langle E_1 E_2^* \rangle}{\sqrt{\langle |E_1|^2 \rangle \langle |E_2|^2 \rangle}} = \frac{3}{(kr)^2}\left[\frac{\sin(kr)}{kr} - \cos(kr)\right] \tag{5.67}$$

where E_1 and E_2 are the longitudinal components of the E-field with distance r as shown in Figure 5.9. The correlation with different distance is also plotted.

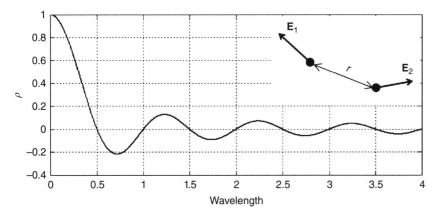

Figure 5.7 Correlation of the total E-field with distance r in wavelength.

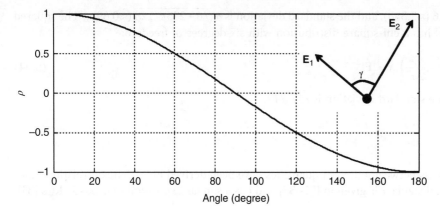

Figure 5.8 Angular correlation with angle γ.

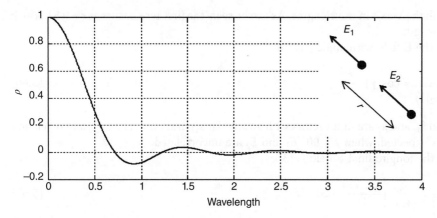

Figure 5.9 Correlation of longitudinal component of E-field with distance r.

The spatial correlation for the traverse E-field is [3]

$$\rho_t(r) = \frac{\langle E_1 E_2^* \rangle}{\sqrt{\langle |E_1|^2 \rangle \langle |E_2|^2 \rangle}} = \frac{3}{2}\left[\frac{\sin(kr)}{kr} - \frac{1}{(kr)^2}\left(\frac{\sin(kr)}{kr} - \cos(kr) \right) \right] \tag{5.68}$$

where E_1 and E_2 are the transverse components of the E-field with distance r. The correlation and the definition of symbols are illustrated in Figure 5.10.

The spatial correlation of the mixed E- and H-fields has been obtained, except the orthogonal transverse components of E and H, other correlations are zero [3, 35].

$$\langle E_x(0)H_x^*(\hat{z}r)\rangle = \langle E_x(0)H_z^*(\hat{z}r)\rangle = \langle E_y(0)H_y^*(\hat{z}r)\rangle = \langle E_y(0)H_z^*(\hat{z}r)\rangle =$$

$$\langle E_z(0)H_x^*(\hat{z}r)\rangle = \langle E_z(0)H_y^*(\hat{z}r)\rangle = \langle E_z(0)H_z^*(\hat{z}r)\rangle = 0 \tag{5.69}$$

$$\rho_{xy}(r) = \frac{\langle E_x H_y^* \rangle}{\sqrt{\langle |E_x|^2 \rangle \langle |H_y|^2 \rangle}} = \frac{3j}{2(kr)^2}[\sin(kr) - kr\cos(kr)] \tag{5.70}$$

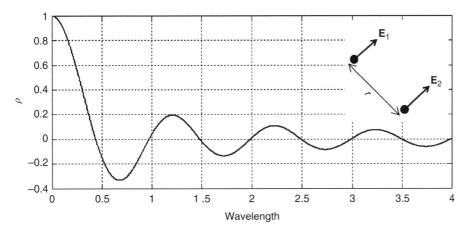

Figure 5.10 Correlation of transverse component of E-field with distance *r*.

The magnitude of $\rho_{xy}(r)$ is plotted in Figure 5.11. It is interesting to see that the correlation is zero at $r = 0$, and it increases with distance first and then decreases.

The correlations of the squared field quantities have been given in [3]. For the square of the longitudinal field component:

$$\rho_{ll}(r) = \frac{\left\langle \left(|E_1|^2 - \left\langle |E_1|^2 \right\rangle \right)\left(|E_2|^2 - \left\langle |E_2|^2 \right\rangle \right) \right\rangle}{\sqrt{\left\langle \left(|E_1|^2 - \left\langle |E_1|^2 \right\rangle \right)^2 \right\rangle \left\langle \left(|E_2|^2 - \left\langle |E_2|^2 \right\rangle \right)^2 \right\rangle}} = \rho_l^2(r) \tag{5.71}$$

the correlation of E_1 and E_2 is shown in Figure 5.9, and $\rho_l(r)$ is defined in (5.67). For the square of the transverse field component:

$$\rho_{tt}(r) = \frac{\left\langle \left(|E_1|^2 - \left\langle |E_1|^2 \right\rangle \right)\left(|E_2|^2 - \left\langle |E_2|^2 \right\rangle \right) \right\rangle}{\sqrt{\left\langle \left(|E_1|^2 - \left\langle |E_1|^2 \right\rangle \right)^2 \right\rangle \left\langle \left(|E_2|^2 - \left\langle |E_2|^2 \right\rangle \right)^2 \right\rangle}} = \rho_t^2(r) \tag{5.72}$$

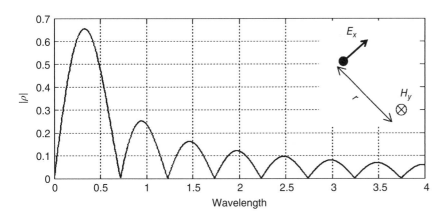

Figure 5.11 Correlation of orthogonal transverse components of E- and H-fields with distance *r*.

The correlation of E_1 and E_2 is illustrated in Figure 5.10, and $\rho_t(r)$ is defined in (5.68). Equation (5.72) has been verified experimentally by using two short monopoles separated with distance r, the received power is proportional to the square of the E-field, and the correlation of the power samples is the correlation of the squared E-field in (5.72). For the square of the total E-field:

$$\rho_{\mathbf{EE}}(r) = \frac{\left\langle \left(|\mathbf{E}_1|^2 - \langle |\mathbf{E}_1|^2 \rangle\right)\left(|\mathbf{E}_2|^2 - \langle |\mathbf{E}_2|^2 \rangle\right)\right\rangle}{\sqrt{\left\langle \left(|\mathbf{E}_1|^2 - \langle |\mathbf{E}_1|^2 \rangle\right)^2\right\rangle \left\langle \left(|\mathbf{E}_2|^2 - \langle |\mathbf{E}_2|^2 \rangle\right)^2\right\rangle}} = \frac{2\rho_{tt}(r) + \rho_{ll}(r)}{3} \tag{5.73}$$

where $\mathbf{E}$ is the total E-field defined in (5.62) and the field points of $\mathbf{E}_1$ and $\mathbf{E}_2$ have a distance of r.

The correlation of electric energy density and magnetic energy density is the same as (5.73) [3]. The electric energy density and the magnetic energy density are defined as

$$W_E(\mathbf{r}) = \frac{\varepsilon}{2}|\mathbf{E}|^2 \text{ and } W_H(\mathbf{r}) = \frac{\mu}{2}|\mathbf{H}|^2 \tag{5.74}$$

The total energy density is defined as

$$W(\mathbf{r}) = W_E(\mathbf{r}) + W_H(\mathbf{r}) \tag{5.75}$$

The correlation of the total energy density is found to be [3, 35]

$$\rho_W(r) = \frac{\langle (W_1 - \langle W_1 \rangle)(W_2 - \langle W_2 \rangle)\rangle}{\sqrt{\langle (W_1 - \langle W_1 \rangle)^2 \rangle \langle (W_2 - \langle W_2 \rangle)^2 \rangle}} = \rho_{\mathbf{EE}}(r) + \frac{2}{3}\left|\rho_{xy}(r)\right|^2 \tag{5.76}$$

where W_1 and W_2 are the total energy densities of two points with distance r. The mean values of the electric energy density, the magnetic energy density, and the total energy density are given by [3]

$$\langle W_E(\mathbf{r}) \rangle = \langle W_H(\mathbf{r}) \rangle = \frac{\varepsilon}{2}E_0^2 \text{ and } \langle W(\mathbf{r}) \rangle = \varepsilon E_0^2 \tag{5.77}$$

where E_0^2 is the mean-square E-field given in (5.45).

5.4.3 Boundary Fields

Because the tangential E-field and normal magnetic field are zero on the boundary of the highly conductive walls in the RC, the field uniformity (FU) of (5.47) cannot be satisfied everywhere. However, by considering the effect of the boundaries, the plane-wave spectrum method can be extended to investigate the field near the planar interface, the right-angle bend and the right-angle corner [3].

Suppose an infinite planar structure is shown in Figure 5.12. The total E-field can be expressed by the sum of the incident and reflected fields:

$$\mathbf{E}^{\text{tot}} = \mathbf{E}^{\text{inc}} + \mathbf{E}^{\text{ref}} \tag{5.78}$$

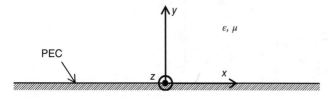

Figure 5.12 Infinite half space.

where $\mathbf{E}^{\text{inc}}$ and $\mathbf{E}^{\text{ref}}$ are the incident and reflected fields, respectively:

$$\mathbf{E}^{\text{inc}}(x,y,z) = \hat{\mathbf{x}}E_x^{\text{inc}}(x,y,z) + \hat{\mathbf{y}}E_y^{\text{inc}}(x,y,z) + \hat{\mathbf{z}}E_z^{\text{inc}}(x,y,z) \tag{5.79}$$

$$\mathbf{E}^{\text{ref}}(x,y,z) = -\hat{\mathbf{x}}E_x^{\text{inc}}(x,-y,z) + \hat{\mathbf{y}}E_y^{\text{inc}}(x,-y,z) - \hat{\mathbf{z}}E_z^{\text{inc}}(x,-y,z) \tag{5.80}$$

At interface $y = 0$, the total E-field is

$$\mathbf{E}^{\text{tot}}(x,0,z) = 2\hat{\mathbf{y}}E_y^{\text{inc}}(x,0,z) \tag{5.81}$$

The tangential E-field is zero and the normal E-field is doubled. Similarly, the normal H-field is zero and the tangential magnetic field is doubled. The mean values of relevant quantities have been derived in [3] by using the plane-wave integral. The mean value of the total E-field is zero, and the normal component of the E-field is enhanced at the boundary while the tangential component of the E-field is zero:

$$\langle \mathbf{E}^{\text{tot}}(x,y,z) \rangle = \langle \mathbf{H}^{\text{tot}}(x,y,z) \rangle = 0 \tag{5.82}$$

$$\left\langle \left| E_y^{\text{tot}}(x,y,z) \right|^2 \right\rangle = \frac{E_0^2}{3}[1 + \rho_l(2y)] \tag{5.83}$$

$$\left\langle \left| E_x^{\text{tot}}(x,y,z) \right|^2 \right\rangle = \frac{E_0^2}{3}[1 - \rho_t(2y)] \tag{5.84}$$

$$\left\langle \left| H_y^{\text{tot}}(x,y,z) \right|^2 \right\rangle = \frac{E_0^2}{3\eta^2}[1 - \rho_l(2y)] \tag{5.85}$$

$$\left\langle \left| H_x^{\text{tot}}(x,y,z) \right|^2 \right\rangle = \frac{E_0^2}{3\eta^2}[1 + \rho_t(2y)] \tag{5.86}$$

where ρ_l and ρ_t are defined in (5.67) and (5.68), respectively. As can be seen, when $y \to \infty$, (5.83)–(5.86) approach the free-space mean-square values in (5.47) and (5.51). Figures 5.13 and 5.14 show the plots of (5.83)–(5.86), the mean-square values are normalised to the values in free space, which are $E_0^2/3$ and $E_0^2/(3\eta^2)$, and '$*$' means the field component can be x or y. Figures 5.13 and 5.14 can provide a guideline in the measurement: the object under test should be placed at least $\lambda/4$ (or $\lambda/2$ more strictly) away from the wall of the RC to have a good FU [4].

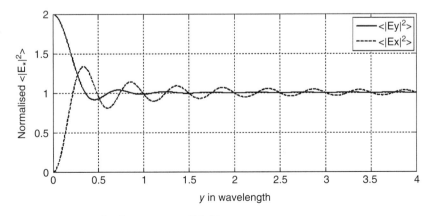

Figure 5.13 Normalised mean-square E-field.

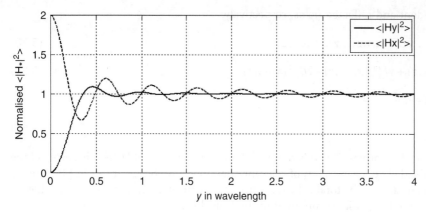

Figure 5.14 Normalised mean-square H-field.

For the right-angle bend shown in Figure 5.15, the following mean values have been derived [3]:

$$\left\langle \left| E_z^{tot}(x,y,z) \right|^2 \right\rangle = \frac{E_0^2}{3}\left[1 - \rho_t(2y) + \rho_t(2x) + \rho_t\left(2\sqrt{x^2 + y^2} \right) \right] \tag{5.87}$$

$$\left\langle \left| E_x^{tot}(x,y,z) \right|^2 \right\rangle = \frac{E_0^2}{3}\left[1 - \rho_t(2y) + \rho_l(2x) - \frac{y^2}{x^2 + y^2}\rho_t\left(2\sqrt{x^2 + y^2} \right) - \frac{x^2}{x^2 + y^2}\rho_l\left(2\sqrt{x^2 + y^2} \right) \right] \tag{5.88}$$

$$\left\langle \left| H_z^{tot}(x,y,z) \right|^2 \right\rangle = \frac{E_0^2}{3\eta^2}\left[1 + \rho_t(2x) + \rho_t(2y) + \rho_t\left(2\sqrt{x^2 + y^2} \right) \right] \tag{5.89}$$

$$\left\langle \left| H_x^{tot}(x,y,z) \right|^2 \right\rangle = \frac{E_0^2}{3\eta^2}\left[1 + \rho_t(2y) - \rho_l(2x) - \frac{y^2}{x^2 + y^2}\rho_t\left(2\sqrt{x^2 + y^2} \right) - \frac{x^2}{x^2 + y^2}\rho_l\left(2\sqrt{x^2 + y^2} \right) \right] \tag{5.90}$$

The normalised mean-square E-field (normalised to $E_0^2/3$) of (5.87) is plotted in Figure 5.16a,b which shows the extracted normalised E-field along the r direction (diagonal direction). As can be seen, the inhomogeneity is enhanced near the corner of the right-angle bend and more than one wavelength is required to be far from the corner to have a uniform mean field. Equation (5.88) is illustrated in Figure 5.17a. By interchanging x and y, the mean-square E-field in y polarisation can be obtained and is shown in Figure 5.17b. The normalised mean-square H-field in (5.89) is shown in Figure 5.18a, and the value along the diagonal direction is extracted and plotted in Figure 5.18b. It is interesting to see that the H-field is four times greater than in free space (far away from the boundary), which also means that the current density is twice that on the planar structure in Figure 5.12. Figure 5.19 shows the mean-square H-field of the x component in (5.90) and the y component (by interchanging x and y).

For the right-angle corner shown in Figure 5.20 the following mean values have been derived [3]; only the z component is given as for other components we only need to interchange the variables.

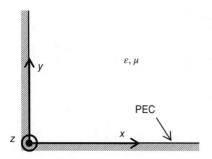

Figure 5.15 Right-angle bend in an RC.

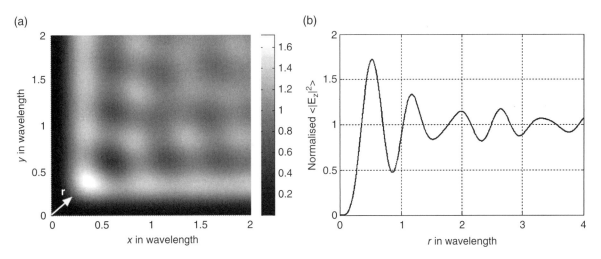

Figure 5.16 (a) Normalised mean-square E-field of (5.87) and (b) values along the diagonal direction.

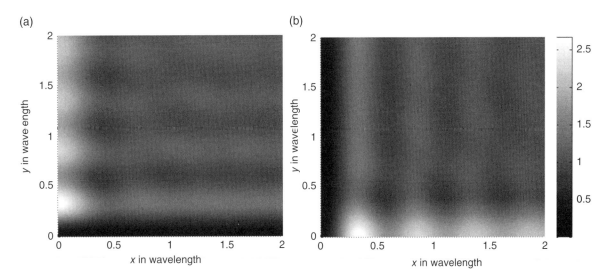

Figure 5.17 (a) Normalised mean-square E-field of (5.88) and (b) normalised $\left\langle \left| E_y^{tot}(x,y,z) \right|^2 \right\rangle$.

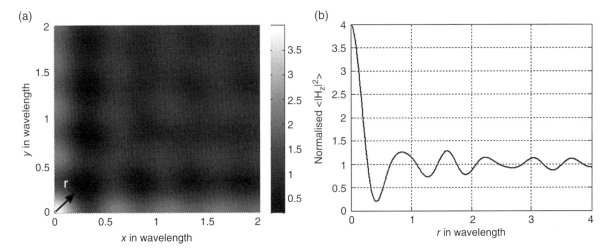

Figure 5.18 (a) Normalised mean-square H-field of (5.89) and (b) values along the diagonal direction.

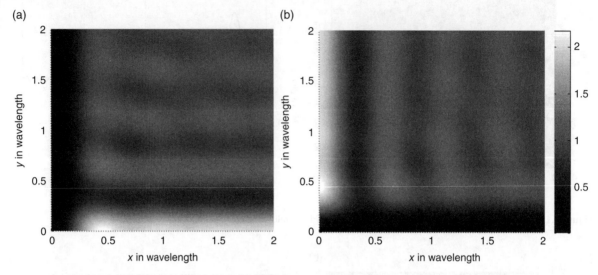

Figure 5.19 (a) Normalised mean-square H-field of (5.90) and (b) normalised $\left\langle \left| H_y^{\text{tot}}(x,y,z) \right|^2 \right\rangle$.

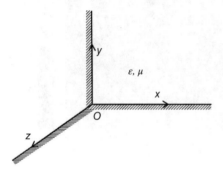

Figure 5.20 Right-angle corner in an RC, *xoy*, *xoz*, and *yoz* planes are PEC.

$$
\begin{aligned}
\left\langle \left| E_z^{\text{tot}}(x,y,z) \right|^2 \right\rangle = \frac{E_0^2}{3} \Bigg[& 1 - \rho_t(2x) - \rho_t(2y) + \rho_t\left(2\sqrt{x^2+y^2}\right) + \rho_l(2z) \\
& - \frac{x^2}{x^2+z^2}\rho_t\left(2\sqrt{x^2+z^2}\right) - \frac{z^2}{x^2+z^2}\rho_l\left(2\sqrt{x^2+z^2}\right) \\
& - \frac{y^2}{y^2+z^2}\rho_t\left(2\sqrt{y^2+z^2}\right) - \frac{z^2}{y^2+z^2}\rho_l\left(2\sqrt{y^2+z^2}\right) \\
& + \frac{x^2+y^2}{x^2+y^2+z^2}\rho_t\left(2\sqrt{x^2+y^2+z^2}\right) \\
& + \frac{z^2}{x^2+y^2+z^2}\rho_l\left(2\sqrt{x^2+y^2+z^2}\right) \Bigg]
\end{aligned}
\tag{5.91}
$$

$$\left\langle \left| H_z^{\text{tot}}(x,y,z) \right|^2 \right\rangle = \frac{E_0^2}{3\eta^2} \Bigg[1 + \rho_t(2x) + \rho_t(2y) + \rho_t\left(2\sqrt{x^2+y^2}\right) - \rho_l(2z)$$

$$- \frac{x^2}{x^2+z^2}\rho_t\left(2\sqrt{x^2+z^2}\right) - \frac{z^2}{x^2+z^2}\rho_l\left(2\sqrt{x^2+z^2}\right)$$

$$- \frac{y^2}{y^2+z^2}\rho_t\left(2\sqrt{y^2+z^2}\right) - \frac{z^2}{y^2+z^2}\rho_l\left(2\sqrt{y^2+z^2}\right) \qquad (5.92)$$

$$- \frac{x^2+y^2}{x^2+y^2+z^2}\rho_t\left(2\sqrt{x^2+y^2+z^2}\right)$$

$$- \frac{z^2}{x^2+y^2+z^2}\rho_l\left(2\sqrt{x^2+y^2+z^2}\right) \Bigg]$$

(a)　　　　　　　　　　　　　　　　(b)

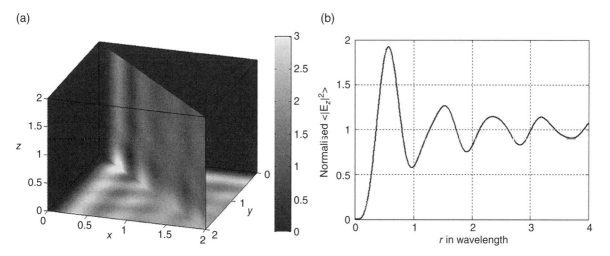

Figure 5.21 (a) Normalised mean-square E-field of (5.91) and (b) values along the diagonal direction.

(a)　　　　　　　　　　　　　　　　(b)

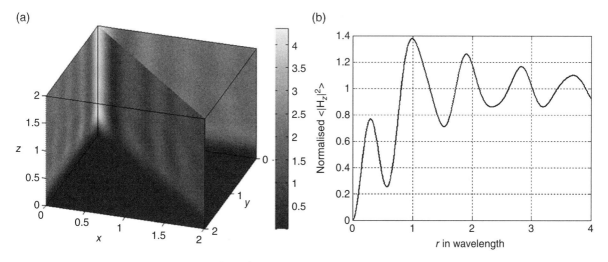

Figure 5.22 (a) Normalised mean-square H-field of (5.92) and (b) values along the diagonal direction.

The normalised mean-square E-field (normalised to $E_0^2/3$) of (5.91) is plotted in Figure 5.21a with a cut plane $x = y$. Figure 5.21b shows the extracted normalised E-field along the diagonal direction ($x = y = z$). The normalised mean-square H-field of (5.92) is illustrated in Figure 5.22a and the values along the diagonal direction are extracted and shown in Figure 5.22b.

It should be noted that although the mean values have a spatial distribution, the PDFs have the same form. The magnitude and the squared magnitude of the E- and H-field components have Rayleigh and exponential PDFs, and the variances are functions of position. However, the magnitude and the squared magnitude of the total E- and H-fields are not chi and chi square distributions [3] because different rectangular components have different distributions near the boundary.

5.4.4 Enhanced Backscattering Effect

The enhanced backscattering effect is a very interesting and universal phenomenon inherent to waves of whatever physical nature: electromagnetics, acoustics, etc. [36]. In the RC, the enhanced backscattering effect is very useful and can be used to characterise the performance of the RC [37] and to measure the radiation efficiency of antennas [38–40].

Suppose two ideal antennas (well-matched, no loss) are positioned in the RC, as shown in Figure 5.23. If the RC is well-stirred, the enhanced backscattering effect tells us that the average power received by the Tx antenna is twice the power received by the Rx antenna, when the Tx and Rx antenna are separated distantly. The average is over different states of the RC (different stirrer positions) and can be written as:

$$\langle |S_{11,s}|^2 \rangle = e_b \langle |S_{21,s}|^2 \rangle = \langle |S_{22,s}|^2 \rangle \tag{5.93}$$

where $S_{*,s} = S_* - \langle S_* \rangle$ means the stirred part of S_*, e_b is defined as the enhanced backscatter constant, for a well-stirred RC, $e_b = 2$.

This effect can be mathematically proved by using geometrical optics [21], plane-wave formulation [3] or scattering matrices [41], and it has been found that the enhanced backscatter constant varies as a function of the distance between the two antennas. To understand this effect easily, we can introduce another antenna,

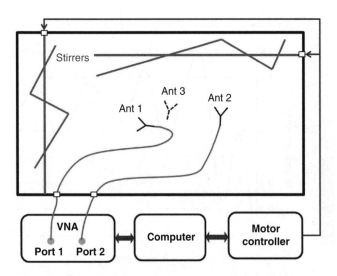

Figure 5.23 Measurement of *S*-parameters at different stirrer positions.

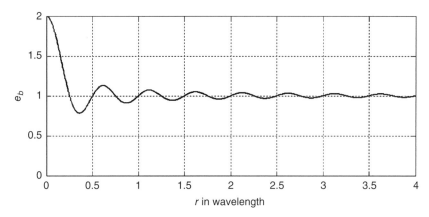

Figure 5.24 Enhanced backscatter constant as a function of distance in wavelength.

antenna 3, and use antenna 2 as a reference. Suppose the distance between antenna 1 and antenna 3 is r, and the distance between antenna 1 and antenna 2 is large, it has been proved that [3]

$$e_b = \frac{\langle |S_{31,s}|^2 \rangle}{\langle |S_{21,s}|^2 \rangle} = 1 + \frac{\sin(2kr)}{2kr} \tag{5.94}$$

where $S_{*,s}$ means the stirred part of S_*. As expected, the enhanced backscattering effect does not appear suddenly, when $r = 0$, antenna 1 and antenna 3 overlap, and $e_b = 2$. When $r \to \infty$, there is no statistical difference between antenna 3 and antenna 2, $e_b = 1$ (no enhanced backscattering effect). (5.94) is plotted in Figure 5.24, when $r > \lambda/4$, e_b is close to 1. A region of enhanced backscatter can be defined with a sphere of radius $\lambda/4$. Outside this region the enhanced backscattering effect is not significant. This also offers a guideline when doing measurement in the RC, the object under test should be placed at least $\lambda/4$ from the Tx antenna, otherwise the average field strength is different to the monitoring antenna (antenna 2) far from the Tx antenna. This condition is easy to satisfy: when the frequency is high, $\lambda/4$ is very close to an antenna and one rarely places other objects so close to an antenna.

It has also been found when the size of the RC is close to or larger than 100λ, e_b deviates from 2 and depends on the radiation patterns of the antennas [37]. How to realise an ideal enhanced backscattering effect at such high frequencies (electrically very large) is still a problem: an intuitive example is that a laser beam in the RC may not be diffused easily.

5.4.5 Loss Mechanism

The Q factor definition is

$$Q = \frac{\omega U}{P_d} \tag{5.95}$$

where U is the energy stored in the cavity and P_d is the power dissipated. The stored energy can be obtained from the average energy density and the volume of the chamber:

$$U = \langle W \rangle V \tag{5.96}$$

From (5.53) we have

$$U = \langle W \rangle V = \varepsilon E_0^2 V \tag{5.97}$$

Thus the Q factor in (5.95) can be related to E_0^2:

$$E_0^2 = \frac{QP_d}{\omega \varepsilon V} \tag{5.98}$$

It has been derived that the average power received by an ideal antenna (perfectly matched, no loss) in the RC can be obtained as [3]

$$\langle P_r \rangle = \frac{1}{2} \frac{E_0^2}{\eta} \frac{\lambda^2}{4\pi} \tag{5.99}$$

which means the average received power is the product of the scalar power density E_0^2/η and the effective area of an isotropic antenna $\lambda^2/(4\pi)$ and a polarisation mismatch factor of 1/2 [3, 42]. Because of the conservation of power, the transmitted power equals the dissipated power in the cavity ($P_t = P_d$). From (5.98) and (5.99), the well-known Hill's equation can be obtained [3]:

$$\langle P_r \rangle = \frac{\lambda^3 Q}{16\pi^2 V} P_t \tag{5.100}$$

The importance of Hill's equation in the RC is just like the Friis equation in free space (AC):

$$P_r = \mathbf{G}_t(\theta,\varphi) \cdot \mathbf{G}_r(\theta,\varphi) \left(\frac{\lambda}{4\pi d}\right)^2 P_t \tag{5.101}$$

where $\mathbf{G}_t(\theta, \varphi)$ and $\mathbf{G}_r(\theta, \varphi)$ are the gains of the Tx antenna and Rx antenna, respectively, polarisations are matched, d is the distance between the Tx and the Rx antennas, and λ is the wavelength. It is interesting to compare the Hill's equation with the Friis equation, in the RC, for a constant transmitting power at a given frequency: the average received power is only related to the Q factor and the volume of the cavity, while in the free space, the angle dependency of the antenna patterns is involved. This angle independency is very useful in many applications; the received power is independent to the posture of the object under test. As can be seen in the measurement section, Hill's equation has been widely used in the electromagnetic compatibility (EMC) and antenna measurement in the RC. One of the important applications is to measure the Q factor of the RC in the frequency domain:

$$Q = \frac{16\pi^2 V}{\lambda^3} \frac{\langle P_r \rangle}{P_t} \tag{5.102}$$

However, one should bear in mind that (5.102) is for ideal antennas; in practice, the loss and mismatch of the Tx antenna and the Rx antenna need to be considered.

The dissipated power loss P_d in (5.95) can be analysed with more detail and can be decomposed into a summation of power loss from four sources [3]:

$$P_d = P_{d1} + P_{d2} + P_{d3} + P_{d4} \tag{5.103}$$

where P_{d1} is the power dissipated in the cavity walls, P_{d2} is the power absorbed in the loading objects within the cavity, P_{d3} is the power lost through aperture leakage and P_{d4} is the power dissipated in the loads of receiving antennas. The total Q can be written as

$$Q^{-1} = Q_1^{-1} + Q_2^{-1} + Q_3^{-1} + Q_4^{-1} \tag{5.104}$$

where

$$Q_1 = \frac{\omega U}{P_{d1}}, Q_2 = \frac{\omega U}{P_{d2}}, Q_3 = \frac{\omega U}{P_{d3}}, \text{and } Q_4 = \frac{\omega U}{P_{d4}} \tag{5.105}$$

When the frequency is high (the RC is electrically large), the contribution from Q_1, Q_2, Q_3, and Q_4 has been derived [3]. The contribution of Q_1 is

$$Q_1 = \frac{3V}{2\mu_r \delta A} \tag{5.106}$$

which has been given in (5.23). The contribution of Q_2 is

$$Q_2 = \frac{2\pi V}{\lambda \langle \sigma_a \rangle} \tag{5.107}$$

where $\langle \sigma_a \rangle$ is the averaged absorption cross section over all incident angles and both (TE and TM) polarisations [3]:

$$\langle \sigma_a \rangle = \frac{1}{8\pi} \iint_{4\pi} (\sigma_{aTE} + \sigma_{aTM}) d\Omega \tag{5.108}$$

and the contribution of $\langle \sigma_a \rangle$ can be superimposed:

$$\langle \sigma_a \rangle = \sum_m \langle \sigma_{am} \rangle \tag{5.109}$$

where $\langle \sigma_{am} \rangle$ is the averaged absorption cross section of the mth absorber. The contribution of Q_3 is

$$Q_3 = \frac{4\pi V}{\lambda \langle \sigma_t \rangle} \tag{5.110}$$

where $\langle \sigma_t \rangle$ is the averaged transmission cross section over 2π steradians and both (TE and TM) polarisations:

$$\langle \sigma_t \rangle = \frac{1}{4\pi} \iint_{2\pi} (\sigma_{tTE} + \sigma_{tTM}) d\Omega \tag{5.111}$$

Because the waves can only propagate from one side of an aperture, a factor of 2 is introduced in (5.110) when comparing with (5.107). The averaged transmission cross section can also be superimposed:

$$\langle \sigma_t \rangle = \sum_m \langle \sigma_{tm} \rangle \tag{5.112}$$

where $\langle \sigma_{tm} \rangle$ is the averaged transmission cross section of the mth aperture. The average transmission cross section of an electrically large aperture has been found to be half of the aperture area [3]:

$$\langle \sigma_t \rangle = \frac{1}{2\pi} \int_0^{2\pi} \int_0^{\pi/2} A \cos\theta \sin\theta d\theta d\varphi = A/2 \tag{5.113}$$

For electrically small apertures,

$$\langle \sigma_t \rangle = Ck^4 \tag{5.114}$$

where C is a constant dependent on aperture size and shape, and independent of frequency, and k is the wavenumber. In the resonant region, when the aperture dimensions are comparable to the wavelength, no simple form can be obtained. Specifically, for a circular aperture, when the aperture is electrically large

$$\langle \sigma_t \rangle = \pi a^2/2 \tag{5.115}$$

where a is the radius of the aperture. When the aperture is electrically small

$$\langle \sigma_t \rangle = \frac{16a^6}{9\pi} k^4 \tag{5.116}$$

In the resonance region, there is no simple expression. However, the circular aperture does not have strong resonance and it can be evaluated using either (5.115) or (5.116), and a crossover wavenumber k_c can be defined when (5.115) equals (5.116)

$$k_c = \left(9\pi^2/32\right)^{1/4}/a \approx 1.29/a \tag{5.117}$$

when $k > k_c$, (5.115) can be used and when $k < k_c$, (5.116) can be used.

The contribution of Q_4 from a receiving antenna can be obtained from (5.99):

$$P_{d4} = \langle P_r \rangle = \frac{1}{2} \frac{E_0^2}{\eta} \frac{\lambda^2}{4\pi} \eta_{tot} \tag{5.118}$$

where η_{tot} is the total efficiency of the antenna, which is defined as the ratio of the total radiated power to the input power and has taken the impedance mismatch and radiation efficiency into account. Because of the reciprocity theorem, the receiving efficiency is the same as the radiation efficiency. Q_4 can be obtained as

$$Q_4 = \frac{16\pi^2 V}{\lambda^3 \eta_{tot}} \tag{5.119}$$

Measurements have confirmed that when the frequency is low, the total loss is dominated by the receiving antenna; when the frequency is high, Q_4 is very large and the contribution from Q_4 is very small [43, 44].

If the excitation is from the outside of the cavity, as shown in Figure 5.25, and the incident angle and polarisation are random and uniformly distributed, by using the power-balance approach, we have [3, 43]

$$S_c = \frac{\langle \sigma_t \rangle \lambda Q}{4\pi V} S_i \tag{5.120}$$

where S_i is the power density outside the cavity (the incident wave) and S_c is the scalar power density inside the cavity shown in (5.55). The shielding effectiveness (SE) can be defined as the ratio of S_i and S_c [43]:

$$SE = 10\log_{10}\left(\frac{4\pi V}{\langle \sigma_t \rangle \lambda Q}\right) \text{ in dB} \tag{5.121}$$

It is interesting to note that if there is no extra loss in the cavity, the only contribution of Q in (5.121) is from the transmission cross section $\langle \sigma_t \rangle$, and by using (5.110) we have $SE = 0\,dB$.

5.4.6 Probability Distribution Functions

Typical PDFs have been given in Section 5.4.1, and in this section more probability/cumulative distributions are given and discussed from a practical measurement point of view.

The squared magnitude of the total E-field can be estimated by using (5.99) [28]:

$$\langle |\mathbf{E}|^2 \rangle = E_0^2 = \frac{8\pi\eta}{\lambda^2 \eta_{tot}} \langle P_r \rangle \tag{5.122}$$

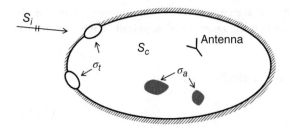

Figure 5.25 Aperture excitation of a cavity.

where E_0^2 is the mean-square value of the total E-field, η_{tot} is the total efficiency of the receiving antenna and $\eta = 120\pi$ is the intrinsic impedance of the medium. The mean-square value of the rectangular component of the E-field can also be obtained [28]:

$$\left\langle |E_x|^2 \right\rangle = \left\langle |E_y|^2 \right\rangle = \left\langle |E_z|^2 \right\rangle = \frac{1}{3} \left\langle |\mathbf{E}|^2 \right\rangle = \frac{8\pi\eta}{3\lambda^2 \eta_{\text{tot}}} \langle P_{\text{r}} \rangle \tag{5.123}$$

From (5.60) it can be found that the mean value of $|E_x|$ is $\sigma\sqrt{\pi/2}$, and from (5.61) we have the mean value of $|E_x|^2$, which is $2\sigma^2$, thus

$$\langle |E_x| \rangle = \frac{\sqrt{\pi}}{2} \sqrt{\left\langle |E_x|^2 \right\rangle} \tag{5.124}$$

By substituting (5.123) into (5.124) and letting $\eta = 120\pi$ (for free space), we have the mean value of the magnitude of the rectangular component of the E-field [28]:

$$\langle |E_x| \rangle = \frac{4\pi\sqrt{5\pi}}{\lambda\sqrt{\eta_{\text{tot}}}} \sqrt{\langle P_{\text{r}} \rangle} \tag{5.125}$$

Equation (5.124) can also be written as

$$\langle |E_x| \rangle = \left\langle \sqrt{|E_x|^2} \right\rangle = \frac{\sqrt{\pi}}{2} \sqrt{\left\langle |E_x|^2 \right\rangle} \tag{5.126}$$

Because $\langle P_{\text{r}} \rangle \propto \langle |E_x|^2 \rangle$, we have

$$\left\langle \sqrt{P_{\text{r}}} \right\rangle = \frac{\sqrt{\pi}}{2} \sqrt{\langle P_{\text{r}} \rangle} \tag{5.127}$$

It should be noted that $\left\langle \sqrt{P_{\text{r}}} \right\rangle \neq \sqrt{\langle P_{\text{r}} \rangle}$, so from (5.127) and (5.125) we have [28]

$$\langle |E_x| \rangle = \frac{8\pi\sqrt{5}}{\lambda\sqrt{\eta_{\text{tot}}}} \left\langle \sqrt{P_{\text{r}}} \right\rangle \tag{5.128}$$

Empirically, (5.128) can be used to estimate the magnitude of the rectangular E-field [28]:

$$|E_x| \approx \frac{8\pi\sqrt{5}}{\lambda\sqrt{\eta_{\text{tot}}}} \sqrt{P_{\text{r}}} \tag{5.129}$$

$$\max(|E_x|) \approx \frac{8\pi\sqrt{5}}{\lambda\sqrt{\eta_{\text{tot}}}} \sqrt{\max(P_{\text{r}})} \tag{5.130}$$

$$\min(|E_x|) \approx \frac{8\pi\sqrt{5}}{\lambda\sqrt{\eta_{\text{tot}}}} \sqrt{\min(P_{\text{r}})} \tag{5.131}$$

It is interesting to note that (5.129) is the same as the result using the antenna factor of an infinite small dipole antenna in free space. If the received power is P_{r}, the E-field can be obtained as

$$|E_x| = AF \times V = \frac{4\sqrt{3}\pi/\sqrt{5}}{\lambda\sqrt{\eta_{\text{tot}}}\sqrt{D}} \times \sqrt{P_{\text{r}} \times 50} = \frac{8\pi\sqrt{5}}{\lambda\sqrt{\eta_{\text{tot}}}} \sqrt{P_{\text{r}}} \tag{5.132}$$

where AF is the antenna factor and $D = 1.5$ is the directivity of the infinite small dipole antenna. However, for the case with $D \neq 1.5$, if the antenna factor is used to calculate the E-field strength, a wrong result will be obtained, even though the field probe has been calibrated in free space. Actually, the antenna factor cannot be used in the RC to obtain the E-field strength from the received power because the received power of the antenna is independent of the pattern of the antenna (antenna factor), which is given in (5.118).

From the mean value of $|\mathbf{E}|$ in (5.63) and the mean value of $|E_x|$ we have [28]

$$\langle|\mathbf{E}|\rangle = \frac{15\sigma\sqrt{2\pi}/16}{\sigma\sqrt{\pi/2}}\langle|E_x|\rangle = \frac{15}{8}\langle|E_x|\rangle \tag{5.133}$$

From (5.125) and (5.128) we can obtain [28]

$$\langle|\mathbf{E}|\rangle = \frac{15\pi\sqrt{5\pi}}{2\lambda\sqrt{\eta_{tot}}}\sqrt{\langle P_r\rangle} = \frac{15\pi\sqrt{5}}{\lambda\sqrt{\eta_{tot}}}\langle\sqrt{P_r}\rangle \tag{5.134}$$

The extreme value distributions can be obtained from the probability distribution of each sample. Suppose the PDF of random variable X is $f(x)$ and the cumulative distribution function (CDF) is $F(x)$. The PDF of the minimum of N independent samples is

$$f_{\min N}(x) = N[1-F(x)]^{N-1}f(x), \text{ where } x = \min(x_1, x_2, \ldots, x_N) \tag{5.135}$$

The PDF of the maximum of N independent samples is

$$f_{\max N}(x) = N[F(x)]^{N-1}f(x), \text{ where } x = \max(x_1, x_2, \ldots, x_N) \tag{5.136}$$

Considering the exponential PDF in (5.61), the CDF can be obtained as

$$CDF(|E_x|^2) = 1 - \exp\left(-\frac{x}{2\sigma^2}\right), x = |E_x|^2 \tag{5.137}$$

The extreme PDFs can be obtained from (5.135) and (5.136), which are also the extreme PDFs of the received power:

$$f_{\min N}(x) = \frac{N}{2\sigma^2}\exp\left(\frac{-xN}{2\sigma^2}\right), x = \min(|E_{x1}|^2, |E_{x2}|^2, \ldots, |E_{xN}|^2) \tag{5.138}$$

The mean value and the standard deviation of (5.138) are both $2\sigma^2/N$ [28]:

$$\left\langle\lfloor|E_x|^2\rfloor_N\right\rangle = std\left(\lfloor|E_x|^2\rfloor_N\right) = 2\sigma^2/N \tag{5.139}$$

where $\lfloor*\rfloor_N$ means the minimum of N independent samples of $*$ and std means the standard deviation.

$$f_{\max N}(x) = \frac{N}{2\sigma^2}\left[1-\exp\left(\frac{-x}{2\sigma^2}\right)\right]^{N-1}\exp\left(\frac{-x}{2\sigma^2}\right), x = \max(|E_{x1}|^2, |E_{x2}|^2, \ldots, |E_{xN}|^2) \tag{5.140}$$

The mean value and the standard deviations are [28]

$$\left\langle\lceil|E_x|^2\rceil_N\right\rangle = 2\sigma^2\sum_{i=1}^{N}\frac{1}{i} \approx 2\sigma^2\left(0.577 + \ln N + \frac{1}{2N}\right), N > 1 \tag{5.141}$$

$$std\left(\lceil|E_x|^2\rceil_N\right) = 2\sigma^2\sqrt{\sum_{i=1}^{N}\frac{1}{i^2}} \approx 2\sigma^2\sqrt{\frac{\pi^2}{6} - \frac{1}{N}}, N > 5 \tag{5.142}$$

where $\lceil*\rceil_N$ means the maximum of N independent samples of $*$. The mean value in (5.141) is normalised to the mean value of $|E_x|^2$ and plotted in Figure 5.26a with different sample number, and the normalised standard deviation is given in Figure 5.26b. The Matlab code used to plot this is given in Appendix A.6. It should be noted that the maximum received power depends on the number of samples, thus the sample number (stirrer position number) should be the same as that in the calibration process if the maximum received power is of interest.

(a)

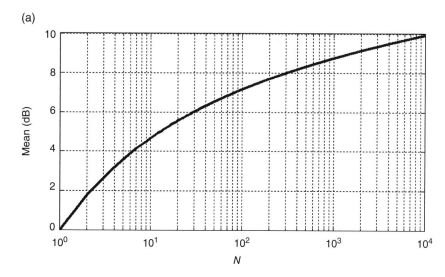

(b)

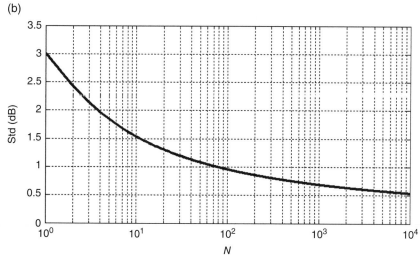

Figure 5.26 (a) Normalised mean value of the maximum received power with different sample number $10\log_{10}\langle \lceil |E_x|^2 \rceil_N \rangle / \langle |E_x|^2 \rangle$ and (b) normalised standard deviation of the maximum received power with different sample number $10\log_{10}[std(\lceil |E_x|^2 \rceil_N) + \langle \lceil |E_x|^2 \rceil_N \rangle] / \langle \lceil |E_x|^2 \rceil_N \rangle$.

Considering the Rayleigh distribution in (5.60), we have

$$f_{\min N}(x) = \frac{Nx}{\sigma^2} \exp\left(\frac{-x^2 N}{2\sigma^2}\right), x = \min(|E_{x1}|, |E_{x2}|, \ldots, |E_{xN}|) \tag{5.143}$$

$$\langle \lfloor |E_x| \rfloor_N \rangle = \sigma\sqrt{\pi/(2N)} \tag{5.144}$$

$$std\left(\lfloor |E_x| \rfloor_N\right) = \sigma\sqrt{2-\pi/2}/\sqrt{N} \tag{5.145}$$

$$f_{\max N}(x) = \frac{Nx}{\sigma^2}\left[1 - \exp\left(\frac{-x^2}{2\sigma^2}\right)\right]^{N-1} \exp\left(\frac{-x^2}{2\sigma^2}\right), x = \max(|E_{x1}|, |E_{x2}|, \ldots, |E_{xN}|) \tag{5.146}$$

There are no analytical expressions for $\langle \lceil |E_x| \rceil_N \rangle$ and $std(\lceil |E_x| \rceil_N)$, and they need to be evaluated numerically from

$$\langle \lceil |E_x| \rceil_N \rangle = \int_0^\infty \frac{Nx^2}{\sigma^2} \left[1 - \exp\left(\frac{-x^2}{2\sigma^2}\right)\right]^{N-1} \exp\left(\frac{-x^2}{2\sigma^2}\right) dx \tag{5.147}$$

and

$$std\left(\lceil |E_x| \rceil_N\right) = \sqrt{\int_0^\infty \left(x - \langle \lceil |E_x| \rceil_N \rangle\right)^2 \frac{Nx^2}{\sigma^2} \left[1 - \exp\left(\frac{-x^2}{2\sigma^2}\right)\right]^{N-1} \exp\left(\frac{-x^2}{2\sigma^2}\right) dx} \tag{5.148}$$

respectively. The mean value in (5.147) is normalised to the mean value of $|E_x|$ and plotted in Figure 5.27a with different sample number, and the normalised standard deviation is given in Figure 5.27b. The Matlab code used to plot this is given in Appendix A.7. It should be noted that around 1 dB difference can be observed in

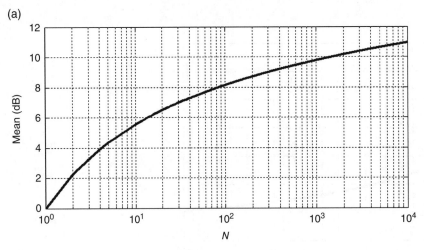

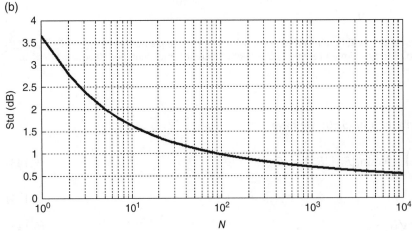

Figure 5.27 (a) Normalised mean value of the maximum rectangular E-field with different sample number $20\log_{10}\langle \lceil |E_x| \rceil_N \rangle / \langle |E_x| \rangle$ and (b) normalised standard deviation of the maximum rectangular E-field with different sample number $20\log_{10}[std(\lceil |E_x| \rceil_N) + \langle \lceil |E_x| \rceil_N \rangle] / \langle \lceil |E_x| \rceil_N \rangle$.

Figures 5.26a and 5.27a when N is large. This should be no surprise because the references they use have 1 dB difference. From (5.60), the mean value of $|E_x|$ is $\sigma\sqrt{\pi/2}$ and from (5.61) the mean value of $|E_x|^2$ is $2\sigma^2$, so the difference is $10\log_{10}2\sigma^2 - 20\log_{10}\sigma\sqrt{\pi/2} = 20\log_{10}(2/\sqrt{\pi}) \approx 1.05\,\mathrm{dB}$.

Similarly, from (5.64) we have the extreme PDFs of the squared magnitude of the total E-field [28]:

$$f_{\min N}(x) = \frac{Nx^2}{16\sigma^6}\exp\left(\frac{-Nx}{2\sigma^2}\right)\left[\sum_{k=0}^{2}\frac{(x/2\sigma^2)^k}{k!}\right]^{N-1}, \; x = \lfloor|\mathbf{E}|^2\rfloor_N \tag{5.149}$$

$$f_{\max N}(x) = \frac{Nx^2}{16\sigma^6}\exp\left(\frac{-x}{2\sigma^2}\right)\left[1-\exp\left(\frac{-x}{2\sigma^2}\right)\sum_{k=0}^{2}\frac{(x/2\sigma^2)}{k!}\right]^{N-1}, \; x = \lceil|\mathbf{E}|^2\rceil_N \tag{5.150}$$

From (5.63) we have the extreme PDFs of the magnitude of the total E-field [28]:

$$f_{\min N}(x) = \frac{Nx^5}{8\sigma^6}\exp\left(\frac{-Nx^2}{2\sigma^2}\right)\left[\sum_{k=0}^{2}\frac{(x^2/2\sigma^2)^k}{k!}\right]^{N-1}, \; x = \lfloor|\mathbf{E}|\rfloor_N \tag{5.151}$$

$$f_{\max N}(x) = \frac{Nx^5}{8\sigma^6}\exp\left(\frac{-x^2}{2\sigma^2}\right)\left[1-\exp\left(\frac{-x^2}{2\sigma^2}\right)\sum_{k=0}^{2}\frac{(x^2/2\sigma^2)}{k!}\right]^{N-1}, \; x = \lceil|\mathbf{E}|\rceil_N \tag{5.152}$$

There are no simple forms of the mean and the standard deviation of (5.149)–(5.152), so numerical integration needs to be performed to evaluate them.

The PDF of the transformed variable can be obtained by using the transformation technique in statistics: for a random variable X with the PDF $f(x)$, in each subset $\left(x_p^{\min}, x_p^{\max}\right), p = 1, 2, \ldots, f(x)$ is increasing or decreasing and differentiable, and the transformation function $Y = y(X)$ will lead to a random variable Y whose PDF $g(y)$ will be

$$g(y) = \sum_p f_p(x(y))\left|\frac{dx_p}{dy}\right| \tag{5.153}$$

where $x(y)$ is the inverse function of $y(x)$. A frequently used transform is $Y = k\log_{10}(X/C)$, where k is normally 10 or 20, and C is a constant that is normally the mean value. If the PDF of X is $f(x)$, by using (5.153) the PDF of Y can be obtained as

$$g(y) = f\left(10^{y/k}C\right)\left|\frac{d\left(10^{y/k}C\right)}{dy}\right| = \frac{f\left(10^{y/k}C\right)10^{y/k}\ln(10)C}{k} \tag{5.154}$$

5.5 Figures of Merit

In Chapter 4, figures of merit for an AC were introduced. There are also many figures of merit for an RC to characterise its performance or quantify how close an RC is to an ideal RC.

5.5.1 Field Uniformity

Like the FU definition for an AC, the FU for an RC also describes how uniform the field is in an RC, but we have to bear in mind that the field in an RC can never be uniform because of the resonance and the standing waves. The word 'uniform' in the RC actually means *statistically* uniform, and for brevity the word 'statistical' is often omitted.

The FU in the RC is defined as the standard deviation from the normalised mean value of the normalised maximum values obtained at each of the eight locations during one rotation of the tuner/stirrer [45]. The FU defined in [45] is based on the maximum magnitude of the E-field measured by using the field probes, but the maximum or average received power (of one rotation of the stirrer) can also be used to obtain the FU [5]. The FU should be related to the application of the RC: if the maximum E-field is of interest, the FU of the maximum E-field/received power should be used; if the average E-field is of interest, the FU of the average E-field/received power should be used. For an ideal RC, because of the statistical uniformity and isotropy given in Section 5.4.1, the standard deviation is zero. The IEC standard [45] has given the minimum sample number (lower boundary) to measure the FU. More samples and number of frequencies can be used in practice to obtain more information and more understanding.

A typical measurement setup is shown in Figure 5.28. Eight field probes are located at the corner of the working volume, and the boundary of the working volume is chosen to be at least 1 m from the walls and $\lambda/4$ from any object. The reason why $\lambda/4$ is used can be understood from Section 5.4: a certain distance is required to have small field correlations and good FU. Each rectangular component of the E-field (E_x, E_y, E_z) is recorded for each stirrer position and each probe position, and the probes do not necessarily need to be oriented along the chamber axes [45]. The minimum sample number of stirrer positions and frequencies are specified in the standard and given in Table 5.1 [45]. f_s is the start frequency in the measurement, which can be smaller than the lowest usable frequency (LUF) so that the dependency of the standard deviation to the frequency can be observed in a wide frequency range. Note that the recommended sample number in the first edition of IEC 61000-4-21 is larger than that in the Table 5.1. This is understandable, as we only need to define the minimum sample number and it can be increased depending on practical/customer requirement.

After a revolution of the stirrer, the maximum magnitude of the rectangular E-field can be obtained and normalised to the input power P_{input} or the net input power $P_{net} = P_{input} - P_{reflected}$ if the reflection is large:

$$\begin{Bmatrix} \widetilde{E}_x \\ \widetilde{E}_y \\ \widetilde{E}_z \end{Bmatrix} = \begin{Bmatrix} E_{\max x} \\ E_{\max y} \\ E_{\max z} \end{Bmatrix} / \sqrt{P_{input}} \tag{5.155}$$

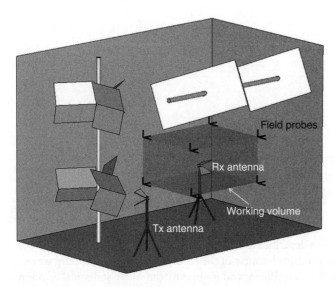

Figure 5.28 Typical FU measurement setup in the RC.

Table 5.1 Sampling requirements [45].

Frequency range	Minimum number of samples[a] required for validation and test[b]	Number of frequencies[c] required for validation
f_s–$3f_s$[d]	12 (50)	20
$3f_s$–$6f_s$	12 (18)	15
$6f_s$–$10f_s$	12 (12)	10
Above $10f_s$	12 (12)	20/decade

[a] The minimum number of tuner steps is 12 for all frequencies. For many chambers the number of tuner steps will need to be increased at lower frequencies. The maximum number of tuner steps is the number of independent samples that a given tuner can produce. This number varies with frequency and needs to be verified when commissioning the chamber. In the event that the chamber fails to meet the uniformity requirement, the number of tuner steps may be increased up to the number of independent tuner samples. The numbers in parenthesis are recommended from IEC 61000-4-21: 2003.
[b] The tuner sequencing used for validation of the chamber shall be the same as for subsequent testing.
[c] Log spaced.
[d] f_s = measurement start frequency.

thus we have eight normalised maximum field (eight probes) for each rectangular component, and the mean value of each rectangular component can be estimated from the average of these eight values:

$$\left\langle \widetilde{E}_x \right\rangle_8 = \frac{1}{8} \sum_{i=1}^{8} \widetilde{E}_{x,i}, \left\langle \widetilde{E}_y \right\rangle_8 = \frac{1}{8} \sum_{i=1}^{8} \widetilde{E}_{y,i}, \left\langle \widetilde{E}_z \right\rangle_8 = \frac{1}{8} \sum_{i=1}^{8} \widetilde{E}_{z,i} \tag{5.156}$$

The mean value of the normalised maximum of all the E-field values can be estimated as:

$$\left\langle \widetilde{E} \right\rangle_{24} = \frac{\left\langle \widetilde{E}_x \right\rangle_8 + \left\langle \widetilde{E}_y \right\rangle_8 + \left\langle \widetilde{E}_z \right\rangle_8}{3} \tag{5.157}$$

The standard deviations of each rectangular component and all components can be obtained:

$$\sigma_x = \sqrt{\frac{\sum_{i=1}^{8} \left(\widetilde{E}_{x,i} - \left\langle \widetilde{E}_x \right\rangle_8 \right)^2}{8-1}} \tag{5.158}$$

$$\sigma_y = \sqrt{\frac{\sum_{i=1}^{8} \left(\widetilde{E}_{y,i} - \left\langle \widetilde{E}_y \right\rangle_8 \right)^2}{8-1}} \tag{5.159}$$

$$\sigma_z = \sqrt{\frac{\sum_{i=1}^{8} \left(\widetilde{E}_{z,i} - \left\langle \widetilde{E}_z \right\rangle_8 \right)^2}{8-1}} \tag{5.160}$$

$$\sigma_{xyz} = \sqrt{\frac{\sum_{i=1}^{8} \sum_{p=x,y,z} \left(\widetilde{E}_{p,i} - \left\langle \widetilde{E} \right\rangle_{24} \right)^2}{24-1}} \tag{5.161}$$

The standard deviation relative to the mean value in dB form can be calculated from the linear value and the mean value:

$$\sigma(\text{dB}) = 20\log_{10}\frac{\sigma(\text{linear}) + \text{mean}}{\text{mean}} \tag{5.162}$$

where the mean value can be $\left\langle\widetilde{E}_x\right\rangle_8$, $\left\langle\widetilde{E}_y\right\rangle_8$, $\left\langle\widetilde{E}_z\right\rangle_8$ and $\left\langle\widetilde{E}\right\rangle_{24}$ for (5.158)–(5.161), respectively. If the average value is of interest, in (5.155) we need to use the average value but not maximum value. If the received power is used, $10\log_{10}$ in (5.162) should be used.

The antenna validation factor (AVF) can be obtained from the average received power of the Rx antenna:

$$AVF = \left\langle\frac{P_{\text{AveRec}}}{P_{\text{input}}}\right\rangle \tag{5.163}$$

The AVF is used to provide a baseline for comparison with a loaded chamber, and the FU is different for different AVF. The FU can be improved when loading absorbers, but the FU will deteriorate when the RC is overloaded [46]. If the RC is loaded, the Q factor decreases, the average mode bandwidth increases, and from (5.29) more modes can be excited and we have more independent sample numbers, thus the FU is improved. If the RC is overloaded, the wave incoming from all directions will be nonuniform (waves from the absorber are small) and the FU will deteriorate. The AVF can be used to monitor the status of the RC when the equipment under test (EUT) is in the RC; this is because when the EUT is inside the RC it may not be convenient to use probes to perform the FU measurement because the working volume is occupied by the EUT. If the FU with different AVFs is already known in the calibration process, when the EUT is inside the RC, the AVF can be used to validate the FU of the RC without repeating the calibration process.

The FU tolerance has been given in [45] and plotted in Figure 5.29. From 80 to 100 MHz the tolerance is 4 dB, from 100 to 400 MHz the tolerance decreases linearly from 4 to 3 dB and above 400 MHz the tolerance is 3 dB. For a good RC, all σ from (5.158)–(5.161) have to be lower than this tolerance. Different standards can have different tolerances.

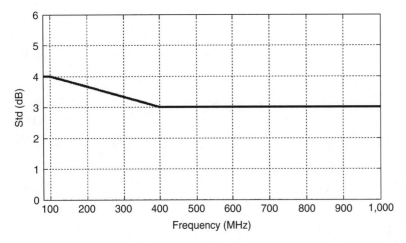

Figure 5.29 Field uniformity tolerance requirements.

5.5.2 Lowest Usable Frequency

The LUF is defined as the frequency at which the chamber meets the requirements of the FU in Figure 5.29 [45]. Since there is no simple equation to calculate the LUF, empirical conclusions have been given:

1) The LUF typically occurs at a frequency slightly above three times the first chamber resonance [45].
2) The LUF is located about five or six times above the first chamber resonance [4].
3) The LUF occurs at a frequency slightly above three to six times the first chamber resonance, $3f_{1st} \leq LUF \leq 7f_{1st}$, where f_{1st} is the first resonant frequency [47].
4) The mode density should be larger than 1 mode/MHz [3].
5) The enclosure has at least 60 possible modes using Weyl's formula (5.14): $N_W(LUF) = 60$ [45].
6) The LUF is three times of the first resonant frequency or $N_W(LUF) \geq 100$ and the mode density should be larger than 1.5 modes/MHz [48, 49].

In practice, the stirrer design and the chamber quality factor also affect the LUF, thus for different RC the LUF can have different empirical values. It should be noted that the definition of LUF can depend on the application of the RC: if the FU is of interest one can follow the definition in the standard [45]; if the statistical distribution is of concern, the LUF can be defined from the rejection rate using hypothesis testing [50] or parameter estimation of a more general distribution (Weibull distribution) [4, 51, 52]. If an exponential power delay profile (PDP) is expected in the time domain, it can also be used to define an LUF [37]. There is no unique LUF definition which satisfies all applications of the RC.

5.5.3 Correlation Coefficient and Independent Sample Number

The independent sample number is an important parameter in the RC measurement. This is because most quantities measured in the RC have statistical meanings; we rarely take only one sample in a statistical environment. Suppose a quantity X measured in the RC has a statistical distribution, and the mean value of X is of interest. If we have an independent sample number of N_{ind}, generally, when N_{ind} is large the standard deviation of the mean value averaged from X_1, X_2, ..., X_{Nind} will be reduced by a factor of $\sqrt{N_{ind}}$ [53]:

$$std\left(\frac{1}{N_{ind}} \sum_{i=1}^{Nind} X_i\right) = \frac{1}{\sqrt{N_{ind}}} std(X) \tag{5.164}$$

If the maximum value is used, the standard deviation also reduces with the increase in N_{ind} (shown in Figure 5.27), but it decreases more slowly than $1/\sqrt{N_{ind}}$ [45]. Generally, if a small uncertainty is required, the independent sample number needs to be increased. The independent sample number from the tuner/stirrer can be obtained from the correlation coefficient of received power at n evenly spaced angular intervals over one tuner rotation [45]. The correlation coefficient ρ can be calculated using the following formula [45]:

$$\rho(\theta) = corr(\mathbf{x}, \mathbf{y}_\theta) = \frac{\frac{1}{n-1} \sum_{i}^{n} (x_i - u_x)(y_{\theta,i} - u_y)}{\sqrt{\frac{\sum_{i=1}^{n}(x_i - u_x)^2}{n-1} \frac{\sum_{i=1}^{n}(y_{\theta,i} - u_y)^2}{n-1}}} \tag{5.165}$$

where n is the number of samples taken over one tuner rotation, x_i is the received power value of each tuner/stirrer position, vector $\mathbf{x} = [x_1, x_2, x_3, ..., x_n]$ contains all received power values for all tuner positions, and $\mathbf{y}_\theta$ is the rotated version of vector $\mathbf{x}$ shifted by angle θ:

$$\begin{aligned}
\mathbf{y}_0 &= x = [x_1, x_2, x_3, ..., x_n] \\
\mathbf{y}_{\theta_1} &= [x_n, x_1, x_2, x_3 ..., x_{n-1}] \\
\mathbf{y}_{\theta_2} &= [x_{n-1}, x_n, x_1, x_2 ..., x_{n-2}] \\
&\quad ... \\
\mathbf{y}_{\theta_{(n-1)}} &= [x_2, x_3, x_4, x, ..., x_n, x_1]
\end{aligned} \tag{5.166}$$

u_x and u_y are the mean values of the received power averaged over all tuner positions:

$$u_x = \frac{1}{n} \sum_{i=1}^{n} x_i = u_y \tag{5.167}$$

If the angle step of the tuner is θ, the total number for one tuner rotation is $n = 360/\theta$, as shown in Figure 5.30.

Using the field strength instead of the received power is also acceptable, but it typically gives similar but slightly different values of ρ [45]. If a vector network analyser is used, x_i can be the measured $|S_{21}|^2$ or $|S_{21}|$.

After the correlation coefficient for different angle shifts is calculated, a threshold of $e^{-1} \approx 0.37$ is chosen to determine the uncorrelated angle. Because the correlation coefficient is also a random variable, for 95% confidence and values of $n \geq 100$, the threshold of the correlation coefficient can be approximated by [45, 54]:

$$\rho_{\text{threshold}} \approx 0.37 \left(1 - \frac{7.22}{n^{0.64}} \right) \tag{5.168}$$

After the angle $\Delta\theta$, $\rho(\Delta\theta) \leq \rho_{\text{threshold}}$, the number of independent samples can be obtained as

$$N_{\text{ind}} = \frac{360}{\Delta\theta} \tag{5.169}$$

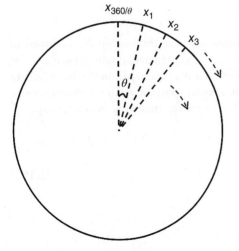

Figure 5.30 Illustration of samples over one tuner rotation.

The PDF of the first order correlation coefficient can be approximated by a normal distribution when $|\rho| \leq 0.5$ [4]:

$$PDF(\rho) \approx Norm(\mu_\rho, \sigma_\rho) = \frac{1}{\sigma_\rho \sqrt{2\pi}} \exp\left[-\frac{1}{2}\left(\frac{\rho - \mu_\rho}{\sigma_\rho}\right)^2\right] \quad \forall |\mu_\rho| \leq 0.5 \tag{5.170}$$

where

$$\rho = corr(\mathbf{x}, \mathbf{y}_{\theta_1}) \tag{5.171}$$

$$\sigma_\rho = \sqrt{\frac{n-1}{n^2}\left(1 - \mu_\rho^2\right)} \tag{5.172}$$

μ_ρ is the expected value (mean value), thus the uncertainty of ρ can also be estimated. A more general expression without the condition $|\rho| \leq 0.5$ in (5.170) has been given in [4, 54].

$$PDF(\rho) = \frac{n-2}{\sqrt{2\pi}} \frac{\Gamma(n-1)}{\Gamma(n-1/2)} \frac{\left(1 - \mu_\rho^2\right)^{\frac{n-1}{2}}\left(1 - \rho^2\right)^{\frac{n-4}{2}}}{\left(1 - \mu_\rho \rho\right)^{n-3/2}} A(\rho) \tag{5.173}$$

where

$$A(\rho) \approx 1 + \frac{1 + \mu_\rho \rho}{4(2n-1)} \tag{5.174}$$

and the function $\Gamma(\cdot)$ is the gamma function:

$$\Gamma(x) = \int_0^\infty t^{x-1} e^{-t} dt \tag{5.175}$$

The upper bound of the independent sample number can be evaluated theoretically. It has been shown that each mode in a rectangular cavity can be decomposed into a superposition of eight plane waves. From (5.4) and (5.5), the wave vectors of the eight plane waves are [4, 5]:

$$\mathbf{k}_1 = k_x\hat{\mathbf{x}} + k_y\hat{\mathbf{y}} + k_z\hat{\mathbf{z}}, \ \mathbf{k}_2 = k_x\hat{\mathbf{x}} + k_y\hat{\mathbf{y}} - k_z\hat{\mathbf{z}},$$
$$\mathbf{k}_3 = k_x\hat{\mathbf{x}} - k_y\hat{\mathbf{y}} + k_z\hat{\mathbf{z}}, \ \mathbf{k}_4 = k_x\hat{\mathbf{x}} - k_y\hat{\mathbf{y}} - k_z\hat{\mathbf{z}},$$
$$\mathbf{k}_5 = -k_x\hat{\mathbf{x}} + k_y\hat{\mathbf{y}} + k_z\hat{\mathbf{z}}, \ \mathbf{k}_6 = -k_x\hat{\mathbf{x}} + k_y\hat{\mathbf{y}} - k_z\hat{\mathbf{z}},$$
$$\mathbf{k}_7 = -k_x\hat{\mathbf{x}} - k_y\hat{\mathbf{y}} + k_z\hat{\mathbf{z}}, \ \mathbf{k}_8 = -k_x\hat{\mathbf{x}} - k_y\hat{\mathbf{y}} - k_z\hat{\mathbf{z}} \tag{5.176}$$

which correspond to the vertices of a cuboid in the spectrum domain shown in Figure 5.31. If only one stirring method (rotating the stirrer) is used, the maximum excited mode number can be obtained from the average

Figure 5.31 Wave vectors of eight independent plane waves.

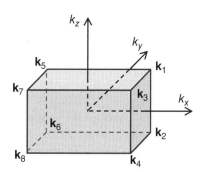

mode bandwidth Δf and the mode density D_w in (5.29). If a frequency stir of a bandwidth B_{FS} is also used, the maximum excited mode number becomes:

$$N_{mod} \le D_w(\Delta f + B_{FS}) \qquad (5.177)$$

thus the upper bound of the independent sample is:

$$N_{ind} \le 8N_{mod} \le 8D_w(\Delta f + B_{FS}) \qquad (5.178)$$

When the stirrer is rotated, the resonant frequencies are not constant but changed, thus a factor of B_{mech} which is defined as the mechanical mode stirring bandwidth is introduced to quantify this effect [13, 14, 53, 55] and the upper bound in (5.178) becomes

$$N_{ind} \le 8D_w(\Delta f + B_{FS} + B_{mech}) \qquad (5.179)$$

Practically, the independent sample number can be written as a product of different contributions, when polarisation, mechanical and frequency stir are enabled [53]:

$$N_{ind} = N_{pol} \cdot N_{mech} \cdot N_{source} \cdot N_{freq} \qquad (5.180)$$

where $N_{pol} = 2$ is the polarisation stir when the Tx antenna is a linear polarisation antenna and excites the RC in two orthogonal polarisations, respectively, N_{mech} is the independent sample number of the stirrer which can be obtained using (5.169), and N_{source} is the independent number of source/platform stir. If the Tx antenna or Rx antenna is moved to a distance larger than half a wavelength (shown in Figure 5.7), the received samples are independent (we assume zero correlation means independent, although it is not rigorous in mathematics). N_{freq} is the independent number of frequency stir, it can be obtained from the bandwidth used in the frequency stir B_{FS} and the average mode bandwidth Δf:

$$N_{freq} = \frac{B_{FS}}{\Delta f} + 1 \qquad (5.181)$$

5.5.4 Field Anisotropy Coefficients and Inhomogeneity Coefficients

To quantify how close an RC is to an ideal RC, the goodness-of-fit test method can be used [56–58], but a predefined acceptance level is required to define. The field anisotropy coefficient is an alternative quantitative approach to quantify the performance of the RC and the statistical distributions of the rectangular components are not required, thus it is a relatively quick way to assess the RC performance and easy to use [45]. The planar and total field anisotropy coefficients $A_{\alpha\beta}$ and A_{tot}, respectively, are defined via their averages as [45]:

$$\langle A_{\alpha\beta} \rangle = \left\langle \frac{|E_\alpha|^2/P_{\alpha i} - |E_\beta|^2/P_{\beta i}}{|E_\alpha|^2/P_{\alpha i} + |E_\beta|^2/P_{\beta i}} \right\rangle = \left\langle \frac{P_\alpha/P_{\alpha i} - P_\beta/P_{\beta i}}{P_\alpha/P_{\alpha i} + P_\beta/P_{\beta i}} \right\rangle \qquad (5.182)$$

$$\langle A_{tot} \rangle = \left\langle \sqrt{\left[A_{xy}^2 + A_{yz}^2 + A_{zx}^2\right]/3} \right\rangle \qquad (5.183)$$

where α or $\beta = x, y, z$ for a given stirrer state (position), $|E_\alpha|$ and $|E_\beta|$ represent the measured magnitude of the field strength of a rectangular component, P_α and P_β represent the received power, P_{*i} means the net input power of each sample, and $\langle \cdot \rangle$ means averaging over all stirrer positions. Ideally, when the RC is isotropic, $\langle A_{xy} \rangle = \langle A_{yz} \rangle = \langle A_{zx} \rangle = \langle A_{tot} \rangle = 0$. Typical numbers of samples used in the anisotropy coefficient measurement and the recommended values for $\langle A_{tot} \rangle$ are listed in Table 5.2.

$A_{\alpha\beta}$ is a random variable and more details on the PDF of $A_{\alpha\beta}$ have been given in [59]. Suppose the real part and the imaginary part of E_α are normally distributed (in (5.57)) and have the same standard deviation σ_α, and E_α, and E_β are statistically independent, then the PDF of $A_{\alpha\beta}$ can be obtained as [59]

Table 5.2 Typical values for total field anisotropy coefficients for 'medium' and 'good' reverberation quality [45].

	N = 10	N = 30	N = 100	N = 300
'Medium' stirring quality	−2.5 dB	−5 dB	−7.5 dB	−10 dB
'Good' stirring quality	−5 dB	−10 dB	−12.5 dB	−15 dB

$$PDF\left(A_{\alpha\beta}\right) = \frac{2\sigma_r}{\left[\sigma_r + 1 + (\sigma_r - 1)A_{\alpha\beta}\right]^2}, \quad -1 \le A_{\alpha\beta} \le 1 \tag{5.184}$$

where $\sigma_r = \sigma_\beta/\sigma_\alpha$. The CDF of $A_{\alpha\beta}$ is [59]:

$$CDF\left(A_{\alpha\beta}\right) = \frac{\left(1 + A_{\alpha\beta}\right)\sigma_r}{\sigma_r + 1 + (\sigma_r - 1)A_{\alpha\beta}}, \quad -1 \le A_{\alpha\beta} \le 1 \tag{5.185}$$

when $\sigma_r = 1$, $PDF(A_{\alpha\beta}) = 1/2$, $A_{\alpha\beta}$ is uniformly distributed. Then the mean and standard deviation of $A_{\alpha\beta}$ are [59]

$$E\left(A_{\alpha\beta}\right) = \frac{1 + 2\sigma_r \ln\sigma_r - \sigma_r^2}{(\sigma_r - 1)^2} \tag{5.186}$$

$$std\left(A_{\alpha\beta}\right) = 2\sqrt{\frac{\sigma_r - \left(2 + \ln^2\sigma_r\right)\sigma_r^2 + \sigma_r^3}{(\sigma_r - 1)^2}} \tag{5.187}$$

when $\sigma_r = 1$, $E(A_{\alpha\beta}) = 0$ and $std\left(A_{\alpha\beta}\right) = 1/\sqrt{3}$. The PDF of A_{tot} has also been given in [59], but it has a more complex form which is not discussed here.

A simplified definition of anisotropy with no statistical characterisation can be defined as [60]:

$$A_{\alpha\beta}^{avg} = \frac{\left\langle |E_\alpha|^2/P_{\alpha i} \right\rangle - \left\langle |E_\beta|^2/P_{\beta i} \right\rangle}{\left\langle |E_\alpha|^2/P_{\alpha i} \right\rangle + \left\langle |E_\beta|^2/P_{\beta i} \right\rangle} \tag{5.188}$$

$$A^{avg} = \sqrt{\left[\left(A_{xy}^{avg}\right)^2 + \left(A_{yz}^{avg}\right)^2 + \left(A_{zx}^{avg}\right)^2\right]/3} \tag{5.189}$$

These definitions are easy to use and only a single value is able to characterise the entire stirring process. The average values across the entire stirring process are needed and there is no need to store all the data at all stirring steps [60].

The definitions of the inhomogeneity coefficients are similar to those of the anisotropy coefficients [45]:

$$\left\langle I_\alpha(r_1, r_2) \right\rangle = \left\langle \frac{|E_\alpha(r_1)|^2/P_{\alpha 1i} - |E_\alpha(r_2)|^2/P_{\alpha 2i}}{|E_\alpha(r_1)|^2/P_{\alpha 1i} + |E_\alpha(r_2)|^2/P_{\alpha 2i}} \right\rangle = \left\langle \frac{P_\alpha(r_1)/P_{\alpha 1i} - P_\alpha(r_2)/P_{\alpha 2i}}{P_\alpha(r_1)/P_{\alpha 1i} + P_\alpha(r_2)/P_{\alpha 2i}} \right\rangle \tag{5.190}$$

$$\left\langle I_{tot}(r_1, r_2) \right\rangle = \left\langle \sqrt{\left[I_x^2 + I_y^2 + I_z^2\right]/3} \right\rangle \tag{5.191}$$

where r_1 and r_2 represent two locations separated by at least one wavelength. It has been found that the anisotropy and homogeneity coefficients are highly correlated statistics. The statistical isotropy implies statistical homogeneity but the converse is not necessarily true, and the assessment of field anisotropy is the more stringent test of the two [45, 61].

5.5.5 Stirring Ratio

The stirring ratio (SR) is defined as the ratio of the maximum amplitude and the minimum amplitude of the data collected during a rotation of 360° of a stirrer [62]:

$$SR\,(\text{dB}) = 10\log_{10}\frac{P_{\max}}{P_{\min}} = 20\log_{10}\frac{|V_{\max}|}{|V_{\min}|} \tag{5.192}$$

where $P_{\max}$ and $P_{\min}$ are the maximum and minimum received powers over a rotation of a stirrer, and $V_{\max}$ and $V_{\min}$ are the corresponding voltages. A low SR means the contribution of direct coupling dominates the received signal or the stirrer is too small to stir the field in the RC. Normally, an efficient mode stirring gives an SR ≥ 20 dB [4].

The SR is easy to use but no statistical distribution is involved, thus the use of the SR is not rigorous. Additional information is normally needed to justify the effectiveness. At the resonant frequency below the LUF, because of resonance, one can easily obtain a very high SR, but the RC may not be usable at this frequency. If the SR keeps high values in a certain bandwidth, it could be more convincing to conclude that the stirrer is effective. One can also apply the hypothesis test to test if the collected samples follow a specific distribution.

5.5.6 *K*-Factor

The *K*-factor (or Rician *K*-factor) is defined as a ratio between the direct power and the stirred power in the RC. From communication channel perspective, when $K = 0$ the channel is a Rayleigh channel (Rayleigh fading), and when $K = \infty$ the channel is a non-fading channel (Gaussian channel). *K*-factor has also been used to characterise the performance of the RC [63] or emulate a controllable communication channel in the RC [3, 64].

The stirred field E_S and unstirred field E_d are illustrated in Figure 5.32. The stirred field fully interacts with the stirrers while the unstirred field is from the direct illumination and the scattered field from the unmoving objects. Thus the E-field at the Rx antenna can be expressed as

$$E_\theta = E_{S\theta} + E_{d\theta} \tag{5.193}$$

Only the θ component is considered; for other components, the expression is the same.

$$E_{S\theta} = E_{S\theta r} + jE_{S\theta i} \tag{5.194}$$

$E_{S\theta r}$ and $E_{S\theta i}$ are the real and imaginary parts which follow the normal (Gaussian) distribution as given in (5.57), where the variances are:

$$\left\langle (E_{S\theta r} - \langle E_{S\theta r}\rangle)^2 \right\rangle = \left\langle E_{S\theta r}^2 \right\rangle = \left\langle E_{S\theta i}^2 \right\rangle = \frac{\eta\lambda QP_t}{12\pi V} = \sigma^2 \tag{5.195}$$

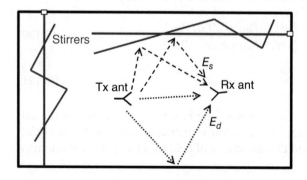

Figure 5.32 Stirred field E_s and unstirred field E_d.

The amplitude of the θ component has a Rice PDF [3, 46, 65]:

$$PDF(|E_\theta|) = \frac{|E_\theta|}{\sigma^2} I_0\left(\frac{|E_{S\theta}||E_{d\theta}|}{2\sigma^2}\right) \exp\left(-\frac{|E_{S\theta}|^2 + |E_{d\theta}|^2}{2\sigma^2}\right) \tag{5.196}$$

where $I_0(\cdot)$ is the modified Bessel function of zero order

$$I_0(x) = \sum_{m=0}^{\infty} \frac{1}{(m!)^2}\left(\frac{x}{2}\right)^{2m} \tag{5.197}$$

From (5.61) we have $\langle|E_{S\theta}|^2\rangle = 2\sigma^2$ and the Rician K-factor is defined as [66]

$$K = \frac{|E_{d\theta}|^2}{2\sigma^2} = \frac{|E_{d\theta}|^2}{\langle|E_{S\theta}|^2\rangle} \tag{5.198}$$

Thus the K-factor describes the power ratio between the unstirred part and the stirred part. When $K \to 0$, (5.196) reduces to the Rayleigh PDF in (5.60).

The K-factor can be controlled using one or two antennas. If one antenna is used, assume the unstirred part is dominated by the direct illumination [64]

$$|E_{d\theta}|^2 = \frac{\eta}{4\pi r^2}P_t D \tag{5.199}$$

where r is the distance between the Tx antenna and the device under test (DUT) and P_t is the transmitted power. D is the directivity of the Tx antenna. We assume the antenna efficiency is high, and the mismatch and the loss of the antenna is ignored. From (5.195) and (5.199), the K-factor can be expressed as [64]

$$K = \frac{|E_{d\theta}|^2}{2\sigma^2} = \frac{3}{2}\frac{V}{\lambda Q}\frac{D}{r^2} \tag{5.200}$$

From (5.200), we know that the K-factor can be controlled by the directivity of the Tx antenna, the distance between the antenna and the DUT, and the quality factor of the chamber.

It is also possible to control the K-factor by using two antennas to have a larger varying range. A second antenna (antenna 2) can be involved but not directed to the DUT, thus waves from antenna 2 contribute the stirred part but not the unstirred part. In this case, σ^2 becomes [64]

$$\sigma^2 = \frac{\eta\lambda Q(P_{t1} + P_{t2})}{12\pi V} \tag{5.201}$$

where P_{t1} and P_{t2} are the transmitted powers of two antennas. In this case, (5.200) becomes [64]

$$K = \frac{3}{2}\frac{V}{\lambda Q}\frac{D_1}{r^2}\frac{P_{t1}}{P_{t1} + P_{t2}} \tag{5.202}$$

where D_1 is the directivity of antenna 1.

In practical measurements, if a network analyser is connected to two antennas (the Tx antenna and the Rx antenna), the K-factor between the two antennas can be calculated from the measured S parameters [64]

$$K = \frac{|\langle S_{21}\rangle^2|}{\langle|S_{21} - \langle S_{21}\rangle|^2\rangle} \tag{5.203}$$

The K-factor can also be treated as a random variable. When the K-factor is very small ($K < -10$ dB), a more accurate unbiased estimator $\langle\hat{K}\rangle$ has been obtained in [67] as

$$\langle\hat{K}\rangle = \frac{N-2}{N-1}\frac{|\langle S_{21}\rangle^2|}{\langle|S_{21} - \langle S_{21}\rangle|^2\rangle} - \frac{1}{N} \tag{5.204}$$

where frequency stir can also be used and N is the independent sample number of S-parameters. It should be noted that when other stirring methods are involved, such as frequency stir or platform (source) stir, an average K-factor can be obtained [68].

5.6 Summary

The fundamentals of the RC have been reviewed in this chapter. The well-known deterministic model was introduced first and followed by the statistical theory, and statistical properties of the fields inside the RC were reviewed. Basic figures of merit of the RC have been introduced. This chapter was mainly focused on theory while measurement procedures for some frequently used parameters such as Q factor and chamber decay time (decay constant) have not been given but will be provided in Chapter 7.

References

1 Harrington, R. F. (1961). *Time-Harmonic Electromagnetic Fields*. New York: McGraw-Hill.

2 Balanis, C. A. (2012). *Advanced Engineering Electromagnetics*, 2e. Wiley.

3 Hill, D. A. (2009). *Electromagnetic Fields in Cavities: Deterministic and Statistical Theories*. Wiley-IEEE Press.

4 Demoulin, B. and Besnier, P. (2011). *Electromagnetic Reverberation Chambers*. Wiley.

5 Boyes, S. and Huang, Y. (2016). *Reverberation Chambers: Theory and Applications to EMC and Antenna Measurements*. Wiley.

6 Carlberg, U., Kildal, P.-S., and Carlsson, J. (2009). Numerical study of position stirring and frequency stirring in a loaded reverberation chamber. *IEEE Transactions on Electromagnetic Compatibility* 51 (1): 12–17.

7 Huang, Y. (1993). The Investigation of Chambers for Electromagnetic Systems. DPhil thesis, University of Oxford.

8 Liu, B. H., Chang, D. C., and Ma, M. T. (1983). *Eigenmodes and the composite quality factor of a reverberating chamber*. US National Bureau of Standards, Technical Note 1066.

9 Weyl, H. (1913). Uber die randwertaufgabe der randwertaufgabe der Strahlungstheorie und asymptotische Spektralgesetze. *Journal für die reine und angewandte Mathematik* 143: 177–202.

10 Strauss, W. A. (2008). *Partial Differential Equations: An Introduction*, 2e. Wiley.

11 Hill, D. A. (1996). A reflection coefficient derivation for the Q of a reverberation chamber. *IEEE Transactions on Electromagnetic Compatibility* 38: 591–592.

12 Rosengren, K. and Kildal, P.-S. (2001). Study of distributions of modes and plane waves in reverberation chambers for the characterization of antennas in a multipath environment. *Microwave and Optical Technology Letters* 30: 386–391.

13 Kildal, P.-S., Chen, X., Orlenius, C. et al. (2012). Characterization of reverberation chambers for OTA measurements of wireless devices: physical formulations of channel matrix and new uncertainty formula. *IEEE Transactions on Antennas and Propagation* 60 (8): 3875–3891.

14 Kildal, P.-S. (2015). *Foundations of Antenna Engineering: A Unified Approach for Line-of-Sight and Multipath*. Artech House.

15 Kwon, D.-H., Burkholder, R. J., and Pathak, P. H. (1998). Ray analysis of electromagnetic field build-up and quality factor of electrically large shielded enclosures. *IEEE Transactions on Electromagnetic Compatibility* 40 (1): 19–26.

16 Amador, E., Lemoine, C., Besnierm, P., and Laisné, A. (2010). Studying the pulse regime in a reverberation chamber with a model based on image theory. IEEE International Symposium on Electromagnetic Compatibility, Fort Lauderdale. pp. 520–525.

17 Amador, E., Lemoine, C., Besnier, P., and Laisné, A. (2010). Reverberation chamber modeling based on image theory: investigation in the pulse regime. *IEEE Transactions on Electromagnetic Compatibility* 52 (4): 778–789.

18 Pirkl, R.J., Ladbury, J.M., and Remley, K.A.. (2011). The reverberation chamber's unstirred field: A validation of the image theory interpretation. IEEE International Symposium on Electromagnetic Compatibility, Long Beach, CA. pp. 670–675.

19 Pirkl, R. J. and Remley, K. A. (2014). Experimental evaluation of the statistical isotropy of a reverberation chamber's plane-wave spectrum. *IEEE Transactions on Electromagnetic Compatibility* 56 (3): 498–509.

20 Choudhury, B., Gouramma, H.S. Jha, R.M., and Bommer, J.P. (2012). Comparison of image method and refined ray tracing method for aircraft cabin application. Proceedings of the 2012 IEEE International Symposium on Antennas and Propagation, Chicago. pp. 1–2.

21 Ladbury, J. and Hill, D.A. (2007). Enhanced backscatter in a reverberation chamber: inside every complex problem is a simple solution struggling to get out. IEEE International Symposium on Electromagnetic Compatibility, Honolulu. pp. 1–5.

22 Erying, C. F. (1930). Reverberation time in dead room. *Journal of the Acoustical Society of America* 1: 217–241.

23 DeLyser, R. R., Holloway, C. L., Johnk, R. T. et al. (1996). Figure of merit for low frequency anechoic chambers based on absorber reflection coefficients. *IEEE Transactions on Electromagnetic Compatibility* 38 (4): 576–584.

24 Holloway, C. L., Shah, H. A., Pirkl, R. J. et al. (2012). Early time behaviour in reverberation chambers and its effect on the relationships between coherence bandwidth, chamber decay time, RMS delay spread, and the chamber buildup time. *IEEE Transactions on Electromagnetic Compatibility* 54 (4): 714–725.

25 Dunens, E. K. and Lambert, R. F. (1977). Impulsive sound-level response statistics in a reverberant enclosure. *Journal of the Acoustical Society of America* 61 (6): 1524–1532.

26 Holloway, C. L., Cotton, M. G., and McKenna, P. (1999). A model for predicting the power delay profile characteristics inside a room. *IEEE Transactions on Vehicular Technology* 48 (4): 1110–1120.

27 Richardson, R. E. (2008). *Reverberant microwave propagation*. Dahlgrem, VA, Tech. Rep. NSWCDD/TR-08/127: Naval Surface Warfare Center.

28 Ladbury, J., Koepke, G., and Camell, D. (1999). *Evaluation of the NASA Langley research centre mode stirred chamber facility*. US National Institute of Standards and Technology, Tech. Note. 1508.

29 Hill, D. A. (1998). *Electromagnetic theory of reverberation chambers*. U.S. National Institute of Standards and Technology Technical Note 1506.

30 Kostas, J. G. and Boverie, B. (1991). Statistical model for a mode-stirred chamber. *IEEE Transactions on Electromagnetic Compatibility* 33 (4): 366–370.

31 Hill, D. A. (1995). Spatial correlation function for fields in a reverberation chamber. *IEEE Transactions on Electromagnetic Compatibility* 37 (1): 138.

32 Lehman, T.H. (1993). A statistical theory of electromagnetic fields in complex cavities, EMP Interaction Note 494.

33 Wolf, E. (1976). New theory of radiative energy transfer in free electromagnetic fields. *Physical Review D* 13: 869–886.

34 Cook, R. K., Waterhouse, R. V., Berendt, R. D. et al. (1955). Measurement of correlation coefficients in reverberant sound fields. *Journal of the Acoustical Society of America* 27: 1072–1077.

35 Hill, D. A. and Ladbury, J. M. (2002). Spatial-correlation functions of fields and energy density in a reverberation chamber. *IEEE Transactions on Electromagnetic Compatibility* 44 (1): 95–101.

36 Barabanenkov, Y. N., Kravtsov, Y. A., Ozrin, V. D., and Saichev, A. I. (1991). Enhanced backscattering: the universal wave phenomenon. *Proceedings of the IEEE* 79 (10): 1367–1370.

37 Dunlap, C.R. (2013). Reverberation chamber characterization using enhanced backscatter coefficient measurements. PhD dissertation, Department of Electrical, Computer and Engineering, University of Colorado, Boulder.

38 Holloway, C. L., Shah, H. A., Pirkl, R. J. et al. (2012). Reverberation chamber techniques for determining the radiation and total efficiency of antennas. *IEEE Transactions on Antennas and Propagation* 60 (4): 1758–1770.

39 Holloway, C.L., Smith, R.S., and Dunlap, C.R. et al. (2012). Validation of a two-antenna reverberation chamber technique for estimating the total and radiation efficiency of antennas. International Symposium on Electromagnetic Compatibility (EMC EUROPE*)*. pp. 1–6, 17–21.

40 Holloway, C.L., Smith, R.S., and Dunlap, C.R. et al. (2012). Validation of a one-antenna reverberation-chamber technique for estimating the total and radiation efficiency of an antenna. IEEE International Symposium on Electromagnetic Compatibility (EMC). pp. 205–209, 6–10.

41 Junqua, I., Degauque, P., Liénard, M., and Issac, F. (2012). On the power dissipated by an antenna in transmit mode or in receive mode in a reverberation chamber. *IEEE Transactions on Electromagnetic Compatibility* 54 (1): 174–180.

42 Tai, C. T. (1961). On the definition of effective aperture of antennas. *IEEE Transactions on Electromagnetic Compatibility* 9: 224–225.

43 Hill, D. A., Ma, M. T., Ondrejka, A. R. et al. (1994). Aperture excitation of electrically large, lossy cavities. *IEEE Transactions on Electromagnetic Compatibility* 36 (3): 169–178.

44 Loughry, T. A. (1991). *Frequency stirring: an alternate approach to mechanical mode stirring for the conduct of electromagnetic susceptibility testing*. Phillips Laboratory, Kirtland Air Force Base, NM Technical Report 91 1036.

45 IEC 61000-4-21 (2011). *Electromagnetic compatibility (EMC) – Part 4–21: Testing and measurement techniques – Reverberation chamber test methods*, 2.0e. IEC Standard.

46 Holloway, C. L., Hill, D. A., Ladbury, J. M., and Koepke, G. (2006). Requirements for an effective reverberation chamber: unloaded or loaded. *IEEE Transactions on Electromagnetic Compatibility* 48 (1): 187–194.

47 Rosnarho, J.–F. and Berre, S. L.. (2016). Reverberation Chambers Handbook, Everything you ever wanted to know about reverberation chambers but never dared to ask, Siepel, ver. 3.

48 Bruns, C. (2005). Three-Dimensional Simulation and Experimental Verification of a Reverberation Chamber. Doctor of Sciences dissertation, Swiss Federal Institute of Technology, Zurich.

49 Li, C., Yang, L., Lu, B. et al. (2016). A reverberation chamber for rodents' exposure to wideband radiofrquency electromagnetic fields with different small-scale fading distributions. *Electromagnetic Biology and Medicine* 35 (1): 30–39.

50 Fall, A. K., Besnier, P., Lemoine, C.et al. (2014). Determining the lowest usable frequency of a frequency-stirred reverberation chamber using modal density. International Symposium on Electromagnetic Compatibility, Gothenburg. pp. 263–268.

51 Orjubin, G., Richalot, E., Mengue, S., and Picon, O. (2006). Statistical model of an undermoded reverberation chamber. *IEEE Transactions on Electromagnetic Compatibility* 48 (1): 248–251.

52 Lemoine, C., Besnier, P., and Drissi, M. (2007). Investigation of reverberation chamber measurements through high-power goodness-of-fit tests. *IEEE Transactions on Electromagnetic Compatibility* 49 (4): 745–755.

53 Pucci, E.M. (2008). Calibration of reverberation chamber for wireless measurements: study of accuracy and characterization of the number of independent samples. Master of Science thesis, Department of Signals and Systems, Chalmers University of Technology, Sweden.

54 Krauthauser, H.G., Winzerling, T., and Nitsch, J. et al. (2005). Statistical interpretation of autocorrelation coefficients for fields in mode-stirred chambers. International Symposium on Electromagnetic Compatibility, Chicago, IL. pp. 550–555.

55 Chen, X., Kildal, P.S., and Carlsson, J. (2014). Characterization and modeling of measurement uncertainty in a reverberation chamber with a rotating mode stirrer. International Symposium on Electromagnetic Compatibility, Gothenburg. pp. 296–300.

56 Arnuat, L.R. and West, P.D. (1999). Electric Field Probe Measurements in the NPL Untuned Stadium Reverberation Chamber. National Physical Laboratory (UK), Technical Report CETM 13.

57 Arnaut, L. R. (2002). Compound exponential distributions for undermoded reverberation chambers. *IEEE Transactions on Electromagnetic Compatibility* 44 (3): 442–457.

58 Lunden, O. and Backstrom, M. (2000). Stirrer efficiency in FOA reverberation chambers. Evaluation of correlation coefficients and chi-squared tests. IEEE International Symposium on Electromagnetic Compatibility. Symposium Record, Washington, DC. Volume 1. pp. 11–16.

59 Arnaut, L. R., Serra, R., and West, P. D. (2017). Statistical anisotropy in imperfect electromagnetic reverberation. *IEEE Transactions on Electromagnetic Compatibility* 59 (1): 3–13.

60 Arnaut, L. R. (1998). Field anisotropy, field inhomogeneity and polarization bias in imperfect reverberation chambers, National Physical Laboratory (UK), Report ID: R981120.

61 Arnaut, L.R. and West, P. D. (1998). Evaluation of the NPL untuned stadium reverberation chamber using mechanical and electronic stirring techniques, National Physical Laboratory (UK), Technical Report CEM 11.

62 Crawford, M.L. and Koepke, G. H. (1986). Design, evaluation, and use of a reverberation chamber for performing electromagnetic susceptibility/vulnerability measurements. NBS Technical Note 1092.

63 Patane, C.S.L. (2010). Reverberation chamber performance and methods for estimating the Rician K-factor. Master of Science thesis, Department of Signals and Systems, Chalmers University of Technology, Sweden.

64 Holloway, C. L., Hill, D. A., Ladbury, J. M. et al. (2006). On the use of reverberation chambers to simulate a Rician radio environment for the testing of wireless devices. *IEEE Transactions on Antennas and Propagation* 54: 3167–3177.

65 Corona, P., Ferrara, G., and Migliaccio, M. (2000). Reverberating chamber electromagnetic field in presence of an unstirred component. *IEEE Transactions on Electromagnetic Compatibility* 42 (2): 111–115.

66 Steele, R. (1974). *Mobile Radio Communications*. New York, NY: IEEE Press.

67 Lemoine, C., Amador, E., and Besnier, P. (2011). On the K-factor estimation for Rician channel simulated in reverberation chamber. *IEEE Transactions on Antennas and Propagation* 59 (3): 1003–1012.

68 Chen, X., Kildal, P. S., and Lai, S. H. (2011). Estimation of average Rician K-factor and average mode bandwidth in loaded reverberation chamber. *IEEE Antennas and Wireless Propagation Letters* 10: 1437–1440.

6

The Design of a Reverberation Chamber

6.1 Introduction

In this chapter, the design of a reverberation chamber (RC) is introduced. If the forward problem is to analyse the performance of a given RC, the inverse problem is to design an RC with given specifications. An analysis problem is normally easier than a design (synthesis) problem: because of the complexity of the statistical electromagnetics, the design of an RC is very challenging, and there is no well-established solution. Like the design of an anechoic chamber (AC), guidelines have been provided to ensure that a margin of safety can be achieved. In this chapter, empirical guidelines for the RC design are introduced first and followed by the theoretical analysis. It can be found that the performance of the RC can be well understood from the time domain, and the design limit of the RC can be obtained theoretically.

6.2 Design Guidelines

The design guidelines for an RC are reviewed from different aspects, and most of them are empirical. Some of these guidelines are dependent, for example, if the working volume of an RC needs to be increased, the whole size of the RC and the size of the stirrers need to be scaled up as well. To minimise direct coupling, one can add extra scatters in the RC or involve more stirring method. Factors affecting the performance of an RC are discussed one by one.

6.2.1 The Shape of the RC

Rectangular shape has been widely used in the RC, but it may not be necessarily the best choice. Degenerate modes exist when m, n, and p are all nonzero in (5.7), and the TE_{mnp} and TM_{mnp} modes have the same resonant frequency. Similar mode degeneracy also exists in the triangular cavity [1]. In a cylindrical cavity, degenerate modes exist because of the radial symmetry of the shape [2, 3]. In a spherical cavity, there are more degenerate modes because of the perfect symmetry [2, 3]. It is expected there will be fewer degenerate modes and each mode has a different resonant frequency, thus for each working frequency the excited mode number is more uniform. It should be noted that when the frequency is high, the shape of the cavity only affects the frequency distribution of the modes and the mode number is determined by the volume and the third order of the frequency according to Weyl's law in (5.14).

For a rectangular RC, from (5.6) we have

$$\frac{k_{mnp}^2}{\pi^2} = \frac{m^2}{a^2} + \frac{n^2}{b^2} + \frac{p^2}{d^2} \tag{6.1}$$

Anechoic and Reverberation Chambers: Theory, Design, and Measurements, First Edition. Qian Xu and Yi Huang.
© 2019 John Wiley & Sons Ltd. Published 2019 by John Wiley & Sons Ltd.

Figure 6.1 RC with nonparallel walls and diffusers.

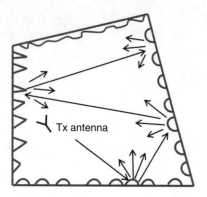

Tx antenna

If the ratios $a^2 : b^2 : d^2$ are rational, there may exist more than one set of $\{m, n, p\}$ values giving the same value of k_{mnp} and the degeneracy of modes arises [4]. Thus the ratio of the dimensions should not be rational to increase the uniformity in distribution of resonant modes:

$$a^2 : b^2 : d^2 \text{ are irrational} \tag{6.2}$$

From (6.2) we can see that a cubic RC should be avoided and a random and asymmetric shape is obviously a better choice [5]. However, when an asymmetric stirrer exists, the symmetry is broken and the degeneracy may not be that bad, and it has been found that a cubic RC does not exhibit worse performance than other rectangular RCs [6].

The RC can also be designed with no walls in parallel to avoid a repeatable propagation path [7] or by enhancing the chaotic phenomenon in the RC [8]. To improve the randomness of the structure, extra scatters/diffusers can be used in a rectangular RC; Schroeder diffusers [9–13] have been widely used in the acoustic RC, similar applications in the EM RC have also been investigated [14–21]. These techniques can be hybridised as shown in Figure 6.1.

The use of diffusors can make the unstirred part (early reflections and scatterings) more uniform; the stirred part contribution is from the stirrers and can be understood more clearly from the time domain, which will be introduced later in this chapter.

Metamaterial structures can also reduce the lowest usable frequency (LUF) and improve the field uniformity (FU), at low frequencies [22–24]; the working principle is different from the diffusers. Metamaterial structures or surfaces are designed to emulate an equivalent boundary condition or equivalent reflection coefficients to support more resonant modes at low frequencies, thus more independent samples can be obtained and the FU is improved.

6.2.2 The Lowest Usable Frequency

The LUF depends on the number of excited modes; the range of LUF is discussed in Section 5.5.2. From the design perspective, at the LUF, we have the mode number

$$N_W(f) = \frac{8\pi V f^3}{3} \frac{f^3}{c^3} \geq 60 \sim 100 \text{ modes} \tag{6.3}$$

and the mode density

$$D_W(f) = 8\pi V \frac{f^2}{c^3} \geq 1 \sim 1.5 \text{ modes/MHz} \tag{6.4}$$

If a rectangular RC is used, the mode number in (6.3) and mode density in (6.4) can be replaced by the mode number and the mode density in the rectangular cavity in (5.12) and (5.13), respectively.

It should be noted that the definition of LUF is from the standard deviation of the maximum electric field (E-field). For different applications the meaning of 'usable' can be different, e.g. in the antenna efficiency measurement without reference antenna, the power delay profile (PDP) is expected to decay exponentially so the decay constant can be extracted, but at low frequencies the PDP does not follow an exponential decay and may not satisfy this condition [25]; the frequency that satisfies the goodness-of-fit test may not be exactly the same as the frequency that satisfies the standard deviation. The 'usable' in this application is different from the 'usable' of maximum E-field.

6.2.3 The Working Volume

The working volume depends on the application of the RC and the edge of the working volume is expected to be at least $\lambda/4$, or $\lambda/2$ more strictly, from the boundary of the RC. The effect of the boundaries has been given in Section 5.4.3. If the working volume is chosen to be larger than $a_w \times b_w \times d_w$, the LUF is chosen to be 80 MHz (the wavelength is 3.75 m), and the dimensions of the RC should be at least

$$a > (a_w + 2), b > (b_w + 2), d > (d_w + 2), \text{unit} : \text{metre} \tag{6.5}$$

where a_w, b_w, and d_w are the dimensions of the designed working volume, and a, b, and d are the dimensions of the RC. Figure 6.2 is given to illustrate the working volume, where a quarter wavelength at 80 MHz is about 1 m ($0.9375 \approx 1$). Note that the dimensions of the stirrers have not been considered; these need to be added to the right-hand side of (6.5).

6.2.4 The Q Factor

The ratio between the unstirred power and the stirred power has been given in (5.200). In the RC we require the stirred part to dominate the response, thus we have [1]

$$K = \frac{3}{2} \frac{V}{\lambda} \frac{D}{Qr^2} << 1 \tag{6.6}$$

where D is the directivity of the Tx antenna, V is the volume of the RC, and λ is the wavelength. In the measurement a quarter-wave distance is normally chosen to separate objects to have a small field correlation coefficient, thus $r = \lambda/4$, and from (6.6) we have

$$Q >> \frac{24VD}{\lambda^3} \tag{6.7}$$

When using (6.7) to estimate the lower bound of the Q factor, r is chosen as $\lambda/4$ and a small D value can be used, e.g. $D \approx 1$ because the Tx antenna is rarely directed to the object under test directly.

Figure 6.2 Working volume in the RC.

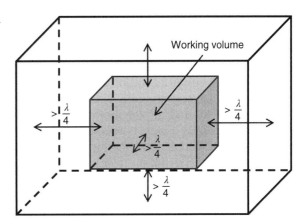

From the consideration of the *K*-factor, we have the constraint in (6.7), but in the electromagnetic compatibility (EMC) test, the maximum E-field strength is normally concerned. From Figure 5.27a, when 100 samples are used, the maximum magnitude of the E-field can be 8 dB (2.5 times) higher than the mean value. From (5.59), (5.60), and (5.98), we have

$$E_0^2 = 6\sigma^2 = 6\left(\frac{\langle |E_x| \rangle}{\sqrt{\pi/2}}\right)^2 = \frac{QP_d}{\omega \varepsilon V} \tag{6.8}$$

The mean of the maximum E-field can be expressed as

$$\langle \lceil |E_x| \rceil_{100} \rangle \approx 2.5 \times \langle |E_x| \rangle = 2.5\sqrt{\frac{QP_d}{\omega \varepsilon V}}\sqrt{\frac{\pi/2}{6}} \tag{6.9}$$

Substituting $\varepsilon = 10^{-9}/(36\pi)$ and $\omega = 2\pi f$ into (6.9), and using the power balance relationship $P_t = P_d$ we have:

$$E_{\text{expect}} = \langle \lceil |E_x| \rceil_{100} \rangle \approx 1.7 \times 10^5 \sqrt{\frac{QP_t}{fV}} \tag{6.10}$$

where E_{expect} is the expected maximum E-field using 100 measured samples. (6.10) is very useful, it links the *Q* factor, the net input power, the working frequency, and the volume of the RC together. Suppose the input power is 1 W, the volume of the RC is 100 m³, the frequency is 1 GHz, and the *Q* factor is 10 000, then the expected maximum E-field can be

$$E_{\text{expect}} \approx 1.7 \times 10^5 \sqrt{\frac{10000 \times 1}{10^9 \times 100}} \approx 54\,\text{V/m} \tag{6.11}$$

An even higher value can be obtained if more independent samples are used in Figure 5.27a, but this would increase the measurement time. Considering the measurement time in practice, $N = 100$ is an acceptable value. The constraint of *Q* factor can also be obtained when other variables defined in (6.10) are given:

$$Q \geq \frac{3.4 \times 10^{-11} E_{\text{expect}}^2 fV}{P_t} \tag{6.12}$$

It should be noted that when the RC is loaded with objects under test, the *Q* factor will reduce; also the reflected power needs to be considered when calculating the net input power. The ratio of the maximum received power of *N* independent samples to the mean value can be approximated by using (5.141):

$$\frac{\langle \lceil |E_x|^2 \rceil_N \rangle}{\langle |E_x|^2 \rangle} \approx 0.577 + \ln N + \frac{1}{2N}, N > 1 \tag{6.13}$$

One should note that if the received power is used, there is 1 dB difference, as can be seen in Figure 5.26a, and this has been discussed in Section 5.4.6.

In practice, (6.7) is easy to satisfy but (6.10) should be considered very carefully. It is not easy to simulate all the loss in the RC and the simulated *Q* may not be reliable. This does not depend on what algorithms or software the designer uses, but depends on how much loss is included in the simulation and the simulated loss could be very different from the loss in practice. Thus, a better way to obtain the *Q* factor is through measurement using a vector network analyser (VNA) or a field probe before using the chamber. From the measured *Q* factor and the expected maximum E-field, the required input power can be estimated and the power amplifier can be configured. Since the power amplifier is normally expensive, it is better to know the Q factor before the purchase of the power amplifier rather than assemble all the parts in one step. The step-by-step procedure can minimise the risk in practice.

The measurement method of the Q factor is not unique; the Q factor can be measured in the frequency domain by using the Hill's equation or in the time domain from the decay constant. In the frequency domain, from the Hill's equation we have [2]

$$Q = \frac{16\pi^2 V}{\lambda^3} \frac{\langle P_r \rangle}{P_t} \tag{6.14}$$

Suppose a network analyser is connected to the Tx antenna and the Rx antenna, and the total efficiency of these two antennas are η_{Txtot} and η_{Rxtot}, respectively. (6.14) becomes

$$Q = \frac{16\pi^2 V}{\lambda^3 \eta_{Txtot} \eta_{Rxtot}} \langle |S_{21}|^2 \rangle \tag{6.15}$$

where $\langle \cdot \rangle$ is the average value for the stirring method used, such as mechanical stir, frequency stir, or source stir. When using (6.15), the total efficiency of antennas needs to be estimated if they are not known: for a log periodic antenna an empirical value 75% can be used, for a horn antenna 90% can be used. These values are approximations and may not be accurate. In the time domain, the Q factor can be obtained from the decay constant of the PDP:

$$Q = \omega \tau \tag{6.16}$$

where the decay constant τ needs to be extracted from the inverse Fourier transform of measured S parameters with a filter function. The measurement details are introduced in Chapter 7, and the total and radiation efficiency of antennas can also be measured without using reference antennas.

There are two types of loss in an RC, the diffused loss and the insertion loss. The Q factor measured in the time domain will not be affected by the insertion loss of cables and antennas, but by the absorption cross section (ACS) of all objects in the RC, while the Q factor measured in the frequency domain needs to be corrected by these insertion losses.

6.2.5 The Stirrer Design

The stirrer design is very important in the RC since stirrers play a vital role in changing the field distribution in the RC. How to design the structure of a stirrer to maximise the field randomness and the equipment under test (EUT) area while minimising the LUF is an import question. A large stirrer certainly gives better performance (good FU and low LUF) but occupies a large space and reduces the working volume, thus there is a trade-off and we need to optimise the size of the stirrer. Empirical conclusions have been given in the international electrotechnical commission (IEC) standard [26]. The stirrers should be adequate to provide the desired (FU), which means the stirrers should be large enough to 'stir' the field. Empirically, the stirrer should be as large as possible, and it has been suggested in [26] that: (i) the stirrers have one dimension that is at least one-quarter wavelength at the LUF, (ii) the stirrers should have one dimension at least three-quarters of the smallest chamber dimension, and (iii) the stirrers should be shaped asymmetrically to avoid a repetitive field pattern in the RC. Mathematically, they are

$$L_{stir} > \frac{1}{4} \lambda_{LUF} \tag{6.17}$$

$$L_{stir} > \frac{3}{4} \min(a, b, d) \tag{6.18}$$

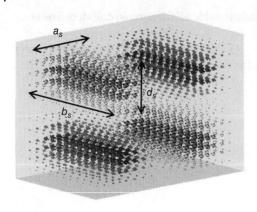

Figure 6.3 Definition of a modal cell, the E-field is represented by vectors.

where L_{stir} is the maximum length of the stirrer, λ_{LUF} is the wavelength at the LUF, and a, b, and d are the dimensions of the RC (if a rectangular RC is used). Experience has shown that the stirred volume should represent approximately 8% of the whole volume of the RC [27]:

$$V_{stir} > V_{RC} 8\% \tag{6.19}$$

The dimensions of the stirrer can also be used to predict the LUF. When the size of the stirrer is comparable to the size of a modal cell, it is expected that the stirrer has enough interactions with the resonance mode and can significantly stir the field at the resonance frequency [28]. The definition of a modal cell is illustrated in Figure 6.3, which is the minimum period of the field pattern of a resonant mode in a cavity. The resonant field pattern in a cavity is repeated (with phase difference) by a modal cell with size a_s, b_s, and d_s. Suppose the boundary of a stirrer is inside a box of dimensions a_s, b_s, and d_s shown in Figure 6.4, and the dimension of the RC are a, b, and d. When the model cell is comparable to the size of the stirrer, it means the stirrer can affect the mode significantly. Under this assumption we have [28]

$$m_s = \text{int}[a/a_s], n_s = \text{int}[b/b_s], p_s = \text{int}[d/d_s] \tag{6.20}$$

where int[·] means the closest integer number, and m_s, n_s, and p_s are the closest mode indices. The LUF can be estimated from (5.7) and (6.20) [28]:

$$f_{LUF} \sim \frac{c}{2}\sqrt{\left(\frac{m_s}{a}\right)^2 + \left(\frac{n_s}{b}\right)^2 + \left(\frac{p_s}{d}\right)^2} \tag{6.21}$$

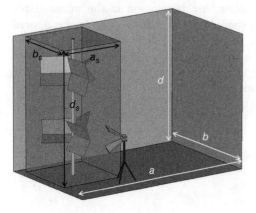

Figure 6.4 A stirrer inside an RC with bounding box $a_s \times b_s \times d_s$.

where $c = 3 \times 10^8$ m/s is the speed of light in the cavity. The RC at the University of Liverpool (UoL) has an LUF of 139–170 MHz with different stirrer designs [29]. The size of the RC is 3.6 m × 5.8 m × 4 m, and $a_s = b_s = 2$ m, $d_s = 4$ m. From (6.20) we have $m_s = \text{int}[3.6/2] = 2$, $n_s = \text{int}[5.8/2] = 3$, and $p_s = \text{int}[4/4] = 1$, and by using (6.19) the LUF can be estimated as [28]

$$f_{LUF} \sim \frac{3 \times 10^8}{2} \sqrt{\left(\frac{2}{3.6}\right)^2 + \left(\frac{3}{5.8}\right)^2 + \left(\frac{1}{4}\right)^2} \approx 120\,\text{MHz} \tag{6.22}$$

As can be seen in (6.22), the estimated LUF is lower than the measured result. Thus, (6.21) can be relaxed to

$$f_{LUF} \approx \kappa \frac{c}{2} \sqrt{\left(\frac{m_s}{a}\right)^2 + \left(\frac{n_s}{b}\right)^2 + \left(\frac{p_s}{d}\right)^2} \tag{6.23}$$

where κ is the relaxation factor. For the RC at the UoL, $\kappa \approx 177/120 \approx 1.5$ could be a good estimated value of first-order approximation.

The detailed structure of the stirrer also affects the RC performance. Numerical methods have been used to optimise the stirrers in the RC. One method is to use the eigenfrequency shift simulated in a spherical cavity [29, 30] to evaluate the performance of a stirrer. This is based on the conclusion that an effective stirrer can shift the eigenfrequencies significantly [31]. The simulation set-up is illustrated in Figure 6.5. A stirrer is placed inside a perfect electric conductor (PEC) spherical cavity, and the resonant frequencies of the cavity can be obtained by using the eigenmode solver. The eigenfrequency shift Δf_{eig} can be obtained as

$$\Delta f_{eig} = \left| f_{empty} - f_{loaded} \right| \tag{6.24}$$

where f_{empty} represents the resonant frequencies of different modes in an empty spherical cavity and f_{loaded} represents the resonant frequencies of different modes in the cavity loaded with a stirrer.

A concept of treating the stirrer as an antenna was introduced and discussed in [29, 30]. To make an antenna efficient (i.e. to make the stirrer efficient) the current path length on the antenna should be comparable with the wavelength of the LUF (typically half-wavelength), thus some cuts were introduced, and this has been demonstrated as an effective way to reduce the LUF without increasing the stirrer size or reduce the EUT area. More details can be found in [29].

A genetic algorithm (GA) combined with the transmission-line matrix (TLM) method was used in [32] to optimise the stirrer design. The average change in angle of the Poynting vector was calculated as the cost function, and a reference stirrer of a simple cube was used. It has been found that increasing the radius of the stirrer has a very significant effect on stirrer performance; this confirms the intuitive conclusion that a large stirrer has a better performance. In [32], the stirrer was simulated in free space but not in an RC, which accelerated the simulation greatly based on the assumption that a stirrer that can scatter waves uniformly in free space should also have a good ability to scatter waves in an RC. Thus, in free space, the simulation time was 1.6 minutes for each plane wave incidence, but in an RC, the simulation time was around 50 minutes. A similar method has also been used in [33] by simulating the first reflection of the stirrer. If no simplification or assumption is made, the simulation of the RC is very time-consuming. The TLM method has also been used to investigate the relationship between the LUF and the maximum working volume, and the trade-off between these two factors was analysed and discussed [34].

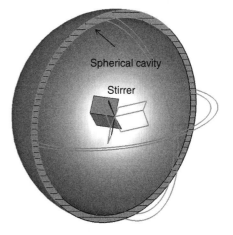

Figure 6.5 Eigenmode simulation model.

Simulations of stirrers on a high-performance computer have been performed by using the finite-difference time domain (FDTD) method in [35, 36]. 512 nodes were used with each node corresponding to a single stirrer position; the total simulation time was about 20 hours. The results showed that the stirrer performance is highly determined by the rotation radius of the stirrer but is not sensitive to the orientation of the rotation axis.

Other ways to stir the field in the RC have been reviewed in Section 1.1.2, and more simulations and experimental investigations on stirrer design can be found in [37–50]. An interesting question is: Is there a best stirrer? The structure of the stirrer can be optimised, but what is the best stirrer in theory? A best stirrer means no matter how the structure of a stirrer is optimised under given constraints, the performance cannot be improved unlimited. However, from the frequency domain it could be very difficult to identify this limit and the time domain investigation could provide better insights. The answer to this question is given in Sections 6.4–6.6.

6.3 Simulation of the RC

In this section, the simulation of an RC is reviewed systematically although it has been discussed partially in the stirrer design. Generally, using the full-wave method to simulate the statistical electromagnetic problems is not easy; the simulation time could be too long to be acceptable for a personal computer (PC). Most of the simulations of an RC are used to verify the performance but rarely used in trial-and-error processes (like the trial-and-error in antenna simulation). In [34], if the problem was solved on a PC with four cores, the simulation time would be $512 \times 20/4 = 2560$ hours, which is 3.5 months! Thus, it is impossible to use a PC to optimise the RC by brute force; smart methods and simplifications are normally required in practice. To save the simulation time and memory requirement, one can lower the upper frequency to reduce the mesh number, use a simplified or equivalent model, use a high frequency approximation method, etc. Typical simulation methods are summarised below.

6.3.1 Monte Carlo Method

The field inside an RC can be described by a superposition of many plane waves coming from random directions, polarisations, and phases. The Monte Carlo method can be a good way to simulate the response in an RC. The Monte Carlo simulation can be considered as a discrete version of the plane-wave integral/spectrum approach discussed in Section 5.4, and the analytical conclusions from the plane-wave integral approach can be verified by using the Monte Carlo simulation [51–53]. The Monte Carlo method has the following advantages [51]:

1) The analytical representation is valid for an ideal RC, but the numerical simulation can be used to describe an imperfect RC and the number of plane waves can be controlled. When the frequency and mode density are not high enough, the Monte Carlo method can still be used to investigate the statistical properties of the field quantities.
2) The simulated sample size can be used to compare with the sample size in measurements, thus the convergence can be verified from simulation.
3) If the analytical closed-form expression cannot be obtained from the plane-wave integral, numerical results can be achieved.
4) User-defined distribution can be used to simulate specific scenarios.

In the Monte Carlo simulation, a single plane wave with linear polarisation can be expressed by

$$\mathbf{E}(\mathbf{r}) = E_0 \left(e_x \hat{\mathbf{x}} + e_y \hat{\mathbf{y}} + e_z \hat{\mathbf{z}} \right) e^{-j\mathbf{k}\cdot\mathbf{r}} e^{j\Phi} \tag{6.25}$$

where $\hat{\mathbf{x}}$, $\hat{\mathbf{y}}$, and $\hat{\mathbf{z}}$ are the unit vectors in axes x, y, and z, respectively, e_x, e_y, and e_z are the scalar portions in the direction of x, y, and z, the direction of incidence is represented by the wave vector $\mathbf{k}$, whose magnitude is the wave number and whose direction is the wave propagation, $\mathbf{r} = x\hat{\mathbf{x}} + y\hat{\mathbf{y}} + z\hat{\mathbf{z}}$ is the field point, and Φ is the phase.

For the waves in an ideal RC, the incident angle is uniformly distributed over the full solid angle of a sphere, the polarisation angle is uniformly distributed in the range of $[0, \pi]$, and the phase Φ is uniformly distributed in the range of $[0, 2\pi]$. The magnitude of the incident waves is assumed to be constant [28]. Each component of the wave vector $\mathbf{k}$ in (6.25) can be written as [51]

$$k_x = k\sin\theta\cos\phi$$

$$k_y = k\sin\theta\sin\phi$$

$$k_z = k\cos\theta \tag{6.26}$$

and the E-field with the polarisation angle α has

$$e_x = \cos\alpha\cos\theta\cos\phi - \sin\alpha\sin\phi$$

$$e_y = \cos\alpha\cos\theta\sin\phi + \sin\alpha\cos\phi$$

$$e_z = -\cos\alpha\sin\theta \tag{6.27}$$

The definitions of the polar angle θ, azimuth angle ϕ, and polarisation angle α are shown in Figure 6.6.

To generate uniformly distributed angles over a sphere, the polar angle and the azimuth angle can be obtained from a uniform distribution. Suppose variables u and v are independent and randomly distributed over $[0, 1]$, the random incident angles θ and ϕ can then be obtained from [51]

$$\theta = \arccos(2u - 1)$$

$$\phi = 2\pi v \tag{6.28}$$

More methods to generate uniformly distributed samples on a sphere can be found in [54].
From (5.98), we have the square of the magnitude of the E-field

$$E_0^2 = \frac{QP_t}{\omega\varepsilon V} \tag{6.29}$$

Figure 6.6 Definitions of θ, ϕ, and α.

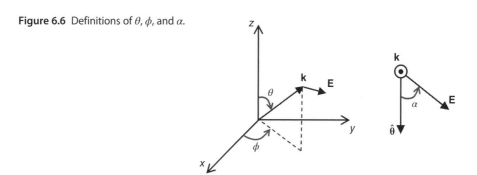

where P_t is the power transmitted into the cavity. This means that the power density of each plane wave need to be normalised, otherwise the more plane waves generated the higher the power density we have. Because

$$P \propto E_0^2 \tag{6.30}$$

for N plane waves, the total power (density) is

$$P_N \propto NE_0^2 \tag{6.31}$$

To keep the total power (density) constant, the magnitude of the E-field needs to be normalised:

$$E_N \propto \frac{E_0}{\sqrt{N}} \tag{6.32}$$

A typical Matlab code snippet is demonstrated in Appendix A.8. One thousand random plane waves were used as shown in Figure 6.7a with $E_0 = 1$ V/m. The Monte Carlo simulation was repeated 1000 times, thus 1000 result samples can be obtained. The simulated cumulative distribution function (CDF) and the theoretical CDF are plotted in Figure 6.7b; as can be seen, a very good agreement was obtained. Note that to verify a distribution a quantified hypothesis test (goodness-of-fit test) needs to be applied. We just show the results here to present a direct understanding. The theoretical CDF can be derived from (5.60), which is

$$CDF(|E_x|) = 1 - \exp\left(-\frac{x^2}{2\sigma^2}\right), \ x = |E_x| \tag{6.33}$$

Here $E_0 = 1$ V/m and $\sigma = E_0/\sqrt{6}$ are used in the code.

The Monte Carlo simulation is normally used to obtain the response of the object under test in an ideal RC; results from Monte Carlo simulation can be used as the theoretical reference when analytical closed-form solution is not available. More applications of the Monte Carlo simulation in the RC can be found in [55–73].

6.3.2 Time Domain Simulation

Time domain methods such as TLM, FDTD, and the finite integration technique (FIT) have been widely used in RC simulation because of advantages such as wideband, memory efficiency, and ease of parallelising [32–36]. The finite wall loss in the RC can be equivalent to the volumetric loss in the cavity [74], and a proper value needs to be chosen, otherwise the energy would never decay in the cavity and the Q factor would be unrealistically high. Insufficient mesh resolution may introduce numerical errors, where the errors can come from the discretisation of structures and numerical dispersion. A similar effect also exists in frequency domain methods. In most cases, errors are not large enough to affect the statistical distributions if the statistical proprieties are of interest. However, in some cases, the mesh error could be an issue, e.g. if the angular correlation coefficient needs to be simulated, when the correlation coefficient is very small, the mesh error could affect the correlation coefficient, which should be treated carefully.

6.3.3 Frequency Domain Simulation

Frequency domain methods such as the finite element method (FEM), the boundary element method (BEM), and the moment method (MoM) have been used in RC simulations. Although frequency domain methods can handle highly resonant structures very efficiently, for wideband simulations a huge number of frequency sample points are necessary. The memory consumption (space complexity) and time complexity could be a problem when the frequency is high. However, it may not be necessary to simulate the RC at high frequencies, as normally the performance at high frequencies becomes better, and the frequencies of interest around the LUF are mostly focused [6]. A summary of full-wave RC simulations and the references are listed in Table 6.1.

(a)

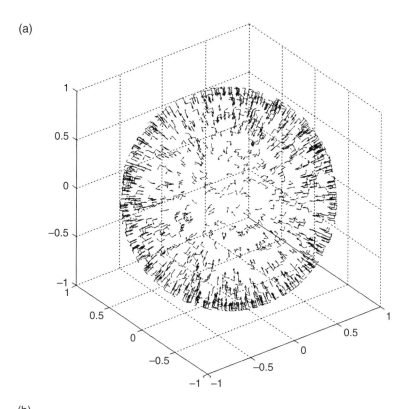

(b)

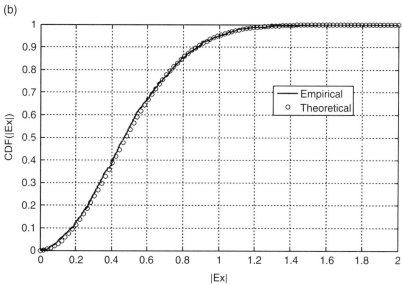

Figure 6.7 Monte Carlo simulation with 1000 plane waves: (a) wave vectors and polarisation vectors and (b) simulated and theoretical CDFs of $|E_x|$.

Table 6.1 Summary of selected full-wave RC simulation.

Authors	Method	References	Year
Harima et al.	FDTD	[75, 76]	1998, 1999
Bai et al.		[77, 78]	1999
Zhang and Song		[79]	2000
Hoijer et al.		[80]	2000
Hoeppe et al.		[81]	2001
Shen et al.		[17]	2002
Kouveliotis et al.		[82]	2003
Moglie et al.		[50, 74, 83–87]	2003, 2004, 2006, 2011, 2012, 2016
Ritter et al.		[88]	2003
Lammers et al.		[89, 90]	2004
Bonnet et al.		[91]	2005
Haffear et al.		[92]	2009
Bosco et al.		[93]	2012
Wang et al.		[94]	2013
Primiani and Moglie		[36]	2014
Bastianelli et al.		[35, 40, 95]	2015, 2016
Gradoni et al.		[96–98]	2013, 2015
Li et al.		[99]	2017
Wu and Chang	TLM	[31]	1989
Petirsch and Schwab		[19]	1999
Clegg et al.		[32, 100]	2002, 2005
Coates et al.		[34, 101, 102]	2002, 2007
Weinzierl et al.		[103–106]	2003, 2004, 2006, 2008
Marvin et al.		[107]	2004
Smartt et al.		[108]	2013

Table 6.1 (Continued)

Authors	Method	References	Year
Asander et al.	IE (BEM, MoM, FMM, MLFMM)	[47]	2002
Leuchtmann et al.		[109]	2003
Bruns et al.		[6, 110]	2005
Gronwald		[111]	2005
Carlberg et al.		[112–114]	2005, 2006, 2009
Karlsson et al.		[115]	2006
Cerri et al.		[116]	2009
Gruber et al.		[117–122]	2013, 2014, 2015
West et al.		[58, 123, 124]	2012, 2013, 2014
Zhao et al.		[125–137]	2008–2016
Borries et al.		[138]	2014
Yang et al.		[139, 140]	2014, 2017
Bunting et al.	FEM	[141–146]	1998–2003
Suriano et al.		[147, 148]	2000, 2001
Zhang and Li		[149]	2002
Huang et al.		[30]	2005
Orjubin et al.		[65, 150–152]	2006, 2007
Zekios et al.		[153–155]	2009, 2010, 2014
Tan et al.		[156]	2016

6.4 Time Domain Characterisation of the RC

In this section, the RC is investigated in the time domain. The time domain behaviour of an RC has been observed in [157, 158] to investigate the statistical isotropy of the RC. The information extracted from the time domain can be combined with that from the frequency domain, and a series of applications become possible, such as measuring the radiation efficiency of antennas without using a reference antenna [159], ACS measurement [160], Q factor extraction, and chamber decay time control [161], etc. A B-scan is a two-dimensional time domain impulse scan that has been widely used in many applications such as radar, medical imaging, non-destructive testing [162], etc. However, a complete B-scan has not yet been implemented in an RC. Some relevant work has been done in [157, 158], which provides important insight from the time domain, and the synthetic-aperture technique was used with nine stirrer positions [157, 158]. In the time domain, the arrived signal with different time and angle (anisotropy of the RC) can be observed directly, which provides important guidelines and insights for the future measurement setup.

The statistical behaviour in the frequency domain has been investigated in Chapter 5 [2], and the statistical distributions of the E-field in the frequency domain are well known. However, there are limited studies in the time domain statistical distribution, which is one of the main contributions to this section. An important definition is the stirrer efficiency, which is hard to quantify in the frequency domain. Actually there is no 'standard' definition of stirrer efficiency. Many efforts have been made to discuss it [32, 163–165], which provide important practical guidelines and experience. It has been found that, for some definitions, the stirrer efficiency (if defined by using K-factor) could be changed by load or antenna positions [166, 167]. Thus, it is related to too many variables and hard to characterise in the frequency domain. In this section, it has been found that the stirrer efficiency can be well defined in the time domain and only related to the equivalent total scattering cross section (TSCS) of the stirrers. The equivalent TSCS of stirrers is determined by the geometric properties of the stirrers (shape, position, etc.) and how the stirrers are moved; it is not sensitive to the load and antenna position. Like the efficiency definition in other applications, the stirrer efficiency defined in the time domain is in the range of 0–100%, which corresponds to an RC that is not stirred and one that is fully stirred, respectively. The definition in the time domain provides a universal way to compare the performance between different RCs or different stirrer designs in one RC. Moreover, the time-gating technique can be used to remove the early-time behaviour in the chamber decay constant extraction [159, 168]. It demonstrates in this section that the time-gating technique can also be used in an RC to filter the unwanted signals to correct the chamber transfer function. By removing the unwanted signals, the stirrer efficiency can be increased virtually without physically changing the stirrers.

In this section, the measurement setup and theory are given first, followed by understanding and discussion of the results where three main aspects are addressed: the statistical behaviour, the stirrer efficiency quantification, and the time-gating technique [169].

6.4.1 Statistical Behaviour in the Time Domain

To gain a direct understanding of the RC in the time domain, it is possible to measure the time domain response directly using an impulse source and an oscilloscope [170]. Another method is to measure the system response in the frequency domain and apply the inverse fast Fourier transform (IFFT) method to obtain the time domain response. Measuring the frequency domain response using a VNA is simpler and can provide a larger dynamic range [171]. The frequency domain measurement method is used here to obtain the time domain response.

The measurement setup is shown in Figure 6.8, where the size of the RC is 3.6 m (W) × 5.8 m (L) × 4 m (H). Two horn antennas are used as antenna 1 (Rohde & Schwarz® HF 906) and antenna 2 (SATIMO SH 2000), and both antennas are well matched from 2 GHz. The rotation platform, stirrers, and VNA are synchronised and controlled by a computer. S-parameters with 10 001 sample points in the frequency range of 2–4 GHz are recorded for different platform angles and stirrer positions; this corresponds to the time domain (after IFFT) resolution of 0.25 ns and duration of 5000 ns. For each platform angle, 100 stirrer positions are used with 3.5°/step. The platform is rotated with 2°/step for one complete revolution to have a good angular resolution (B-scan). Thus we have 180 × 100 = 18 000 set of S-parameters in total.

The time domain response can be obtained from the IFFT (or inverse fourier transform (IFT)) of the measured S-parameters, and a 10th order elliptic band pass filter is applied to reduce the ripples caused by a rectangular window [172], more discussion on the use of different filters is given in Appendix D for a detailed understanding. Here the pass band is set as 2.4–3.6 GHz. We denote the E-field at the receiving antenna as

$$E(t, \theta, n) = \text{IFFT}\left[\widetilde{S}(f, \theta, n)\right] \tag{6.34}$$

where $\widetilde{S}$ represents the filtered S-parameters that depend on the rotation angle of antenna 1 (θ), frequency (f), and stirrer position (n). The plots of $\text{IFFT}\left[\widetilde{S}_{11}(f, \theta, n)\right]$ and $\text{IFFT}\left[\widetilde{S}_{21}(f, \theta, n)\right]$ give monostatic and bistatic B-scan maps, respectively. Although the E-field obtained from the IFFT of S-parameters is not the actual E-field in space, it does not affect our study; we can consider it as the *equivalent* E-field and it has been shown

(a)

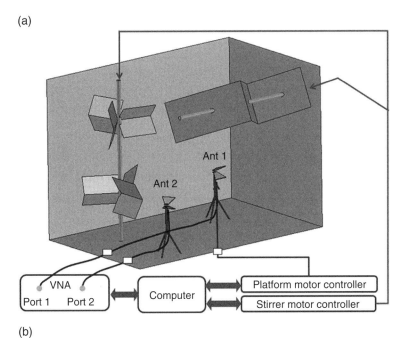

(b)

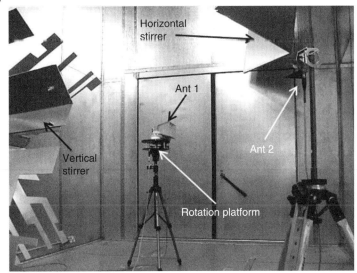

Figure 6.8 B-scan measurement setup in an RC: (a) schematic measurement setup and (b) measurement setup in the RC at the University of Liverpool.

that the statistical behaviour of the received voltage and E-field are the same [2]. The gain of antenna 1 in the measured frequency range is around 10 dBi (half-power beamwidth $\approx 60°$), thus a good angular resolution can be obtained. Obviously, rotating an omnidirectional (low gain) antenna is meaningless.

Since the measured S-parameter includes the unstirred part $\widetilde{S}_{\mathrm{us}}$ and the stirred part $\widetilde{S}_{\mathrm{s}}$ [159]:

$$\widetilde{S} = \widetilde{S}_{\mathrm{us}} + \widetilde{S}_{\mathrm{s}} = \left\langle \widetilde{S} \right\rangle + \widetilde{S}_{\mathrm{s}} \tag{6.35}$$

where $\langle \cdot \rangle$ means the average using any stirring method (e.g. mechanical stir, frequency stir, source stir, etc.). Applying the IFFT to both sides of (6.35) gives

$$\text{IFFT}\left(\widetilde{S}\right) = \text{IFFT}\left(\left\langle \widetilde{S} \right\rangle\right) + \text{IFFT}\left(\widetilde{S}_s\right) \tag{6.36}$$

where $\text{IFFT}\left(\widetilde{S}\right)$ is the total time domain response and $\text{IFFT}\left(\left\langle \widetilde{S} \right\rangle\right)$ is the unstirred part in the time domain. For an ideal/well-stirred RC, $\text{IFFT}\left(\left\langle \widetilde{S} \right\rangle\right)$ is the free-space S-parameter and $\left\langle \widetilde{S} \right\rangle = \widetilde{S}_{\text{FreeSpace}}$ [173], and the measurement of free-space S-parameter of an antenna in an RC is introduced in Chapter 7. As can be seen later, when the RC is not ideal (stirrer efficiency is not 100%), the unstirred part $\left\langle \widetilde{S} \right\rangle$ includes not only the free-space response, but also the contribution of the equivalent TSCS of stirrers (the moving objects in the RC). It is not rigorous to consider $\left\langle \widetilde{S} \right\rangle$ as just the unstirred part of the free-space response in the RC.

If we consider the statistical behaviour of the impulse response in the time domain, the E-field can be regarded as a non-stationary stochastic process. For a specific time, because the space waves superimpose with each other randomly in large numbers, by applying the Lindberg central limit theorem, the rectangular E-field follows normal distribution at each specific time. If the early-time response and the unstirred part are ignored, the probability density function (PDF) can be expressed as

$$\text{PDF}[E_x(t)] = \text{PDF}\left[E_y(t)\right] = \text{PDF}[E_z(t)] = \frac{1}{\sigma\sqrt{2\pi}}\exp\left(-\frac{x^2}{2\sigma^2}\right), \ x = E_x, E_y \text{ or } E_y \tag{6.37}$$

with a mean value of zero, and the standard deviation σ is a time dependent variable. An expression for $\sigma(t)$ needs to be known. If we consider the power decay $P(t)$, it decays exponentially, which can be expressed as:

$$\langle P(t) \rangle \propto \left\langle E(t)^2 \right\rangle = P_0 \exp\left(-\frac{t}{\tau_{\text{RC}}}\right) \tag{6.38}$$

where τ_{RC} is the decay constant of the RC and P_0 is a constant which determines the initial power. From (6.37), the PDF of $E(t)^2$ which follows chi-square distribution with one degree of freedom can be obtained as:

$$\text{PDF}\left[E_*(t)^2\right] = \frac{1}{\sigma\sqrt{2\pi x}}\exp\left(-\frac{x^2}{2\sigma^2}\right), \ x = E_x^2, E_y^2 \text{ or } E_z^2 \tag{6.39}$$

From (6.39), the mean value can be obtained as

$$\left\langle E_*(t)^2 \right\rangle = \sigma^2 \tag{6.40}$$

If we compare (6.38) and (6.40), the time dependent σ can be obtained as

$$\sigma(t) = \sqrt{P_0 \exp\left(-\frac{t}{\tau_{\text{RC}}}\right)} \tag{6.41}$$

Thus the statistical behaviour of the impulse response E-field is well characterised. Note that τ_{RC} is frequency dependent and can be considered as the average value in the frequency range of the excitation impulse.

To have an intuitive understanding, CST Microwave Studio is used to simulate the time domain response in the RC. The simulation model is shown in Figure 6.9, where the dimensions of the RC are the same as the RC at UoL, an antenna is places at the centre of the RC, and a probe is used to record the E-field in the time domain. The excited signal is a modulated Gaussian impulse, as shown in Figure 6.10a; the frequency range is set as 200 MHz–1 GHz. After around 1 hour simulation on a workstation (central processing unit (CPU): Intel Xeon

E5-1660 v3), the E-fields in the time domain recorded by the probe are shown in Figure 6.10c–d. As expected in (6.41), the standard deviation of each field component decays exponentially. Also from Figure 6.10b, the energy in the calculation domain (whole chamber) decays exponentially if the early-time built-up response is ignored.

The E-field distributions at different times are plotted in Figure 6.11. The early-time distributions at 10, 15, 20, and 25 ns are illustrated in Figure 6.11a–d, respectively. It can be seen clearly that the E-field at early-time is not well diffused compared to the E-field in Figure 6.11e–f where no clear wave fronts can be observed and field is diffused randomly.

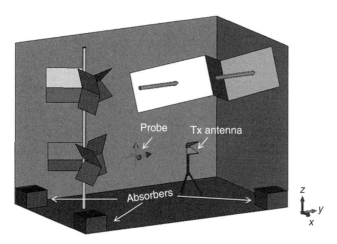

Figure 6.9 RC simulation model in the CST Microwave Studio.

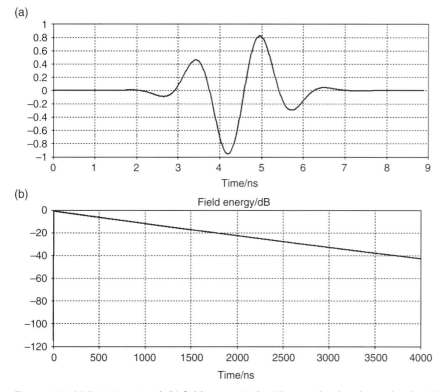

Figure 6.10 (a) Excitation signal, (b) field energy in the RC, normalised to the peak value, (c) $E_x(t)$, (d) $E_y(t)$, and (e) $E_z(t)$.

(c)

(d)

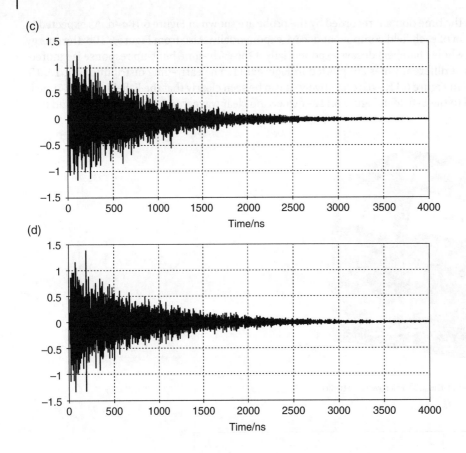

(e)

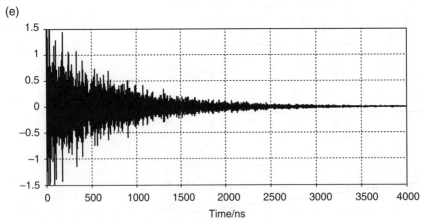

Figure 6.10 (Continued)

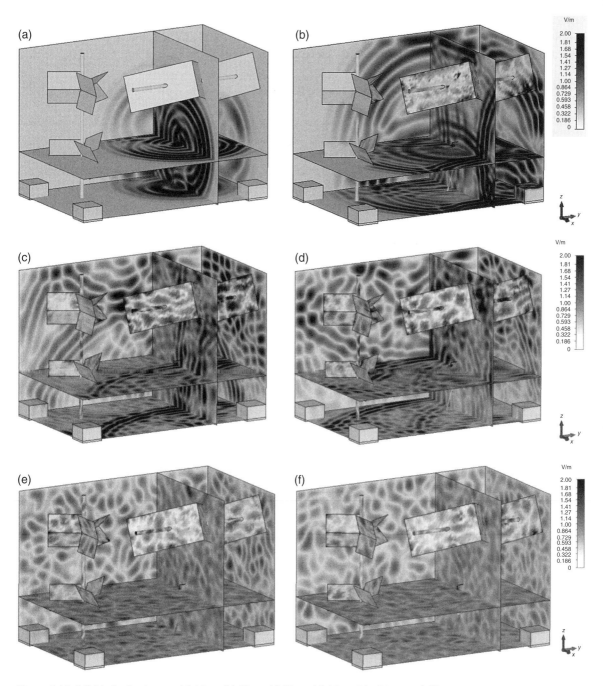

Figure 6.11 E-field distributions at (a) 10 ns, (b) 15 ns, (c) 20 ns, (d) 25 ns, (e) 400 ns, and (f) 500 ns.

6.4.2 Stirrer Efficiency Based on Total Scattering Cross Section

If we check the TSCS measurement in [174] (also in Chapter 7), the stirrers can be considered as an object under test, and this has been used to characterise the TSCS of stirrers in simulation [175, 176]. It should be noted that, in the TSCS measurement, the object under test is required to be moved freely in space. However, in practice, the stirrers are rotating around fixed axes (cannot be moved freely), thus the measured TSCS is actually the *equivalent* TSCS. Since only the equivalent TSCS plays a part in the measurement, we are not interested in the real value TSCS of stirrers. We denote the equivalent TSCS as $\widetilde{TSCS}$, thus the $\widetilde{TSCS}$ of stirrers can be obtained from the difference of the decay speed of $\langle E(t)^2 \rangle$ and $\langle E(t) \rangle^2$ [174–176]:

$$\langle E(t)^2 \rangle = P_0 \exp\left(-\frac{t}{\tau_{RC}}\right) \tag{6.42}$$

$$\langle E(t) \rangle^2 = P_0 \exp\left[-t\left(\frac{1}{\tau_{RC}} + \frac{1}{\tau_s}\right)\right] \tag{6.43}$$

The least-squares fit can be applied to extract the chamber decay time τ_{RC} and the scattering damping time τ_s [174], and the $\widetilde{TSCS}$ of stirrers can be obtained as [174]

$$\widetilde{TSCS} = \frac{V}{\tau_s c_0} \tag{6.44}$$

Thus $\tau_s = V / \left(\widetilde{TSCS} \times c_0\right)$, where V is the volume of the RC and c_0 is the speed of light. If we check (6.42)–(6.44) carefully, it can be found that the contribution of τ_s is independent of the load of the chamber because the load of the chamber has been included in τ_{RC}. The contribution of τ_s can be extracted by

$$\frac{\langle E(t) \rangle^2}{\langle E(t)^2 \rangle} = \exp\left(-\frac{t}{\tau_s}\right) \tag{6.45}$$

If we check (6.36), it can be found that τ_s and $\widetilde{TSCS}$ actually describe how fast the unstirred part decays compared with the total signal strength (PDP). If we define the 'stirrer efficiency' η_s as the residual of the ratio in (6.45) caused by the scattering damping time τ_s, we have

$$\eta_s = 1 - \exp\left(-\frac{t_0}{\tau_s}\right) \tag{6.46}$$

where t_0 is a typical/reference time (similar to the concept of the typical physical dimension in [170, 177]). The stirrer efficiency is in the range of 0–100%. It can be seen that when a very small stirrer is used, the stirrer efficiency is small, as there is no significant difference between (6.42) and (6.43), $\tau_{RC}^{-1} \approx \tau_{RC}^{-1} + \tau_s^{-1}$, $\tau_s \rightarrow +\infty$, and $\eta_s \rightarrow 0\%$. When the RC is well stirred, $\langle E(t) \rangle^2$ decays to zero very quickly and $\tau_{RC}^{-1} + \tau_s^{-1} \rightarrow +\infty$, $\tau_s \rightarrow 0$, and $\eta_s \rightarrow 100\%$.

An appropriate typical/reference time t_0 needs to be chosen to determine η_s. Assuming the shape of the RC is a cube, if the wave is allowed to interact with all the walls inside the RC at least twice (the wave follows the dotted line in Figure 6.12), the travel time is $12\sqrt[3]{V}/c_0$. As can be seen, when $\widetilde{TSCS} = V^{2/3}/4$ (a quarter of the surface area of the face in Figure 6.12), $\eta_s = 1 - e^{-3} = 95.0\%$. This is a reasonable value to our knowledge. If we have such a large sphere with radius $r = V^{1/3}/\sqrt{8\pi}$ which can move freely in the RC, the stirrer efficiency should be high (the TSCS of the sphere is $2\pi r^2 = V^{2/3}/4$ when the electrical size is large [174]). Normally, the stirrers are rotating around fixed axes; the equivalent TSCS is smaller than the real TSCS of the stirrers.

It should be noted that theoretically there are infinite ways to map $\widetilde{TSCS}/V^{2/3}$ to the range of 0–100%. If we use a variable α to control the allowed travel time and let

$$t_0 = \alpha \sqrt[3]{V}/c_0 \tag{6.47}$$

η_s with different α values are shown in Figure 6.13. As can be seen, when α is small, high stirrer efficiency becomes hard to achieve, which means that a very short time is allowed for the waves to travel to become random. In practice, we need an appropriate value to have an intuitive understanding and it is not reasonable to have a too small α (high η_s can never be achieved) or too large α (always give a high η_s). Here, $\alpha = 12$ is recommended. It is also possible to use the mean-free time of $4V/Ac_0$ from (5.32), but there is a potential problem: the inner surface area A actually cannot affect the stirrer efficiency and the TSCS cannot be changed by increasing the surface are of the RC. Thus a typical time defined in (6.47) is more reasonable or the mean-free time can be redefined by using (6.47).

By using $\alpha = 12$, η_s in (6.46) becomes

$$\eta_s = 1 - \exp\left(-\frac{12\sqrt[3]{V}}{c_0 \tau_s} \right) \qquad (6.48)$$

It can also be written in the TSCS form using (6.44):

$$\eta_s = 1 - \exp\left(-\frac{12\widetilde{TSCS}}{V^{2/3}} \right) \qquad (6.49)$$

which is only related to the ratio between the equivalent TSCS of the stirrers and the surface area of the RC. Thus we have a definition of stirrer efficiency which is only related to the $\widetilde{TSCS}$ of stirrers and is not sensitive to the load and the antenna positions in

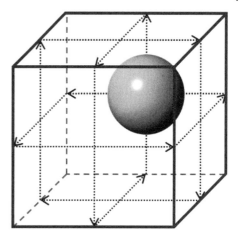

Figure 6.12 An intuitive explanation of the wave interaction with the boundaries in a cubic RC. A sphere with radius $r = V^{1/3}/\sqrt{8\pi}$ is also shown.

the RC. The stirrer efficiency can be understood from the time domain response in (6.48) or from the $\widetilde{TSCS}$ of stirrers in (6.49), and they are physically equivalent.

In the measurement shown in Figure 6.8, antenna 1 is rotated first but the stirrers are kept fixed, so there is no variable n in (6.34). The top view of the measurement setup inside the chamber is shown in Figure 6.14a, and typical filtered S-parameters S_{21} and S_{11} are shown in Figure 6.14b. Then antenna 1 is rotated with $2°$/step, and the IFFT is applied to all filtered S-parameters. We denote the bistatic and monostatic time domain responses as

$$E_{21}(t,\theta)^2 = \left\{ \text{IFFT}\left[\widetilde{S}_{21}(f,\theta) \right] \right\}^2 \qquad (6.50)$$

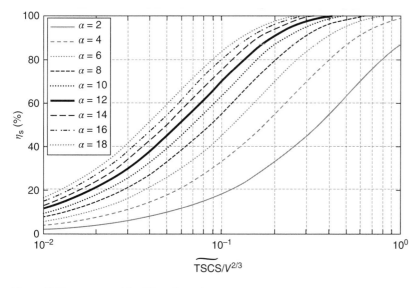

Figure 6.13 η_s curves with different α values.

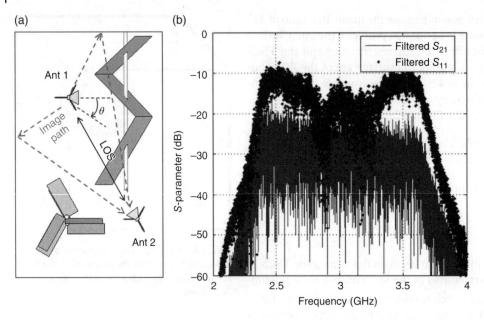

Figure 6.14 (a) Top view of B-scan measurement setup inside an RC and (b) typical filtered *S*-parameters.

and

$$E_{11}(t,\theta)^2 = \left\{ \text{IFFT}\left[\widetilde{S}_{11}(f,\theta) \right] \right\}^2 \tag{6.51}$$

respectively.

The bistatic and monostatic B-scan power maps in the range of 0–50 ns are shown in Figure 6.15. In Figure 6.15a, the line-of-sight (LoS) component, which can be seen clearly, arrived first ($\theta \approx 45°$, the distance is 3 m between antenna 1 and antenna 2, the travelling time is 10 ns), followed by signals from the image sources ($\theta \approx 135°$, $\theta \approx 280°$), which are also significant; In Figure 6.15b, reflections from walls and corners are easily identified (for a monostatic map, the time needs to be divided by 2 when calculating the distance). Note the reflected wave from $\theta = 90°$ is diffused because of the vertical stirrer. The concentric circles in the centre are the reflection from the antenna itself, which is independent of rotation angle. The bistatic and monostatic B-scan power maps in the range of 0–500 ns are given in Figure 6.16. As expected, the field is diffused as it travels in the RC.

To investigate the angle dependency of the stirrer efficiency, 100 stirrer positions are used for each angle of antenna 1. $\langle E_{21}(t,\theta)^2 \rangle$ and $\langle E_{11}(t,\theta)^2 \rangle$ are shown in Figure 6.17, while the unstirred parts $\langle E_{21}(t,\theta) \rangle^2$ and $\langle E_{11}(t,\theta) \rangle^2$ are shown in Figure 6.18. $\langle E_{21}(t,\theta)^2 \rangle$ is actually the PDP, as expected, (6.43) decays faster than (6.42), and the early-time response from the walls cannot be cancelled since the waves have not interacted with the stirrers yet. $\langle E_{21}(t)^2 \rangle$ and $\langle E_{21}(t) \rangle^2$ for a fixed angle are shown in Figure 6.19. It can be seen from Figure 6.19 that, because the value of $\left(\tau_{\text{RC}}^{-1} + \tau_{\text{s}}^{-1} \right)$ in (6.43) is not infinite, η_{s} is not 100% and $\text{IFFT}\left(\langle \widetilde{S} \rangle \right) \neq \text{IFFT}\left(\widetilde{S}_{\text{FreeSpace}} \right)$. This explains the difference between the free-space *S*-parameters measured in the AC and RC [173]. Only when the stirrer efficiency is high are these two parameters close. The difference between $\langle \widetilde{S} \rangle$ and $\widetilde{S}_{\text{FreeSpace}}$ is due to the unstirred part, which can be observed in the time domain; it includes not only the free-space response but also the contribution of the $\widetilde{TSCS}$ (decays exponentially). In practice, if η_{s} is high, the difference between $\langle \widetilde{S} \rangle$ and $\widetilde{S}_{\text{FreeSpace}}$ is small, and $\widetilde{S}_{\text{FreeSpace}}$ can still be measured in the RC.

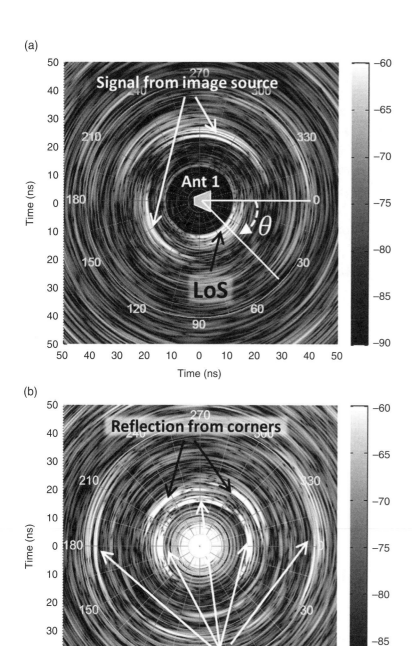

Figure 6.15 B-scan in the range of 0–50 ns with stirrers fixed: (a) bistatic map, $E_{21}(t, \theta)^2$ in dB (all the figures in this section share the same θ definition given in Figure 6.14a) and (b) monostatic map, $E_{11}(t, \theta)^2$ in dB.

(a)

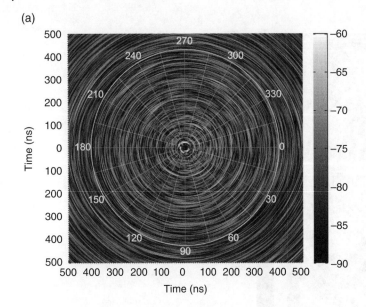

(b)

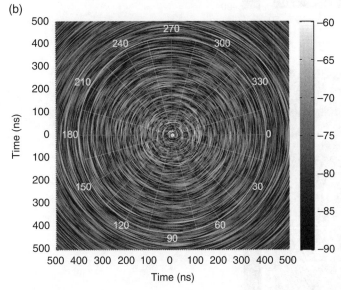

Figure 6.16 B-scan in the range of 0–500 ns with stirrers fixed: (a) bistatic map, $E_{21}(t, \theta)^2$ in dB, and (b) monostatic map, $E_{11}(t, \theta)^2$ in dB.

The least-squares fit is applied to extract τ_s in (6.43). As shown in Figure 6.19a, the slopes of $\langle E_{21}(t)^2 \rangle$ and $\langle E_{21}(t) \rangle^2$ in dB are k_1 and k_2, respectively. From (6.42) and (6.43), we have

$$\tau_{RC}^{-1} = -k_1 \ln 10/10 \tag{6.52}$$

$$\tau_{RC}^{-1} + \tau_s^{-1} = -k_2 \ln 10/10 \tag{6.53}$$

τ_s can be obtained as

$$\tau_s = \frac{10}{(k_1 - k_2) \ln 10} \tag{6.54}$$

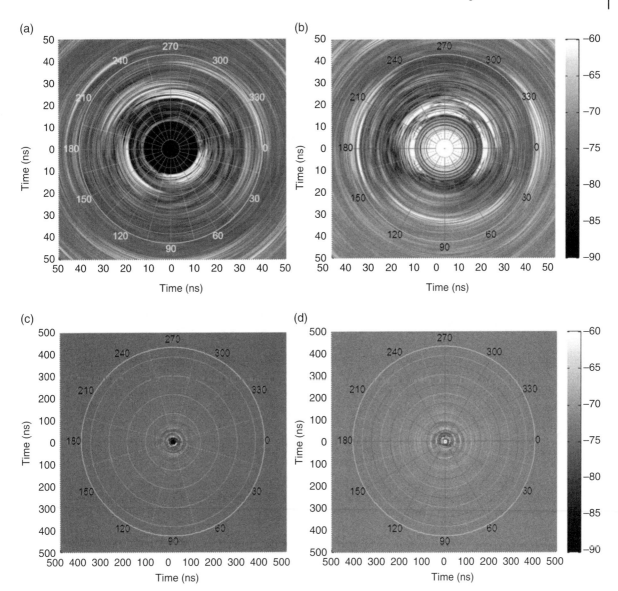

Figure 6.17 PDP plot: (a) $\langle E_{21}(t,\theta)^2 \rangle$ in the range of 0–50 ns, (b) $\langle E_{11}(t,\theta)^2 \rangle$ in the range of 0–50 ns, (c) $\langle E_{21}(t,\theta)^2 \rangle$ in the range of 0–500 ns, and (d) $\langle E_{11}(t,\theta)^2 \rangle$ in the range of 0–500 ns.

or directly from the least-squares fit using the ratio between $\langle E_{21}(t)^2 \rangle$ and $\langle E_{21}(t) \rangle^2$ in (6.45). In Figure 6.19b, $\langle E_{21}(t) \rangle$ and E_{21} are also shown with time in log scale and magnitude in linear scale. As can be seen, at the beginning the time domain response is dominated by the free-space response (first arrived waves), and because of the contribution of $\widetilde{TSCS}$, $\langle E_{21}(t) \rangle$ decays faster than $E_{21}(t)$ in about 100 ns, where the decay speed is determined by τ_s in (6.45). For an ideal RC we have $\text{IFFT}\left(\langle \widetilde{S} \rangle\right) = \text{IFFT}\left(\widetilde{S}_{\text{FreeSpace}}\right)$, which means $\langle E_{21}(t) \rangle$ decays so fast that $\tau_s \rightarrow 0$.

In order to investigate the load effect on the stirrer efficiency, we have repeated the whole measurement with the RC loaded with radio absorbing materials. Loaded and unloaded τ_{RC} and η_s with different rotation angles

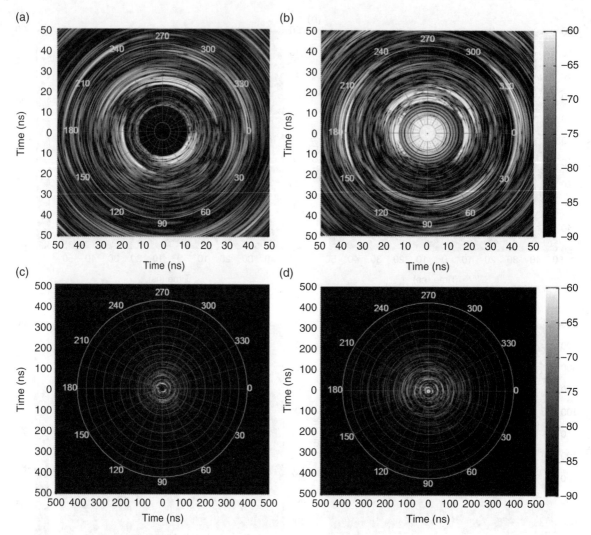

Figure 6.18 Unstirred time-domain response: (a) $\langle E_{21}(t,\theta)\rangle^2$ in the range of 0–50 ns, (b) $\langle E_{11}(t,\theta)\rangle^2$ in the range of 0–50 ns, (c) $\langle E_{21}(t,\theta)\rangle^2$ in the range of 0–500 ns, and (d) $\langle E_{11}(t,\theta)\rangle^2$ in the range of 0–500 ns.

are shown in Figure 6.20. Obviously, τ_{RC} is not sensitive to antenna positions, so we have $\tau_{RCload} = 993$ ns and $\tau_{RCunload} = 1726$ ns. Although τ_{RC} are different, we still have the same η_s, and η_s remains insensitive to the rotation angles (95.8 ± 2%) as expected. As discussed previously, η_s depends only on the equivalent TSCS of stirrers.

The measurement scenarios and measurement times are summarised in Table 6.2. We have measured these in a bandwidth of 1.2 GHz (2.4–3.6 GHz) with 180 angle samples of antenna 1 (Figure 6.20b). Since τ_{RC} and τ_s are frequency dependent, η_s is also frequency dependent. If we use a smaller bandwidth (200 MHz) and sweep the centre frequency, a frequency dependency of stirrer efficiency can be observed (like the extraction of τ_{RC} in [159]). Further measurement scenarios include rotating only the horizontal stirrer, only the vertical stirrer and both stirrers (Table 6.3). Because of the limitation of the maximum sample number of the VNA, the frequency range of 0.2–4.1 GHz is divided into three bands with 10 001 sample points in each band, thus we have a total of 30 003 sample points. Finally, we remove half of the vertical stirrer (Figure 6.21) and repeat the measurement with 10 random antenna positions, one of which is LoS to check if there is any relation between the K-factor and η_s.

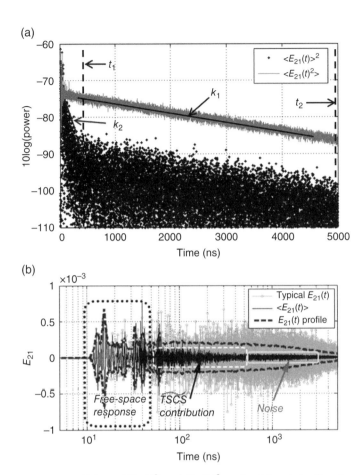

Figure 6.19 (a) Typical $\langle E_{21}(t)^2 \rangle$ and $\langle E_{21}(t) \rangle^2$ at a fixed θ angle in dB scale, and (b) typical $E_{21}(t)$, $\langle E_{21}(t) \rangle$, and E_{21} profile (square root of the PDP) plot in linear scale. The dominated responses in different time ranges are also marked.

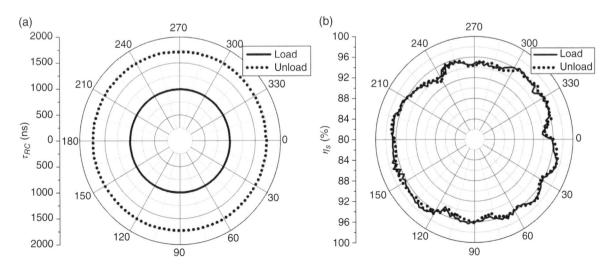

Figure 6.20 (a) Loaded and unloaded τ_{RC} and (b) η_s with different rotation angles.

Table 6.2 Measurement scenarios with 180 antenna angles.

Scenario	Stirrer position no.	Platform position no.	Load/unload	Measurement time (h)
1	1	180	Unload	4
2	100	180	Unload	402
3	100	180	Load	402

Table 6.3 Further measurement scenarios with different stirrer configurations.

Stirrer	Stirrer position no.	Load/unload	Antenna position no.	Frequency range
Only H	360	Load	1	0.2–4.1 GHz (30 003 points)
Only V	360	Load	1	0.2–4.1 GHz (30 003 points)
H and V	360	Load	1	0.2–4.1 GHz (30 003 points)
Small V[a]	360	Load	10	1.9–4.1 GHz (10 001 points)

[a] Small V means half of the vertical stirrer is removed, as shown in Figure 6.21.

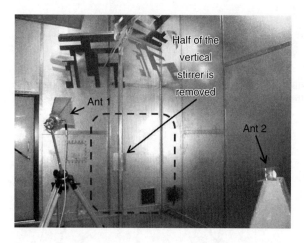

Figure 6.21 Measurement with half of the vertical stirrer.

The results are given in Figures 6.22 and 6.23, and K-factors with random antenna positions are given in Figure 6.24. As expected, when both stirrers are used we have the highest stirrer efficiency and a smaller stirrer gives small stirrer efficiency. Although the K-factors have a large variation (~15 dB), they do not affect the stirrer efficiency. The LoS component only affects the initial response in a few nanoseconds, but the decay speed in a few 100 ns (Figure 6.25) is determined by the equivalent TSCS of the stirrers and is not sensitive to the antenna position.

If we check the mean value of η_s, it can be found that the mean value of η_s of the half vertical stirrer is around 55%, which corresponds to the $\widetilde{TSCS}$ of $0.0665V^{2/3}$ in (6.49). Because the TSCS can be superimposed under the dilute approximation [174, 178], if a whole vertical stirrer is used, the $\widetilde{TSCS}$ is roughly $2 \times 0.0665V^{2/3}$, which

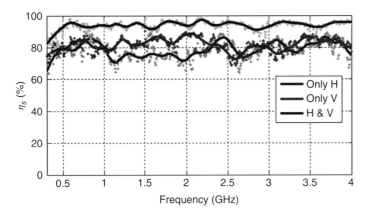

Figure 6.22 Stirrer efficiency with different stirrer rotation. The light dots show the measured result and the solid line is the smoothed result.

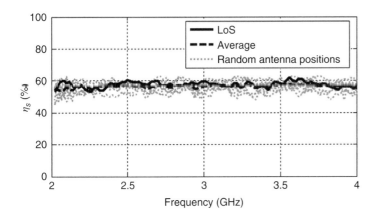

Figure 6.23 Stirrer efficiency with half of the vertical stirrer (10 random antenna positions are used with one LoS).

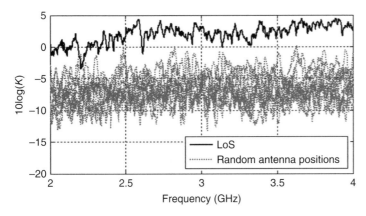

Figure 6.24 *K*-factor in dB with half of the vertical stirrer (10 random antenna positions are used with one LoS).

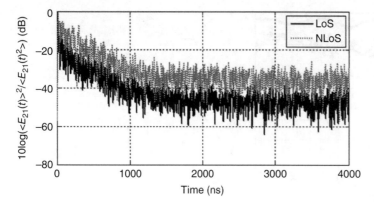

Figure 6.25 $\langle E(t)\rangle^2/\langle E(t)^2\rangle$ in dB with half of the vertical stirrer, a comparison between LoS and non-line-of-sight (NLoS).

corresponds to a stirrer efficiency of 80%, which agrees well with Figure 6.22. We also compare stirrer efficiency between horizontal and vertical stirrers in Figure 6.22. The $\widetilde{TSCS}$ of the two stirrers is around $2 \times 2 \times 0.0665 V^{2/3}$, which corresponds to a stirrer efficiency of 96%, which agrees well with the H and V in Figure 6.22. Thus from (6.49) we have

$$\eta_{\mathrm{stot}} = 1 - \prod_{i=1}^{M}(1-\eta_{si}) \tag{6.55}$$

where η_{si} is the stirrer efficiency with only the ith stirrer and η_{stot} is the stirrer efficiency when all M stirrers are moving together.

An important question is: how can the stirrer efficiency be increased? From (6.49) we know that by increasing the stirring surface area (the minimum surface area wrapping the rotating stirrer), the stirrer efficiency can be increased. This conforms with existing conclusions, but in the previous work, the 'stirring volume' (the minimum occupied volume when a stirrer is rotating) was used empirically [32, 35, 36]. It should be noted that it is actually the stirring surface that affects the performance, not the stirring volume. This can be understood from the average absorption coefficient measurement in Chapter 7, as TSCS and ACS are dual quantities. When a stirrer is placed at the corner, the stirring volume is the same but the stirring surface is different.

From (6.55), we can conclude that a big stirrer or many stirrers working together is obviously better, but the trade-off is a reduction in the test volume. The source stir is a proven method to improve the performance of the RC [179, 180]. Actually, if we consider the moving antenna as the stationary coordinate frame, rotating the antenna is actually rotating the whole RC, which obviously has a large $\widetilde{TSCS}$. We have also tried to aim the antenna to the stirrer (in 10 random antenna positions in Figure 6.21) because there are always leaky waves (from side lobes, back lobes, scattering from rotation axes, and stationary objects) which do not fully interact with the moving stirrer, thus waves from the antenna have not fully interacted with the stirrer except for the extreme case (source stir). Note the TSCS contribution dominates the time domain response in a few hundred seconds. In this time scale the leaky wave dominates the unstirred part. The stirrer and the antenna can otherwise be considered as an integrated big antenna and the moving of the stirrer becomes the moving of the source (source stir). To verify this, we use the data from scenario 1 in Table 6.2 to calculate the stirrer efficiency (keep the stirrers steady and rotate the antenna) and a nearly 100% stirrer efficiency is obtained in the frequency range of 2.4–3.6 GHz. The ratio between $\langle E(t)\rangle^2$ and $\langle E(t)^2\rangle$ is shown in Figure 6.26; it drops down to the noise level in 100 ns.

The stirrer efficiency defined here is highly related to the equivalent TSCS of stirrers (the moving object in the RC), which is not sensitive to the antenna position and the load of the RC, thus it provides a general way to compare the performance of different RCs or one chamber with different stirrers. The proposed definition is intuitive and can be understood from another point of view: a small stirrer means the wave needs to travel a

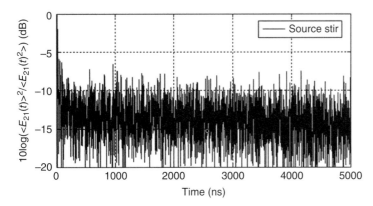

Figure 6.26 $\langle E(t)\rangle^2/\langle E(t)^2\rangle$ in dB using the source stir method.

relatively long time and interact with stirrers more times to be random enough (a slow $\langle E(t)\rangle^2$ decay), while a big stirrer means the wave can become random quickly (a fast $\langle E(t)\rangle^2$ decay). Because the TSCS can be super-imposed, which means the stirrer efficiency is predictable, this is important for the RC design, as the designer can evaluate how large a stirrer needs to be (or how many stirrers are needed) based on the existing design using (6.55).

It is necessary to clarify the difference between the stirrer efficiency and the FU discussed in Chapter 5. The FU is determined by the measured maximum (or average) E-field in the frequency domain [26], which includes both the unstirred part and the stirred part. Stirrer efficiency describes how fast the unstirred part decays compared with the total response or how fast the total response is dominated by the stirred part. It also describes the difference between $\langle \widetilde{S} \rangle$ and $\widetilde{S}_{\text{FreeSpace}}$: as discussed before, $\langle \widetilde{S} \rangle$ includes not only the free-space response (in a few nanoseconds), but also the contribution of the $\widetilde{TSCS}$ of moving objects (in a few hundred nanoseconds). We need $\langle E(t)\rangle^2$ to decay very quickly make sure the unstirred part is dominated by the free-space response (a few nanoseconds), that is when $\eta_s \to 100\%$, $\langle \widetilde{S} \rangle \to \widetilde{S}_{\text{FreeSpace}}$. To have a good RC, we need the stirred part to be as uniform as possible and the unstirred part to be as small as possible. The use of diffusers [9–13] on the specular reflection wall actually diffuses the unstirred part to improve the FU, but the stationary diffusers cannot change the $\widetilde{TSCS}$ of the stirrers and thus cannot change the stirrer efficiency (the decay speed of $\langle E(t)\rangle^2$). However, if only the FU is used as the standard it is already a good RC, unless for some special applications (e.g. measuring the free-space S-parameters in an RC) we need $\text{IFFT}\left(\langle \widetilde{S} \rangle\right) = \text{IFFT}\left(\widetilde{S}_{\text{FreeSpace}}\right) \left(\langle \widetilde{S} \rangle = \widetilde{S}_{\text{FreeSpace}}\right)$.

High stirrer efficiency does not necessarily mean a good FU (e.g. a scenario with a high directivity antenna direct to the test region) and a good FU does not necessarily mean high stirrer efficiency (both the stirred part and the unstirred part are uniform but the unstirred part decays slowly in the time domain). Normally, if the stirrer efficiency is high and the K-factor is small, the field in the test region is dominated by the stirred part; if the stirred part is uniform, a good FU is obtained (except in some special cases when the stirred part can also be non-uniform when the field is close to the boundary or an absorber [181]).

6.4.3 Time-Gating Technique

In the frequency domain, it is well known that the chamber transfer function T can be corrected by removing the unstirred part of S-parameters [159]

$$T_{\text{CFD}} = \left\langle \left| S_{21} - \langle S_{21} \rangle \right|^2 \right\rangle \tag{6.56}$$

where T_{CFD} denotes that the chamber transfer function is corrected in the frequency domain. Correspondingly, in the time domain, if we check (6.42) and (6.43), because of τ_s, $(\tau_{RC}^{-1} + \tau_s^{-1}) > \tau_{RC}^{-1}$, and the unstirred part $\langle E(t) \rangle^2$ decays faster than $\langle E(t)^2 \rangle$. This is easy to understand as the longer the wave travels, the more times it interacts with stirrers. Filtering the signals in the time domain can also remove the unstirred part

$$T_{CTD} = \left\langle \left| FFT\left\{ TG\left[IFFT\left(\tilde{S}_{21} \right) \right] \right\} \right|^2 \right\rangle \tag{6.57}$$

where TG means the time-gating operation, FFT denotes the fast Fourier transform, and T_{CTD} is the chamber transfer function corrected in the time domain. The philosophy is similar to the reflectivity measurement of radio absorbing material [182] in Chapter 2: to measure the S-parameter in the frequency domain, transfer it to the time domain, then apply the time domain truncation to select the wanted signals, and, finally, transfer the selected signals back to the frequency domain.

From the B-scan measurement data, the time-gating technique can be investigated. Suppose the time gate is from t_1 to t_2, for the loaded and unloaded RC, the power range must be the same, which means that we have the same initial and dissipated powers for the loaded and unloaded RC during the time gate (energy conservation). Otherwise, T_{CTD} in different scenarios cannot be compared. If the chamber build-up time is ignored [172], from (6.42) we have:

$$P_0 \exp\left(-\frac{t_{1unload}}{\tau_{RCunload}} \right) = P_0 \exp\left(-\frac{t_{1load}}{\tau_{RCload}} \right) \tag{6.58}$$

$$P_0 \exp\left(-\frac{t_{2unload}}{\tau_{RCunload}} \right) = P_0 \exp\left(-\frac{t_{2load}}{\tau_{RCload}} \right) \tag{6.59}$$

thus

$$\frac{t_{1unload}}{\tau_{RCunload}} = \frac{t_{1load}}{\tau_{RCload}} \tag{6.60}$$

$$\frac{t_{2unload}}{\tau_{RCunload}} = \frac{t_{2load}}{\tau_{RCload}} \tag{6.61}$$

As can be seen in Figure 6.19a, after 500 ns, the unstirred part is quite small, and we use $t_{1unload} = 500$ ns, $t_{2unload} = 5000$ ns for the unloaded RC and $t_{1load} = 288$ ns, $t_{2load} = 2876$ ns for the loaded RC. The results are shown in Figure 6.27 and a frequency stir with the nearest 100 frequency points is used. The uncorrected chamber transfer function $\langle |S_{21}|^2 \rangle$ is given first in Figure 6.27a, and because of the unstirred part it shows angle dependency. The angle dependency of the K-factor in dB is also shown in Figure 6.27b. At each rotation angle the K-factor is calculated using the unbiased estimator [164]:

$$K = \frac{N-2}{N-1} \left\langle \frac{|\langle S_{21} \rangle|^2}{|S_{21} - \langle S_{21} \rangle|^2} \right\rangle - \frac{1}{N} \tag{6.62}$$

where $N = 100$ with a frequency stir with 100 nearest frequencies. The corrected chamber transfer function using (6.56) and (6.57) is given in Figure 6.27c,d. As expected, no angle dependency is observed, which means both methods can remove the unstirred component. A comparison between T_{CFD} and T_{CTD} is given in Figure 6.28. Because part of the time domain signal is filtered by using time gating, this results in $T_{CTD} < T_{CFD}$. This is not an issue because normally we are interested in the relative T rather than the absolute T, and it can be seen in Figure 6.28b that there is very good agreement between $T_{CFD} - T_{CFDload}$ and $T_{CTD} - T_{CTDload}$. The small deviation (<0.2 dB) could be due to the ignorance of the different chamber build-up time in (6.60) and (6.61), which leads to slightly different total energy for the loaded and unloaded RC in the time gate.

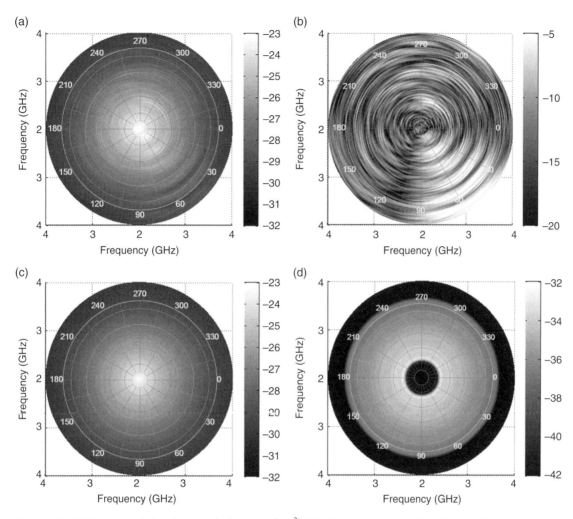

Figure 6.27 (a) Uncorrected chamber transfer function $\langle |S_{21}|^2 \rangle$, (b) K-factor, (c) T_{CFD}, and (d) T_{CTD}, the RC is unloaded, all in dB.

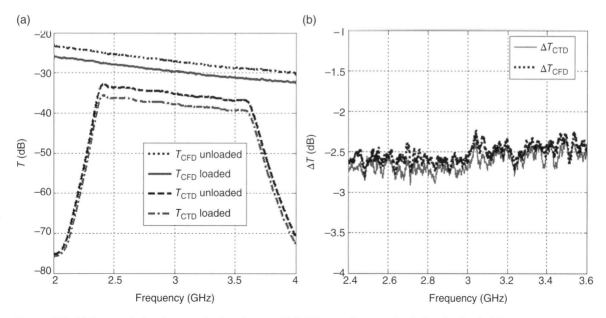

Figure 6.28 (a) Corrected chamber transfer functions and (b) difference between loaded and unloaded T.

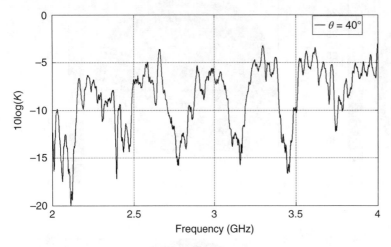

Figure 6.29 *K*-factor in dB at $\theta = 40°$, the RC is unloaded.

It is interesting to note that even though two antennas are positioned in LoS, the *K*-factor (smoothed with frequency stir) can still be very small at some specific frequencies (Figure 6.29). Remember, the unstirred part is not only from the LoS, it can be also from the walls (specular reflection [158]) and other structures (also the contribution of $\widetilde{TSCS}$). When the waves from these sources cancel each other at some frequencies, a small *K*-factor is observed.

The time-gating technique in the RC provides an alternative method to remove the unstirred part in the time domain. It should be noted that when using this time-gating technique to compare the transfer functions in the unloaded and loaded RCs, the start time and the end time of the time gate are different. The time-gating technique can be used to improve the stirrer efficiency: by filtering the early time response, a uniform chamber transfer function can be obtained. For antenna measurements, the time domain response can be easily truncated; for other measurements if the time domain response cannot be easily separated (radiated susceptibility or radiated emission measurement), physically high stirrer efficiency is still required.

6.5 Duality Principle in the RC

It is well known that in classical electromagnetics if variables in two equations occupy identical positions, they are known as dual quantities. The well-known dual equations for fields induced by electric current sources **J** and magnetic current sources **M** are given in Table 6.4 [3]; correspondingly, the dual quantities can be obtained by interchange of variables and are listed in Table 6.5. Similarly, dual quantities exist in statistical electromagnetics, the scattering and absorbing phenomenon in the RC can be related, and relevant quantities are dual [183].

To investigate this duality, we first introduce an equivalent boundary from the full wave point of view and then use a ray model (multipath model [170, 184]) to complete the analysis. A typical measurement scenario is shown in Figure 6.30a. Suppose the vertical stirrer is rotating, then there is an equivalent boundary (stirring surface) which covers the stirrer when it is rotating. When the stirrer is rotated, from the full wave point of view, we just change the boundary condition of each small piece on the equivalent surface. Different boundary condition configurations correspond to different stirrer positions. It should be noted that the equivalent boundary can be chosen arbitrarily as long as it covers the rotating stirrer, but the limit can be considered

Table 6.4 Dual equations for electric and magnetic current sources.

Electric sources (J ≠ 0, M = 0)	Magnetic sources (J = 0, M ≠ 0)
$\nabla \times \mathbf{E} = -j\omega\mu\,\mathbf{H}$	$\nabla \times \mathbf{H} = j\omega\varepsilon\,\mathbf{E}$
$\nabla \times \mathbf{H} = \mathbf{J} + j\omega\varepsilon\,\mathbf{E}$	$-\nabla \times \mathbf{E} = \mathbf{M} + j\omega\mu\,\mathbf{H}$
$\nabla^2\mathbf{A} + k^2\mathbf{A} = -\mu\,\mathbf{J}$	$\nabla^2\mathbf{F} + k^2\mathbf{F} = -\varepsilon\,\mathbf{M}$
$\mathbf{A} = \dfrac{\mu}{4\pi}\displaystyle\iiint_V \mathbf{J}\dfrac{e^{-jkR}}{R}dV'$	$\mathbf{F} = \dfrac{\varepsilon}{4\pi}\displaystyle\iiint_V \mathbf{M}\dfrac{e^{-jkR}}{R}dV'$
$\mathbf{H} = \frac{1}{\mu}\nabla \times \mathbf{A}$	$\mathbf{E} = -\frac{1}{\varepsilon}\nabla \times \mathbf{F}$
$\mathbf{E} = -j\omega\mathbf{A} - \dfrac{j}{\omega\mu\varepsilon}\nabla(\nabla\cdot\mathbf{A})$	$\mathbf{H} = -j\omega\mathbf{F} - \dfrac{j}{\omega\mu\varepsilon}\nabla(\nabla\cdot\mathbf{F})$

Table 6.5 Dual quantities for electric and magnetic current sources.

Electric sources (J ≠ 0, M = 0)	Magnetic sources (J = 0, M ≠ 0)
$\mathbf{E}$ (electric field intensity)	$\mathbf{H}$
$\mathbf{H}$ (magnetic field intensity)	$-\mathbf{E}$
$\mathbf{J}$ (electric current density)	$\mathbf{M}$ (magnetic current density)
$\mathbf{A}$ (magnetic vector potential)	$\mathbf{F}$ (electric vector potential)
ε (permittivity)	μ
μ (permeability)	ε
k (wave number)	k

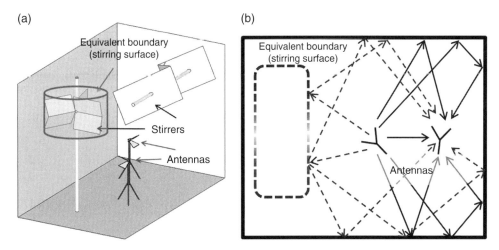

Figure 6.30 (a) Measurement scenario and the equivalent boundary (stirring surface) and (b) equivalent boundary and all interaction paths between two antennas. The dashed lines represent the stirrer path and the solid lines represent the unstirred part.

as the smallest surface that covers the stirrer when it is rotating. As can be seen later, when the stirrer is close to the corner (wall) of the RC, the equivalent boundary shown in Figure 6.30a may not be the smallest surface as the effect of the wall needs to be considered. Actually we can obtain the equivalent stirring surface area numerically without knowing how to wrap the rotating stirrer. We analyse two scenarios: an RC with a stirring boundary and an RC with an absorbing boundary. It can be seen that, interestingly, these two scenarios are dual.

The frequency response of the transmission coefficient (S_{21}) can be considered as a superposition of all scattering and reflection paths between two antennas in the RC, as shown in Figure 6.30b. The dashed lines represent the paths that interact with the stirrer (stirred part) while the solid lines represent the paths that do not interact with the stirrer (unstirred part). Mathematically, the frequency response at each frequency can be expressed as

$$S_{21} = \sum_i H_i(j\omega)E_i = \sum_s H_s(j\omega)E_s + \sum_u H_u(j\omega)E_u \tag{6.63}$$

where $H_s(j\omega)$ and $H_u(j\omega)$ represent the transfer functions of each stirred path and unstirred path between the transmitting (Tx) and receiving (Rx) antenna, respectively, and E_s and E_u are the excitation amplitudes of each path determined by the radiation power and pattern of the transmitting antenna. The reason why $H_s(j\omega)$ and $H_u(j\omega)$ are introduced is that, when averaging, the stirred parts will cancel each other and only the unstirred part will be left. To simplify the derivation we can consider the stirrer as a lossless reflector, the magnitude of the transfer function of each path does not change, and the rotation of the stirrer becomes the stirring of the phase of each path. Thus for each stirrer position (boundary configuration) and each frequency we have

$$S_{21,1} = \sum_s H_s(j\omega)e^{j\phi_{s,1}}E_s + \sum_u H_u(j\omega)E_u$$

$$S_{21,2} = \sum_s H_s(j\omega)e^{j\phi_{s,2}}E_s + \sum_u H_u(j\omega)E_u$$

$$\cdots$$

$$S_{21,n} = \sum_s H_s(j\omega)e^{j\phi_{s,n}}E_s + \sum_u H_u(j\omega)E_u \tag{6.64}$$

where n represents each stirrer position. The averaged S_{21} with N boundary configurations can be obtained as

$$\langle S_{21}\rangle_N = \sum_s \left(H_s(j\omega)E_s\frac{1}{N}\sum_{n=1}^N e^{j\phi_{s,n}}\right) + \sum_u H_u(j\omega)E_u \tag{6.65}$$

For a well-stirred RC, the reflected phases are random and the distribution of the phase $\phi_{s,n}$ should be uniform, thus we have

$$\lim_{N\to\infty}\frac{1}{N}\sum_{n=1}^N e^{j\phi_{s,n}} = 0 \tag{6.66}$$

Eq. (6.65) becomes

$$\lim_{N\to\infty}\langle S_{21}\rangle_N = \sum_u H_u(j\omega)E_u \tag{6.67}$$

which is the contribution from the unstirred part. Actually the assumption 'the magnitude of the transfer function of each path does not change' is not necessary if the PDF of the transfer function is centrosymmetric on the complex plane, which is

$$\lim_{N\to\infty}\frac{1}{N}\sum_{n=1}^N H_{s,n}(j\omega) = 0 \tag{6.68}$$

and we have the same result as (6.67). This is easy to understand: the stirred parts cancel each other for an ideal stirrer.

If we replace the equivalent boundary by a perfect absorber/perfect matched layer (PML), as shown in Figure 6.31, all the waves that interact with the PML will be totally absorbed, and (6.63) becomes

$$S_{21} = \sum_u H_u(j\omega) E_u \qquad (6.69)$$

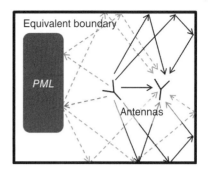

Figure 6.31 The equivalent surface is replaced by a PML. Waves with light dashed lines are perfectly absorbed.

As can be seen, (6.69) and (6.67) are the same. In the time domain the decay of $[\text{IFFT}\langle S_{21}\rangle]^2$ follows $\exp[-t(1/\tau_{RC} + 1/\tau_s)]$ in (6.43) [174], where τ_{RC} is the chamber decay time and τ_s is the scattering damping time. By correcting the loss of the RC (including the loss of the stirrer), the scattering damping time can be obtained from the decay speed of $C(t)$ in (6.45) [174–176]:

$$C(t) = [\text{IFFT}\langle S_{21}\rangle]^2 / \langle [\text{IFFT}(S_{21})]^2 \rangle = \exp\left(-\frac{t}{\tau_s}\right) \qquad (6.70)$$

Since the chamber loss is corrected in (6.70), we can consider the walls of the RC in Figures 6.30b and 6.31 as being lossless, thus the decay speed of $[\text{IFFT}\langle S_{21}\rangle]^2$ in (6.70) determines τ_s in Figure 6.30b and the decay speed of $[\text{IFFT}(S_{21})]^2$ in (6.69) determines τ_{RC} in Figure 6.31. Since (6.69) and (6.70) are the same, τ_s in the scenario of Figure 6.30b equals τ_{RC} in the scenario of Figure 6.31. The dual quantities can be obtained and are shown in Table 6.6 [183].

6.6 The Limit of ACS and TSCS

It has been proven that, for an electrically large perfect absorber, the maximum ACS is a quarter of the overall surface area of the object and independent of the shape of the object [10]. From Table 6.6, it is easy to conclude that the theoretical limit of the TSCS of an electrically large perfect diffuse wave scatterer is a quarter of the stirring surface area.

We have shown that the equivalent TSCS can be used to quantify the stirrer efficiency of an RC. By comparing the measured $\widetilde{TSCS}$ to the theoretical limit, the performance of a stirrer can be quantified and how close it is to an ideal stirrer can be determined. This means no matter how the structure of the stirrer is optimised, the performance of a stirrer can only approach this limit but cannot exceed it. Measurements have been performed to verify this.

The measurement setup is shown in Figure 6.32. Two horn antennas were used (Rohde & Schwarz HF 906 and SATIMO SH 2000) and 10 001 samples of S-parameters in the range of 9.8–10.2 GHz were collected at each stirrer position. The vertical stirrer was rotated with 360 stirrer positions (1°/step); since we were

Table 6.6 Dual quantities in the RC.

In the scenario of Figure 6.30b	In the scenario of Figure 6.31
$\lim_{N\to\infty} \langle S_{21}\rangle_N$	S_{21}
τ_s	τ_{RC}
$\widetilde{TSCS} = V/(c_0\tau_s)$	$\widetilde{ACS} = V/(c_0\tau_{RC})$

(a)　　　　　　　　　　　　　(b)

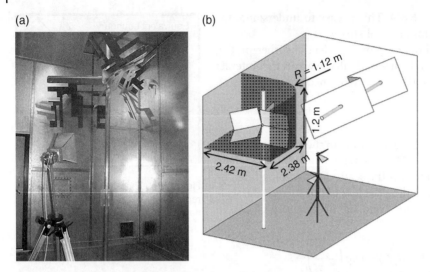

Figure 6.32 TSCS measurement setup in the RC. The size of the RC is 3.6 m (*W*) × 5.8 m (*L*) × 4 m (*H*). The equivalent boundary is shown in (b) with dots, and the surface area is 10.6 m².

measuring the TSCS of the vertical stirrer, the horizontal stirrer was fixed. After all the *S*-parameters were collected, a 10th-order elliptic band pass filter with 10 GHz centre frequency and 200 MHz bandwidth was used to filter the *S*-parameters. The time domain impulse response can be obtained from the IFFT of the filtered *S*-parameters. The extraction procedure of τ_s has been given in Section 6.4.2. τ_s can be extracted from the least-squares fit of the logarithm of $C(t)$ in (6.70), as shown in Figure 6.33; thus the measured TSCS can be obtained using $TSCS = V/(c_0\tau_s)$ [174]. $C(t)$ with different degree steps are also shown. It can be seen that smaller degree steps (higher sample numbers) can reduce the least-squares fit error, and more available signals can be used for the least-squares fit.

To obtain an averaged TSCS, we randomly changed the positions of antennas and the horizontal stirrer, and repeated the measurement 10 times. The measured TSCS values of 10 random configurations are shown in Figure 6.34a. The limit of the TSCS is also given. By removing one paddle we can have two different stirrers with the same stirring surface, shown in Figure 6.34b. Because the stirrer was close to the corner of the RC, the

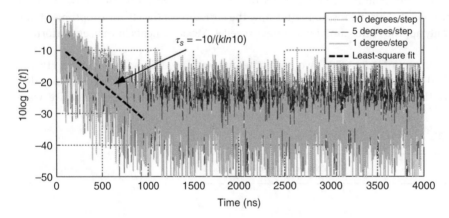

Figure 6.33 Measured $10\log[C(t)]$ with different step degrees. The *S*-parameters are filtered using an elliptical filter, $\tau_s = -10/(k\ln 10)$, where *k* is the slope of the least-squares fit line. Signals after 1000 ns are the noise floor and are not used for the least-squares fit.

(a)

(b)

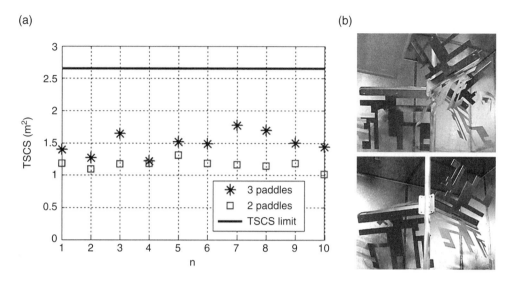

Figure 6.34 (a) Measured TSCS at 10 random RC configurations ($n = 1, 2, \ldots, 10$) and the TSCS limit (solid line), and (b) stirrers with three and two paddles.

use of a cylinder as the equivalent boundary (stirring surface) would give an overestimated value. A better choice is shown in Figure 6.32b; only surfaces that do not overlap with the walls of the RC contribute to the equivalent boundary. This may not be the smallest equivalent boundary. How to obtain the theoretical maximum TSCS (or ACS) of an arbitrary object is introduced in Chapter 7 [185]. The dimensions are given in Figure 6.32b and the maximum TSCS can be calculated as $10.6/4 = 2.65\,\mathrm{m}^2$. As expected, all the measured TSCS values are smaller than the theoretical limit, and the ratio of $4TSCS/A_\mathrm{s}$ (A_s is the area of the stirring surface) can be used to characterise how close the stirrer is to a perfect diffuse wave scatterer. We use the average value to calculate the ratio. As can be seen, the removal of one paddle degrades the performance of the stirrer as expected. At 10 GHz, the stirrer with three paddles is about 56% and the one with two paddles is about 44%, thus we can quantify the effectiveness of the stirrer.

It should be noted that although the derivation is for one stirrer, there is no difference for the case of an arbitrary number of stirrers, the derivation procedures are the same. Like the ACS, the TSCS can also be superimposed [174]. It can be found that the measured averaged S-parameters in a perfect RC and perfect AC are the same, as shown in Figure 6.35a. The averaged multipath signals (from each stirrer position configuration)

(a)

(b)

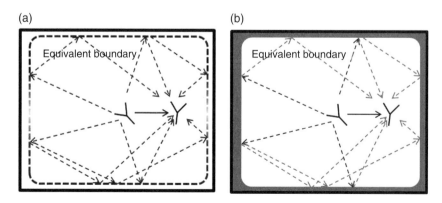

Figure 6.35 (a) A perfect RC and (b) a perfect AC wave. The dashed lines in (b) are perfectly absorbed.

cancel each other out. The averaged S-parameter is just the LoS path, which corresponds to a perfect AC in Figure 6.35b. In this case, the measured $\langle S \rangle$ in an RC equals the measured S-parameter in an AC (free-space response). We have shown that the TSCS of a stirrer actually describes how close the measured averaged S-parameter in an RC ($\langle S \rangle$) and the free-space response in an AC ($S_{\mathrm{FreeSpace}}$) are. Especially when source stir is introduced [173, 179, 180], if we consider the rotating source as the origin of the frame of reference, rotating the source is actually rotating the whole RC, which has been proven to give a better performance [179, 180].

The TSCS of a stirrer can also be used to characterise how close the RC is to a perfect RC. Suppose the measured TSCS of a stirrer is $\widetilde{TSCS}$, thus the surface area of the equivalent perfect diffuse wave scatterer is $4\widetilde{TSCS}$. Suppose the stationary surface area (ground floor, walls, ceiling, etc.) is A_{RC}, we can use $4\widetilde{TSCS}/\left(4\widetilde{TSCS}+A_{\mathrm{RC}}\right)$ to describe how large the stirring surface area is compared to the overall surface area (including the stirring surface) in the RC. Equation (6.49) can also be understood in the same way.

An interesting problem is that, in [174], a sphere was moved freely in the RC and the TSCS was found to be $2\pi a^2$ (electrically large), and a is the radius of the sphere, which is half of the surface area of the sphere. But here, the maximum TSCS is $A_{\mathrm{s}}/4$, a free moving object has a larger equivalent TSCS, and how to obtain the theoretical limit of the TSCS of a free moving object could be an interesting problem [183].

6.7 Design Example

A typical RC design is shown in Figure 6.36, and the dimensions are given in the figure. Two stirrers can be controlled independently. The vertical stirrer has two parts and the radius of rotation is 0.9 m. The horizontal part has a length of 3.7 m and the radius of rotation is 0.7 m. The LUF has been measured in [29]: for a normal vertical paddle, the LUF is around 177 MHz; for an optimised vertical paddle, the LUF is around 139 MHz. The Q factor and the decay constant are given in measurements in Chapter 7.

A control system is required to synchronise the stirrers and the radio frequency (RF) measurement system. A general schematic plot is shown in Figure 6.37. The device under test (DUT) can be an antenna or any other device and the RF measurement system depends on the instruments: it can be a radiated power measurement system, a radiation immunity measurement system, a channel emulator etc. The stirrers can be controlled in both mode-tuned (stepped tuner rotation) and mode-stirred (continuous tuner rotation) ways [26]. The working mode of the RC depends on the type of measurement, e.g. in the diversity gain measurement, the mode-tuned method is normally required: the stirrers need to be stationary to collect responses from a receiver

Figure 6.36 A typical RC design at UoL. Unit: metre.

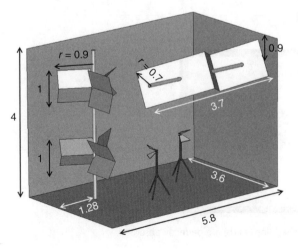

Figure 6.37 A typical RC control system.

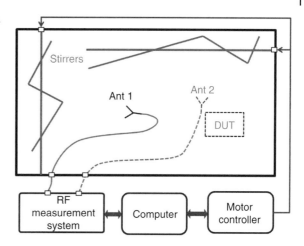

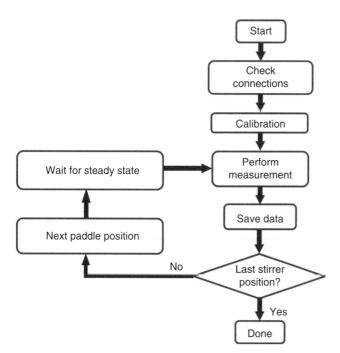

Figure 6.38 A typical measurement flow diagram.

with multiple antennas otherwise the received signals are not in the same environment and are decorrelated. A typical measurement flow is given in Figure 6.38, after the stirrer is rotated we need to wait a few seconds until the electromagnetic (EM) state in the RC is steady, and then the measurement is triggered by the computer. Finally, the measurement data at each stirrer position is recorded on the computer. After the measurement is finished, the data can be analysed by using Matlab or user-defined algorithms. Detailed steps for different types of measurements are given in Chapter 7.

It should be noted that the volume of the RC can be scaled up or down according to the required LUF. We can predict the LUF from existing measurement results in the RC at UoL [29]. By using the RC at UoL as a reference, another in-house developed compact RC measurement system at the Nanjing University of Aeronautics and Astronautics (NUAA) is reported in Appendix E. As expected, the measured LUF agrees well with the predicted value.

6.8 Summary

The design guidelines for the RC have been reviewed. Design constraints have been discussed, including the RC shape, the LUF, the working volume, the Q factor, and the stirrer design. Simulation methods such as Monte Carlo, FDTD, TLM, integral equation (IE), and FEM have been reviewed. The detailed study of the RC in the time domain has been conducted and the statistical behaviour of the fields in the time domain has been investigated. It has been found that the received power of the impulse response follows chi-square distribution with one degree of freedom; the stirrer efficiency has been defined and quantified based on the equivalent TSCS of stirrers. This definition is not sensitive to the antenna position and load in an RC, and can be regarded as an inherent value once the RC is built. The quantified stirrer efficiency provides a universal way to evaluate the stirrer design and the chamber performance. The stirrer efficiency equation has been derived in (6.55), which makes the stirrer efficiency predictable. The time-gating technique in an RC has been proposed which provides an alternative method to eliminate the early time response and correct the chamber transfer function; this could potentially be used to increase the stirrer efficiency without changing the stirrers.

Dual quantities in the RC have been given and by applying the duality between the TSCS and the ACS, the limit of the TSCS has been obtained, which is a quarter of the stirring surface, thus the limit of the stirrer efficiency can be well defined. This means no matter how the structure of the stirrer is optimised, its performance cannot exceed the theoretical limit. The TSCS can be used to quantify the performance of the stirrer only or the performance of the RC. It has been found that it is actually the stirring surface that determines the performance, not the stirring volume.

References

1 Huang, Y. (1999). Conducting triangular chambers for EMC measurements. *Measurement Science and Technology* 10: 121–124.

2 Hill, D.A. (2009). *Electromagnetic Fields in Cavities: Deterministic and Statistical Theories*. Wiley-IEEE Press.

3 Balanis, C.A. (2012). *Advanced Engineering Electromagnetics*, 2e. Wiley.

4 Liu, B.H., Chang, D.C., and Ma, M.T. (1066). *Eigenmodes and the Composite Quality Factor of a Reverberating Chamber*, 1983. US National Bureau of Standards, Technical Note.

5 Leferink, F. B. J. (2010). Using reverberation chambers for EM measurements. *18th International Conference on Software, Telecommunications and Computer Networks*. Split, Dubrovnik. pp. 1–5.

6 Bruns, C. (2005). Three-Dimensional Simulation and Experimental Verification of a Reverberation Chamber. Doctor of Sciences Dissertation, Swiss Federal Institute of Technology Zurich.

7 Leferink, F.B.J. (1998). High field strength in a large volume: the intrinsic reverberation chamber. *IEEE EMC International Symposium on Electromagnetic Compatibility, Symposium Record (Cat. No.98CH36253)*. Denver, CO. vol. 1 pp. 24–27.

8 Selemani, K., Gros, J.B., Richalot, E. et al. (2015). Comparison of reverberation chamber shapes inspired from chaotic cavities. *IEEE Transactions on Electromagnetic Compatibility* 57 (1): 3–11.

9 Schroeder, M.R. (1975). Diffuse sound reflection by maximum-length sequences. *Journal of the Acoustical Society of America* 57 (1): 149–150.

10 Cox, T. and P. D'Antonio. (2005). Thirty years since "Diffuse Sound Reflection by Maximum-Length Sequences": Where are we now?. Forum Acusticum, Budapest. pp. 2129–2134.

11 Cox, T. and D'Antonio, P. (2003). Schroeder diffusers: a review. *Building Acoustics* 10 (1): 1–32.

12 Cox, T. and D'Antonio, P. (2016). *Acoustic Absorbers and Diffusers – Theory, Design and Application*, 3e. CRC Press.

13 Bradley, D., Trapet, M.M., Adelgren, J., and Vorlander, M. (2014). Effect of boundary diffusers in a reverberation chamber: standardized diffuse field quantifiers. *Journal of the Acoustical Society of America* 134 (4): 1898–1906.

14 Karadimou, E. (2013). Investigation of Wave Diffusers to Scatter the Unstirred Radiation of an Antenna in a Reverberation Chamber, Master Thesis, Department of Electronics, University of York.

15 Marvin, A.C. and Karadimou, E. (2013). The use of wave diffusers to reduce the contribution of specular wall reflections to the unstirred energy in a reverberation chamber, IEEE International Symposium on Electromagnetic Compatibility. Denver, CO. pp. 227–231.

16 Rhee, E. and Rhee, J.-G.(2006). Comparison of field uniformity characteristics in a triangular reverberation chamber with QRS diffusers, 17th International Zurich Symposium on Electromagnetic Compatibility. Singapore. pp. 489–492.

17 Shen, B., Ren, W., Gao, B., and S. Yang. (2002). The analysis of several diffusers in a reverberation chamber by FDTD method, 3rd International Conference on Microwave and Millimeter Wave Technology (ICMMT). pp. 911–914.

18 Yun, J., Rhee, J., and Chung, S. (2001). An improvement of field uniformity of reverberation chamber by the variance of diffuser volume ratio, *Asia-Pacific Microwave Conference (APMC)*. Taipei, 2001, vol. 3. pp. 1123–1126.

19 Petirsch, W. and Schwab, A.J. (1999). Investigation of the field uniformity of a mode-stirred chamber using diffusers based on acoustic theory. *IEEE Transactions on Electromagnetic Compatibility* 41 (4): 446–451.

20 Arnaut, L.R. (2003). Comments on: investigation of the field uniformity of a mode-stirred chamber using diffusors based on acoustic theory. *IEEE Transactions on Electromagnetic Compatibility* 45 (1): 146–147.

21 He, P., Jiang, Q., Zhou, X. et al. (2009). Influence of reverberation chamber field uniformity using acoustic diffusor. *Journal of Microwaves* 25 (2).

22 Wanderlinder, L.F., Lemaire, D., and Seetharamdoo, D. (2016). Experimental analysis for metamaterials used to lower the LUF of a reverberation chamber, International Symposium on Electromagnetic Compatibility – EMC EUROPE. Wroclaw. pp. 240–244.

23 M.I. Andries, D. Seetharamdoo, and P. Besnier (2014) .Analytical modal analysis to evaluate the contribution of metamaterials to the improvement of reverberation chambers. *International Symposium on Electromagnetic Compatibility*. Gothenburg. pp. 883–888.

24 Seetharamdoo, D. and Coccato, I.M. (2011). Investigation on the use of metamaterials to lower the operating frequency of reverberation chamber, 10th International Symposium on Electromagnetic Compatibility. York. pp. 680–685.

25 Dunlap, C.R. (2013). Reverberation chamber characterization using enhanced backscatter coefficient measurements, PhD dissertation, Department of Electrical, Computer and Engineering, University of Colorado Boulder, Boulder, CO.

26 IEC 61000-4-21. (2011). Electromagnetic compatibility (EMC) – Part 4–21: Testing and measurement techniques – Reverberation chamber test methods, IEC Standard, Ed 2.0, 2011–01.

27 Rosnarho, J.-F. and Berre, S.L. (2016). Reverberation Chambers Handbook, Everything you ever wanted to know about reverberation chambers but never dared to ask. Siepel France, ver. 3.

28 Demoulin, B. and Besnier, P. (2011). *Electromagnetic Reverberation Chambers*. Wiley.

29 Boyes, S. and Huang, Y. (2016). *Reverberation Chambers: Theory and Applications to EMC and Antenna Measurements*. Wiley.

30 Huang, Y., Zhang, J., and Liu, P.(2005). A novel method to examine the effectiveness of a stirrer, International Symposium on Electromagnetic Compatibility. Chicago, IL. pp. 556–561.

31 Wu, D.I. and Chang, D.C. (1989). The effect of an electrically large stirrer in a mode-stirred chamber. *IEEE Transactions on Electromagnetic Compatibility* 31 (2): 164–169.

32 Clegg, J., Marvin, A.C., Dawson, J.F., and Porter, S.J. (2005). Optimization of stirrer designs in a reverberation chamber. *IEEE Transactions on Electromagnetic Compatibility* 47 (4): 824–832.

33 Manicke, A., Pasche, K., and Krauthäuser, H.G. (2014). Evaluation of stirrer efficiency by means of first reflection, International Symposium on Electromagnetic Compatibility. Gothenburg. pp. 290–295.

34 Coates, A. and Duffy, A.P. (2007). Maximum working volume and minimum working frequency tradeoff in a reverberation chamber. *IEEE Transactions on Electromagnetic Compatibility* 49 (3): 719–722.

35 Bastianelli, L., Primiani, V.M., and Moglie, F. (2015). Stirrer efficiency as a function of its axis orientation. *IEEE Transactions on Electromagnetic Compatibility* 57 (6): 1732–1735.

36 Primiani, V.M. and Moglie, F. (2014). Reverberation chamber performance varying the position of the stirrer rotation axis. *IEEE Transactions on Electromagnetic Compatibility* 56 (2): 486–489.

37 Yuan, Z., He, J., and Chen, S. (2006). Evaluation of the split stirrer in reverberation chamber, 4th Asia-Pacific Conference on Environmental Electromagnetics. Dalian. pp. 454–457.

38 Vosoogh, A., Khaleghi, A., and Keyghobad, K. (2011). Design and simulation of a mode stirred reverberation chamber for IEC 61000-4-21, Loughborough Antennas & Propagation Conference. Loughborough. pp. 1–4.

39 Xiaoqiang, L., Guanghui, W., Yongqiang, Z., and Chenghuai, Z., (2008). Effects of stirrer on the field uniformity at low frequency in a reverberation chamber and its simulation, International Symposium on Computer Science and Computational Technology. Shanghai. pp. 517–519.

40 Bastianelli, L., Moglie, F., and Primiani, V.M. (2016). Evaluation of stirrer efficiency varying the volume of the reverberation chamber, IEEE International Symposium on Electromagnetic Compatibility. Ottawa, ON. pp. 19–24.

41 Tan, H., Huang, X., and Zheng, S. (2013). Simulation on field uniformity of reverberation chamber with different stirrer schemas, 5th IEEE International Symposium on Microwave, Antenna, Propagation and EMC Technologies for Wireless Communications. Chengdu. pp. 563–567.

42 Hong, J.I. and Huh, C.S. (2010). Optimization of stirrer with various parameters in reverberation chamber. *Progress in Electromagnetics Research* 104: 15–30.

43 Lunden, O. and Backstrom, M.(2003). A factorial designed experiment for evaluation of mode-stirrers in reverberation chambers, IEEE International Symposium on Electromagnetic Compatibility. Istanbul. vol. 1. pp. 465–468.

44 Wellander, N., Lunden, O., and Backstrom, M. (2007). Experimental investigation and mathematical modeling of design parameters for efficient stirrers in mode-stirred reverberation chambers. *IEEE Transactions on Electromagnetic Compatibility* 49 (1): 94–103.

45 Huang, Y., Abumustafa, N., Wang, Q.G., and Zhu, X. (2006). Comparison of two stirrer designs for a new reverberation chamber, 4th Asia-Pacific Conference on Environmental Electromagnetics. Dalian. pp. 450–453.

46 Ubin, A., Vogt-Ardatjew, R., and Leferink, F. (2015). Statistical analysis of three different stirrer designs in a reverberation chamber, Asia-Pacific Symposium on Electromagnetic Compatibility (APEMC). Taipei. pp. 604–607.

47 Asander, H.J., Eriksson, G., Jansson, L., and Akermark, H. (2002). Field uniformity analysis of a mode stirred reverberation chamber using high resolution computational modeling, IEEE International Symposium on Electromagnetic Compatibility. Minneapolis, MN. vol. 1. pp. 285–290.

48 Boyes, S.J., Huang, Y., and Khiabani, N.(2011). Improved Rayleigh field statistic in reverberation chambers from modified mechanical stirring paddles, Loughborough Antennas & Propagation Conference. Loughborough. pp. 1–4.

49 Zhu, S. and Zhu, Y. (2013). Stirrer swing experiment in reverberation chambers, 5th IEEE International Symposium on Microwave, Antenna, Propagation and EMC Technologies for Wireless Communications. Chengdu. pp. 642–645.

50 Moglie, F. and Primiani, V.M. (2012). Numerical analysis of a new location for the working volume inside a reverberation chamber. *IEEE Transactions on Electromagnetic Compatibility* 54 (2): 238–245.

51 Magdowski, M., Tkachenko, S.V., and Vick, R. (2011). Coupling of stochastic electromagnetic fields to a transmission line in a reverberation chamber. *IEEE Transactions on Electromagnetic Compatibility* 53 (2): 308–317.

52 Ladbury, J. M. (1999). Monte Carlo simulation of reverberation chambers, Proceedings of the Digital Avionics Systems Conference. vol. 2. St Louis, MO. October 1999, pp. 10.C.1-1–10.C.1-8.

53 Musso, L., Berat, V., Canavero, F., and Demoulin, B. (2002). A plane wave Monte Carlo simulation method for reverberation chambers, Proceedings of the International Symposium on Electromagnetic Compatability. pp. 1–6.

54 Weisstein, E.W., Sphere Point Picking, From MathWorld – A Wolfram Web Resource. http://mathworld. wolfram.com/SpherePointPicking.html.

55 Amador, E., Lemoine, C., and Besnier, P. (2012). A probabilistic approach to susceptibility measurement in a reverberation chamber, Asia-Pacific Symposium on Electromagnetic Compatibility. Singapore. pp. 761–764.

56 Gradoni, G., Antonsen, T.M., Anlage, S.M., and Ott, E. (2015). A statistical model for the excitation of cavities through apertures. *IEEE Transactions on Electromagnetic Compatibility* 57 (5): 1049–1061.

57 Hallbjorner, P. (2007). Accuracy in reverberation chamber antenna correlation measurements, International workshop on Antenna Technology: Small and Smart Antennas Metamaterials and Applications. Cambridge. pp. 170–173.

58 West, J.C., Bunting, C.F., and Rajamani, V. (2012). Accurate and efficient numerical simulation of the random environment within an ideal reverberation chamber. *IEEE Transactions on Electromagnetic Compatibility* 54 (1): 167–173.

59 Rehammar, R., Skårbratt, A., and Lötbäck-Patané, C. (2015). Antenna measurements in reverberation chambers and their relation to Monte Carlo integration methods, International Symposium on Antennas and Propagation (ISAP). Hobart, Tasmania. pp. 1–3.

60 Andries, M.I., Besnier, P., and Lemoine, C. (2013). Diversity gain for Rician and Rayleigh environments in reverberation chamber, International Symposium on Electromagnetic Compatibility. Brugge. pp. 502–507.

61 F. Diouf, F. Paladian, M. Fogli, C. Chauviere and P. Bonnet (2007). Emission in reverberation chamber: numerical evaluation of the total power radiated by a wire with a stochastic collocation method. *18th International Zurich Symposium on Electromagnetic Compatibility*. Munich. pp. 99–102.

62 Serra, R. and Canavero, F. (2009). Field statistics in a one-dimensional reverberation chamber. *Comptes Rendus Physique* 10: 31–41.

63 Junqua, I., Parmantier, J.P., and Degauque, P. (2010). Field-to-wire coupling in an electrically large cavity: a semianalytic solution. *IEEE Transactions on Electromagnetic Compatibility* 52 (4): 1034–1040.

64 Zhang, X., Robinson, M.P., Flintoft, I.D., and Dawson, J.F. (2016). Inverse Fourier transform technique of measuring averaged absorption cross section in the reverberation chamber and Monte Carlo study of its uncertainty, International Symposium on Electromagnetic Compatibility – EMC EUROPE. Wroclaw. pp. 263–267.

65 Orjubin, G. (2007). Maximum field inside a reverberation chamber modeled by the generalized extreme value distribution. *IEEE Transactions on Electromagnetic Compatibility* 49 (1): 104–113.

66 Bonnet, P., Diouf, F., Chauviere, C. et al. (2009). Numerical simulation of a reverberation chamber with a stochastic collocation method. *Comptes Rendus Physique* 10: 54–64.

67 Wilson, P.F., Hill, D.A., and Holloway, C.L. (2002). On determining the maximum emissions from electrically large sources. *IEEE Transactions on Electromagnetic Compatibility* 44 (1): 79–86.

68 Garcia-Fernandez, M.A., Decroze, C., Carsenat, D. et al. On the relationship between field amplitude distribution, its maxima distribution, and field uniformity inside a mode-stirred reverberation chamber. *International Journal of Antennas and Propagation* 2012, 483287: 1–7.

69 Amador, E., Lemoine, C., and Besnier, P. (2012). Optimization of immunity testings in a mode tuned reverberation chamber with Monte Carlo simulations, ESA Workshop on Aerospace EMC. Venice. pp. 1–6.

70 Wang, C.M., Remley, K.A., Kirk, A.T. et al. (2014). Parameter estimation and uncertainty evaluation in a low Rician K-factor reverberation-chamber environment. *IEEE Transactions on Electromagnetic Compatibility* 56 (5): 1002–1012.

71 Y. Li, X. Zhao, L. Yan, K. Huang and H. Zhou (2016). Probabilistic-statistical model based on mode expansion of the EM field of a reverberation chamber and its Monte Carlo simulation. *Asia-Pacific International Symposium on Electromagnetic Compatibility (APEMC)*. Shenzhen, pp. 779–781.

72 Flintoft, I., Marvin, A., and Dawson, L. (2008). Statistical response of nonlinear equipment in a reverberation chamber, IEEE International Symposium on Electromagnetic Compatibility. Detroit, MI. pp. 1–6.

73 Monsef, F. (2012). Why a reverberation chamber works at low modal overlap. *IEEE Transactions on Electromagnetic Compatibility* 54 (6): 1314–1317.

74 Moglie, F., Bastianelli, L., and Primiani, V.M. (2016). Reliable finite-difference time-domain simulations of reverberation chambers by using equivalent volumetric losses. *IEEE Transactions on Electromagnetic Compatibility* 58 (3): 653–660.

75 Harima, K. and Yamanaka, Y. (1999). FDTD analysis on the effect of stirrers in a reverberation chamber, Proceedings of the International Symposium on Electromagnetic Compatibility, Tokyo. IEICE. pp. 223–229.

76 Harima, K. (1998). FDTD analysis of electromagnetic fields in a reverberation chamber. *IEICE Transactions on Communications* E81-B (10): 1946–1950.

77 Bai, L., Wang, L., Wang, B., and Song, J. (1999). Reverberation chamber modeling using FDTD, Proceedings of the IEEE International Symposium on Electromagnetic Compatibility. Piscataway, NJ, IEEE. pp. 7–11.

78 Bai, L., Wang, L., Wang, B., and J. Song, (1999). Effects of paddle configurations on the uniformity of the reverberation chamber, Proceedings of the IEEE International Symposium on Electromagnetic Compatibility. Piscataway, NJ, IEEE. pp. 12–16.

79 Zhang, D. and Song, J. (2000). Impact of stirrers' position on the properties of a reverberation chamber with two stirrers, Proceedings of the IEEE International Symposium on Electromagnetic Compatibility. vol. 1. Piscataway, NJ, IEEE. pp. 7–10.

80 Hoijer, M., Andersson, A.M., Lunden, O., and M. Backstrom (2000). Numerical simulations as a tool for optimizing the geometrical design of reverberation chambers, Proceedings of the IEEE International Symposium on Electromagnetic Compatibility. vol. 1. Piscataway, NJ, IEEE. pp. 1–6.

81 Hoeppe, F., Gineste, P.-N., and Demoulin, B. (2001). Numerical predictions applied to mode stirred reverberation chambers, Proceedings of the 2001 Reverberation Chamber, Anechoic Chamber and OATS Users Meeting, Seattle, WA.

82 Kouveliotis, N.K., Trakadas, P.T., and Capsalis, C.N. (2003). DTD modeling of a vibrating intrinsic reverberation chamber. *Progress in Electromagnetics Research* 39: 47–59.

83 Moglie, F. (2004). Convergence of the reverberation chambers to the equilibrium analyzed with the finite-difference time-domain algorithm. *IEEE Transactions on Electromagnetic Compatibility* 46 (3): 469–476.

84 Moglie, F. and Pastore, A. (2004). FDTD analysis of reverberating chambers, Proceedings of the International Symposium on Electromagnetic Compatibility, Eindhoven, The Netherlands, Technische Universiteit Eindhoven. pp. 6–11.

85 Moglie, F. (2003). Finite difference, time domain analysis convergence of reverberation chambers, Proceedings of the 15th International Zurich Symposium and Technical Exhibition on Electromagnetic Compatibility, Zurich, Switzerland, Swiss Federal Institute of Technolology. Zurich. pp. 223–228.

86 Moglie, F. and Primiani, V.M. (2011). Reverberation chambers: Full 3D FDTD simulations and measurements of independent positions of the stirrers, IEEE International Symposium on Electromagnetic Compatibility. Long Beach, CA. pp. 226–230.

87 Moglie, F. and Pastore, A.P. (2006). FDTD analysis of plane wave superposition to simulate susceptibility tests in reverberation chambers. *IEEE Transactions on Electromagnetic Compatibility* 48 (1): 195–202.

88 Ritter, J. and Rothenhausler, M.(2003). Mode stirring chambers for full size aircraft tests: Concept- and design-studies, Proceedings of the European Microwave Conference. London, Horizon House Publishing Ltd.

89 Lammers, T.M. (2004). Numerical analysis of mode stirred chambers and their loaded and unloaded configurations, MSc thesis, University of Colorado, Boulder, CO.

90 Lammers, T.M., Holloway, C.L., and Ladbury, J. (2004). The effects of loading configurations on the performance of reverberation chambers, Proceedings of the International Symposium on Electromagnetic Compatibility. Eindhoven, The Netherlands, Technische Universiteit Eindhoven. pp. 727–732.

91 Bonnet, P., Vernet, R., Girard, S., and Paladian, F. (2005). FDTD modelling of reverberation chamber. *Electronics Letters* 41 (20): 1101–1102.

92 El Haffar, M., Reineix, A., Guiffaut, C., and Adardour, A. (2009). Reverberation chamber modeling using the FDTD method, International Conference on Advances in Computational Tools for Engineering Applications. Zouk Mosbeh. pp. 151–156.

93 Bosco, G., Picciani, C., Primiani, V.M., and Moglie, F. (2012). Numerical and experimental analysis of the performance of a reduced surface stirrer for reverberation chambers, *IEEE International Symposium on Electromagnetic Compatibility*. Pittsburgh, PA. pp. 156–161.

94 Wang, S., Wu, Z. and Cui, Y. (2013). FDTD simulation of field performance in reverberation chamber excited by two excitation antennas, *7th International Conference on Applied Electrostatics, Journal of Physics: Conference Series 418* 012006.

95 Bastianelli, L., Primiani, V.M. and Moglie, F. (2016). Effect of loss distribution on uncorrelated spatial points and frequency steps in reverberation chambers, International Symposium on Electromagnetic Compatibility – EMC EUROPE. Wroclaw. pp. 211–216.

96 Gradoni, G., Primiani, V.M., and Moglie, F. (2013). Reverberation chamber as a multivariate process: FDTD evaluation of correlation matrix and independent positions. *Progress in Electromagnetic Research* 133: 217–234.

97 Gradoni, G., Primiani, V.M., and Moglie, F. (2013). Carousel stirrer efficiency evaluation by a volumetric lattice-based correlation matrix, IEEE International Symposium on Electromagnetic Compatibility. Denver, CO. pp. 819–824.

98 Gradoni, G., Bastianelli, L., Primiani, V.M., and Moglie, F. (2015). Uncorrelated frequency steps in a reverberation chamber: A multivariate approach. *IEEE International Symposium on Electromagnetic Compatibility (EMC)*. Dresden. pp. 558–562.

99 Li, W., Yue, C., and Elsherbeni, A. (2017). A fast finite-difference time domain simulation method for the source-stirring reverberation chamber. *International Journal of Antennas and Propagation* 2017, 8715020: 1–8.

100 Clegg, J., Marvin, A.C., and J.F. Dawson (2002). Optimisation of stirrer designs in a mode stirred chamber using TLM, Proceedings of the 2002 URSI XXVIIth General Assembly, Maastricht (NL). presentation 1158, Ghent, URSI.

101 Coates, A.R., Duffy, A.P., Hodge, K.G., and Willis, A.J. (2002). Validation of mode-stirred reverberation chamber modelling, Proceedings of the International Symposium on Electromagnetic Compatibility. Sorrento, Milan, AEI. pp. 35–40.

102 Coates, A., Sasse, H.G., Coleby, D.E. et al. (2007). Validation of a three-dimensional transmission line matrix (TLM) model implementation of a mode-stirred reverberation chamber. *IEEE Transactions on Electromagnetic Compatibility* 49 (4): 734–744.

103 Weinzierl, D., Raizer, A., Kost, A., and Ferreira, G.S. (2003). Simulation of a mode stirred chamber excited by wires using the TLM method. *COMPEL – The international journal for computation and mathematics in electrical and electronic engineering* 22 (3): 770–778.

104 Weinzierl, D., Raizer, A., and Kost, A. (2004). Investigation of exciting fields in an alternative mode stirred chamber, Proceedings of the International Symposium on Electromagnetic Compatibility. Eindhoven, Technische Universiteit Eindhoven. pp. 723–727.

105 Weinzierl, D. (2006). Improvement of field distribution in a reverberation chamber by phase shift of exciting wires, calculated by TLM, 17th International Zurich Symposium on Electromagnetic Compatibility. Singapore. pp. 180–183.

106 Weinzierl, D., Sartori, C.A.F., Perotoni, M.B. et al. (2008). Numerical evaluation of noncanonical reverberation chamber configurations. *IEEE Transactions on Magnetics* 44 (6): 1458–1461.

107 Marvin, A.C., Dawson, J.F., and Clegg, J. (2004). Stirrer optimisation for reverberation chambers, Proceedings of the International Symposium on Electromagnetic Compatibility. Eindhoven, Technische Universiteit Eindhoven. pp. 330–335.

108 Smartt, C., Christopoulos, C., Sewell, P., and Greedy, S. (2013). Efficient modelling of a reverberation chamber environment in the time domain, International Symposium on Electromagnetic Compatibility. Brugge. pp. 354–358.

109 Leuchtmann, P., Bruns, C., and Vahldieck, R. (2003). Broadband method of moment simulation and measurement of a medium-sized reverberation chamber. *IEEE Symposium on Electromagnetic Compatibility* 2: 844–849.

110 Bruns, C. and Vahldieck, R. (2005). A closer look at reverberation chambers – 3-D simulation and experimental verification. *IEEE Transactions on Electromagnetic Compatibility* 47 (3): 612–626.

111 Gronwald, F. (2005). Calculation of mutual antenna coupling within rectangular enclosures. *IEEE Transactions on Electromagnetic Compatibility* 47 (4): 1021–1025.

112 Carlberg, U., Kildal, P.S., and Carlsson, J. (2005). Study of antennas in reverberation chamber using method of moments with cavity Green's function calculated by Ewald summation. *IEEE Transactions on Electromagnetic Compatibility* 47 (4): 805–814.

113 Carlberg, U., Kildal, P.S., and Carlsson, J. (2009). Numerical study of position stirring and frequency stirring in a loaded reverberation chamber. *IEEE Transactions on Electromagnetic Compatibility* 51 (1): 12–17.

114 Carlberg, U., Kildal, P.S., and Kishk, A.A. (2006). Fast numerical model of reverberation chambers with metal stirrers using moment method and cavity Green's function calculated by Ewald summation, IEEE Antennas and Propagation Society International Symposium. Albuquerque, NM pp. 2827–2830.

115 Karlsson, K., Carlsson, J., and Kildal, P.S. (2006). Reverberation chamber for antenna measurements: modeling using method of moments, spectral domain techniques, and asymptote extraction. *IEEE Transactions on Antennas and Propagation* 54 (11): 3106–3113.

116 Cerri, G., Primiani, V.M., Monteverde, C., and Russo, P. (2009). A theoretical feasibility study of a source stirring reverberation chamber. *IEEE Transactions on Electromagnetic Compatibility* 51 (1): 3–11.

117 Gruber, M.E., Adrian, S.B., and Eibert, T.F. (2013). A finite element boundary integral formulation using cavity Green's function and spectral domain factorization for simulation of reverberation chambers, International Conference on Electromagnetics in Advanced Applications (ICEAA). Torino. pp. 460–463.

118 Gruber, M.E. and Eibert, T.F. (2014). A hybrid Ewald-spectral representation of the rectangular cavity Green's function, International Symposium on Electromagnetic Compatibility. Gothenburg. pp. 906–909.

119 Gruber, M.E. and Eibert, T.F. (2014). A cavity Green's function boundary element method with spectral domain acceleration for modeling of reverberation chambers. *IEEE Transactions on Electromagnetic Compatibility* 56 (6): 1466–1473.

120 Gruber, M.E., Dengler, T.M., and Knaak, A. (2015). Analysis of a simultaneously clockwise and counterclockwise rotated mode stirrer in a reverberation chamber, IEEE International Symposium on Electromagnetic Compatibility (EMC). Dresden. pp. 402–405.

121 Gruber, M.E. and Eibert, T.F. (2015). A hybrid Ewald-spectral cavity Green's function boundary element method with spectral domain acceleration for modeling of over-moded cavities. *IEEE Transactions on Antennas and Propagation* 63 (6): 2627–2635.

122 Gruber, M.E. and Eibert, T.F. (2015). A cavity Green's function boundary element method for the modeling of reverberation chambers: validation against measurements, IEEE International Symposium on Electromagnetic Compatibility (EMC). Dresden. pp. 563–566.

123 West, J.C., Rajamani, V. and Bunting, C.F. (2014). Simulation of stirred fields within a reverberation chamber using a refined spectral-domain-factorization moment method, IEEE International Symposium on Electromagnetic Compatibility (EMC). Raleigh, NC. pp. 781–786.

124 West, J.C., Rajamani, V. and Bunting, C.F. (2013). Practical considerations for the evaluation of the 3-D Green's function in a rectangular cavity moment method at high frequency, IEEE International Symposium on Electromagnetic Compatibility. Denver, CO. pp. 813–818.

125 Zhao, H. and Shen, Z. (2015). Fast wideband analysis of reverberation chambers using hybrid discrete singular convolution-method of moments and adaptive frequency sampling. *IEEE Transactions on Magnetics* 51 (3): 1–4.

126 Zhao, H. and Shen, Z. (2009). Hybrid DSC-MoM analysis of two-dimensional transverse magnetic reverberation chamber, IEEE Antennas and Propagation Society International Symposium. Charleston. pp. 1–4.

127 Zhao, H., Hu, J., and Chen, Z. (2016). An alternative solution method for hybrid discrete singular convolution-method of moments modeling of reverberation chambers, Asia-Pacific International Symposium on Electromagnetic Compatibility (APEMC). Shenzhen. pp. 988–990.

128 Zhao, H. and Shen, Z. (2010). Hybrid discrete singular convolution-method of moments analysis of a 2-D transverse magnetic reverberation chamber. *IEEE Transactions on Electromagnetic Compatibility* 52 (3): 612–619.

129 Zhao, H. and Shen, Z. (2008). Modal-expansion analysis of a monopole in reverberation chamber, Asia-Pacific Microwave Conference. Macau. pp. 1–4.

130 Zhao, H., Shen, Z. and Li, E. (2012). Hybrid numerical modelling of reverberation chambers, *Asia-Pacific Symposium on Electromagnetic Compatibility*, Singapore. pp. 777–780.

131 Zhao, H. and Shen, Z. (2012). Memory-efficient modeling of reverberation chambers using hybrid recursive update discrete singular convolution-method of moments. *IEEE Transactions on Antennas and Propagation* 60 (6): 2781–2789.

132 Zhao, H. and Shen, Z. (2011). Efficient modeling of three-dimensional reverberation chambers using hybrid discrete singular convolution-method of moments. *IEEE Transactions on Antennas and Propagation* 59 (8): 2943–2953.

133 Zhao, H. and Shen, Z. (2011). Recursive update-discrete singular convolution method for modeling highly resonant structures, IEEE International Symposium on Antennas and Propagation (APSURSI). Spokane. pp. 309–312.

134 Zhao H. and Shen, Z. (2011). Fast and accurate prediction of reverberation chambers' resonant frequencies using time-domain integral equation and matrix pencil method, 8th International Conference on Information, Communications & Signal Processing. Singapore. pp. 1–4.

135 Zhao, H. (2013). MLFMM-accelerated integral equation modeling of reverberation chambers. *IEEE Antennas and Propagation Magazine* 55 (5): 299–308.

136 Zhao, H., Li, E., and Shen, Z. (2012). Modeling of reverberation chamber using FMM-accelerated hybrid integral equation, IEEE Asia-Pacific Conference on Antennas and Propagation. Singapore. pp. 213–214.

137 Zhao, H. and Hu, J. (2016). Efficient solution of electromagnetic problems based on integral equation, IEEE International Conference on Computational Electromagnetics (ICCEM). Guangzhou. pp. 109–110.

138 Borries, O., Hansen, P.C., Meincke, P. (2014). Integral equation modeling of reverberation chambers using high-order basis functions, Proceedings of the AMTA 36th Annual Meeting & Symposium. Tucson.

139 Yang, K. and Yilmaz, A.E. (2014). An FFT-accelerated integral-equation solver for analyzing scattering in rectangular cavities. *IEEE Transactions on Microwave Theory and Techniques* 62 (9): 1930–1942.

140 Yang, K., Ning, C., and Yin, W.Y. (2017). Characterization of near-field coupling effects from complicated three-dimensional structures in rectangular cavities using fast integral equation method. *IEEE Transactions on Electromagnetic Compatibility* 59 (2): 639–645.

141 Bunting, C.F., Moeller, K.J., Reddy, C.J., and Scearce, S.A. (1998). Finite element analysis of reverberation chambers: a two-dimensional study at cutoff, IEEE EMC International Symposium on Electromagnetic Compatibility. Denver, CO. vol. 1. pp. 208–212.

142 Bunting, C.F. (1999). Two-dimensional finite element analysis of reverberation chambers: the inclusion of a source and additional aspects of analysis, IEEE International Symposium on Electromagnetic Compatibility. Seattle, WA. vol. 1. pp. 219–224.

143 Bunting, C.F., Moeller, K.J., and Reddy, C.J. (1999). A two-dimensional finite-element analysis of reverberation chambers. *IEEE Transactions on Electromagnetic Compatibility* 41 (4): 280–289.

144 Bunting, C.F. (2001). Shielding effectiveness in a reverberation chamber using finite element techniques, International Symposium on Electromagnetic Compatibility. Montreal, Quebec. vol. 2. pp. 740–745.

145 Bunting, C.F. (2002). Statistical characterization and the simulation of a reverberation chamber using finite-element techniques. *IEEE Transactions on Electromagnetic Compatibility* 44 (1): 214–221.

146 Bunting, C.F. (2003). Shielding effectiveness in a two-dimensional reverberation chamber using finite-element techniques. *IEEE Transactions on Electromagnetic Compatibility* 45 (3): 548–552.

147 Suriano, C.R., Thiele, G.A., and Suriano, J.R. (2000). Low frequency behavior of a reverberation chamber with monopole antenna. IEEE International Symposium on Electromagnetic Compatibility, Symposium Record. Washington, DC. vol. 2. pp. 645–650.

148 Suriano, C., Thiele, G.A. and Suriano, J.R. (2001). Predicting low frequency behavior of arbitrary reverberation chamber configurations, IEEE EMC International Symposium on Electromagnetic Compatibility. Montreal, Quebec. vol. 2. pp. 757–761.

149 Zhang, D. and Li, E. (2002). Characterization of a reverberation chamber by 3D finite element method, 3rd International Symposium on Electromagnetic Compatibility. pp. 394–396.

150 Orjubin, G., Richalot, E., Mengue, S., and Picon, O. (2006). Statistical model of an undermoded reverberation chamber. *IEEE Transactions on Electromagnetic Compatibility* 48 (1): 248–251.

151 Orjubin, G., Richalot, E., Picon, O., and Legrand, O. (2007). Chaoticity of a reverberation chamber assessed from the analysis of modal distributions obtained by FEM. *IEEE Transactions on Electromagnetic Compatibility* 49 (4): 762–771.

152 Orjubin, G., Richalot, E., Mengue, S. et al. (2007). On the FEM modal approach for a reverberation chamber analysis. *IEEE Transactions on Electromagnetic Compatibility* 49 (1): 76–85.

153 Zekios, C.L., Allilomes, P.C., Lavranos, C.S., and Kyriacou, G.A. (2009). A three dimensional finite element eigenanalysis of reverberation chambers, International Symposium on Electromagnetic Compatibility – EMC Europe. Athens. pp. 1–4.

154 Zekios, C.L., Allilomes, P.C., and Kyriacou, G.A. (2010). Eigenfunction expansion for the analysis of closed cavities, Loughborough Antennas & Propagation Conference, Loughborough. pp. 537–540.

155 Zekios, C.L., Allilomes, P.C., Chryssomallis, M.T., and Kyriacou, G.A. (2014). Finite element based eigenanalysis for the study of electrically large lossy cavities and reverberation chambers. *Progress in Electromagnetics Research B* 61: 269–296.

156 Tan, H., Fang, C., and Huang, M. (2016). Numerical simulation of field uniformity of reverberation chamber, 11th International Symposium on Antennas, Propagation and EM Theory (ISAPE). Guilin. pp. 481–484.

157 Pirkl, R.J. and Remley, K.A. (2014). Experimental evaluation of the statistical isotropy of a reverberation chamber's plane-wave spectrum. *IEEE Transactions on Electromagnetic Compatibility* 46 (3): 498–509.

158 Pirkl, R.J., Ladbury, J.M., and Remley, K.A. (2011). The reverberation chamber's unstirred field: A validation of the image theory interpretation, IEEE International Symposium on EMC. pp. 670–675, Long Beach.

159 Holloway, C.L., Shah, H.A., Pirkl, R.J. et al. (2012). Reverberation chamber techniques for determining the radiation and total efficiency of antennas. *IEEE Transactions on Antennas and Propagation* 60 (4): 1758–1770.

160 Carlberg, U., Kildal, P.-S., Wolfgang, A. et al. (2004). Calculated and measured absorption cross sections of lossy objects in reverberation chamber. *IEEE Transactions on Electromagnetic Compatibility* 46 (2): 146–154.

161 Genender, E., Holloway, C.L., Remley, K.A. et al. (2010). Simulating the multipath channel with a reverberation chamber: application to bit error rate measurements. *IEEE Transactions on Electromagnetic Compatibility* 52 (4): 766–777.

162 Daniels, D.J. (2004). *Ground Penetration Radar*, 2e. The IEE Press.

163 Lunden, O. and Backstrom, M., Stirrer efficiency in FOA reverberation chambers. Evaluation of correlation coefficients and chi-squared tests, IEEE International Symposium on EMC, pp. 11–16, Washington DC, 2000.

164 Lemoine, C., Amador, E., and Besnier, P. (2011). Mode-stirring efficiency of reverberation chambers based on Rician K-factor. *Electronic Letters* 47 (20).

165 Lemonie, C., Besnier, P., and Drissi, M. (2008). Evaluation of frequency and mechanical stirring efficiency in a reverberation chamber, International Symposium on Electromagnetic Compatibility – EMC Europe. Hamburg. pp. 1–6.

166 Holloway, C.L., Hill, D.A., Ladbury, J.M., and Koepke, G. (2006). Requirements for an effective reverberation chamber: unloaded or loaded. *IEEE Transactions on Electromagnetic Compatibility* 48 (1): 187–194.

167 Holloway, C.L., Hill, D.A., Ladbury, J.M. et al. (2006). On the use of reverberation chambers to simulate a Rician radio environment for the testing of wireless devices. *IEEE Transactions on Antennas and Propagation* 54 (11): 3167–3177.

168 Burger, W.T.C., Holloway, C.L., and Remley, K.A. (2013).Proximity and orientation influence on Q-factor with respect to large-form factor loades in a reverberation chamber, International Symposium on Electromagnetic Compatibility. Brugge. pp. 369–374

169 Xu, Q., Huang, Y., Xing, L. et al. (2016). B-scan in a reverberation chamber. *IEEE Transactions on Antennas and Propagation* 64 (5): 1740–1750.

170 Amador, E., Lemonie, C., Besnier, P., and Laisné, A. (2010). Reverberation chamber modeling based on image theory: investigation in the pulse regime. *IEEE Transactions on Electromagnetic Compatibility* 52 (4): 778–789.

171 Shah, S.M.H.A. (2011). Wireless channel characterization of the reverberation chamber at NIST, M.Sc. thesis, Department of Signals and System, Chalmers University of Technology, Gothenburg.

172 Holloway, C.L., Shah, H.A., Pirkl, R.J. et al. (2012). Early time behavior in reverberation chambers and its effect on the relationships between coherence bandwidth, chamber decay time, RMS delay spread, and the chamber buildup time. *IEEE Transactions on Electromagnetic Compatibility* 54 (4): 717–725.

173 Kildal, P.-S., Carlsson, C., and Yang, J. (2001). Measurement of free-space impedances of small antennas in reverberation chambers. *Microwave and Optical Technology Letters* 32 (2): 112–115.

174 Lerosey, G. and de Rosny, J. (2007). Scattering cross section measurement in reverberation chamber. *IEEE Transactions on Electromagnetic Compatibility* 52 (2): 280–284.

175 Lallechere, S., Baba, I.E., Bonnet, P., and Paladian, F. (2011). Total scattering cross section improvements from electromagnetic reverberation chambers modelling and stochastic formalism, 5th European Conference on Antennas and Propagation (EUCAP). Rome. pp. 81–85.

176 Baba, I.E., Lallechere, S., Bonnet, P. (2012). Computing total scattering cross section from 3-D reverberation chambers time modelling, *Asia-Pacific Symposium on Electromagnetic Compatibility (APEMC)*, Singapore, May.

177 Amador, E., Lemoine, C., and Besnier, P. (2011). An empirical statistical detection of non-Ideal field distribution in a reverberation chamber confirmed by a simple numerical model based on image theory. *Annales des Telecommunications* 66: 445–455.

178 Ishimaru, A. (1978, ch. 14). *Wave Propagation and Scattering in Random Media*, vol. 2, 253–294. New York, Academic.

179 Huang, Y. and Edwards, D.J. (1992). A novel reverberating chamber: source-stirred chamber, IEE 8th International Conference on Electromagnetic Compatibility. Edinburghpp. 120–124.

180 Rosengren, K., Kildal, P.-S., Carlsson, C., and Carlsson, J. (2001). Characterization of antennas for mobile and wireless terminals in reverberation chambers: improved accuracy by platform stirring. *Microwave and Optical Technology Letters* 39 (6): 391–397.

181 Toorn, J.A.D., Remley, K.A., Holloway, C.L. et al. (2015). Proximity-effect test for lossy wireless-device measurements in reverberation chambers. *IET Science, Measurement & Technology* 9 (5): 540–546.

182 IEEE. (1998). IEEE recommended practice for radio-frequency (RF) absorber evaluation in the range of 30 MHz to 5 GHz, IEEE Standard 1128–1998, April.

183 Xu, Q., Huang, Y., Xing, L. et al. (2016). The limit of the total scattering cross section of electrically large stirrers in a reverberation chamber. *IEEE Transactions on Electromagnetic Compatibility* 58 (2): 623–626.

184 Ladbury, J.M. and Hill, D.A. (2007). Enhanced backscatter in a reverberation chamber: Inside every complex problem is a simple solution struggling to get out, IEEE International Symposium on Electromagnetic Compatibility. pp. 1–5, July 9–13.

185 Xu, Q., Huang, Y., Xing, L. et al. (2016). Average absorption coefficient measurement of arbitrarily shaped electrically large objects in a reverberation chamber. *IEEE Transactions on Electromagnetic Compatibility* 58 (6): 1776–1779.

7

Applications in the Reverberation Chamber

7.1 Introduction

In this chapter, typical measurements in the reverberation chamber (RC) are introduced. Instead of repeating the measurement steps in the standard (International Electrotechnical Commission (IEC) 61000-4-21), the measurement principles are the main focus. Thus one can conduct accurate and confident measurement even without following the standard. Based on different properties and features of the RC, many measurements can be performed to obtain a range of applications. In the past few years, the applications of the RC have been extended from electromagnetic compatibility (EMC) to other applications, including antenna measurements, channel emulation, Doppler effect, bioelectromangetics, material measurements, etc. In addition to the conventional EMC measurements, some new measurements which have not been included in the standard are also introduced and discussed in this chapter.

7.2 Q Factor and Decay Constant

The Q factor and the decay constant τ_{RC} are two most important parameters in the RC measurement. If one parameter is known, the other one can be easily obtained by using $Q = \omega\tau_{RC}$. They can be measured in the frequency domain (FD) (the power-ratio method) or in the time domain (TD) (the time-constant method) [1].

In the FD, by using the Hill's equation in (5.100), and considering the mismatch of the Tx and Rx antennas, we have:

$$Q = \frac{16\pi^2 V}{\lambda^3} \frac{\langle P_r \rangle}{P_t} \tag{7.1}$$

$$P_{Rx} = P_r \eta_{RxTot} \tag{7.2}$$

$$P_t = P_{in} \eta_{TxTot} \tag{7.3}$$

where P_{Rx} is the average received power at the port of the receiving (Rx) antenna, η_{RxTot} is the total efficiency (including radiation efficiency and mismatch) of Rx antenna, P_{in} is the input power at the port of the transmitting (Tx) antenna, and η_{TxTot} is the total efficiency of the transmitting antenna. Note that when using Hill's equation, the antenna efficiency and the cable loss need to be corrected. A typical measurement setup is illustrated in Figure 7.1. Suppose the Tx antenna is antenna 1 and the Rx antenna is antenna 2, the Q factor can be measured by using a signal generator and a spectrum analyser (or power meter) in Figure 7.1a, or by using a vector network analyser (VNA) in Figure 7.1b. A computer is used to control the measurement and record the data for each stirrer position.

Anechoic and Reverberation Chambers: Theory, Design, and Measurements, First Edition. Qian Xu and Yi Huang.
© 2019 John Wiley & Sons Ltd. Published 2019 by John Wiley & Sons Ltd.

(a)

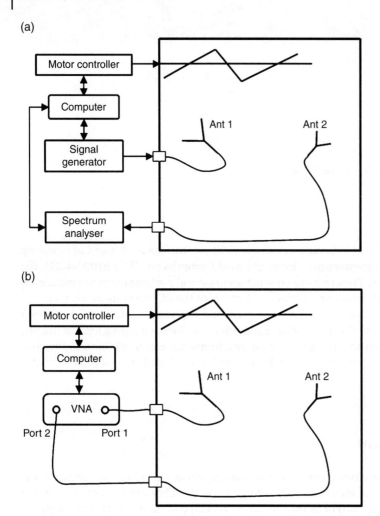

(b)

Figure 7.1 *Q* factor measurement in the FD: (a) measurement with a signal generator and a spectrum analyser (or power meter), and (b) measurement with a VNA.

More specifically, if the reference planes of the VNA in Figure 7.1b are calibrated to the input of the two antennas, the loss of the cables are corrected through the VNA calibration, so (7.1) can be written as

$$Q = \frac{16\pi^2 V}{\lambda^3} \frac{\langle |S_{21} - \langle S_{21} \rangle|^2 \rangle}{\left(1 - |\langle S_{11} \rangle|^2\right)\left(1 - |\langle S_{22} \rangle|^2\right)\eta_{1\text{rad}}\eta_{2\text{rad}}} \tag{7.4}$$

where $\eta_{1\text{rad}}$ and $\eta_{2\text{rad}}$ are the radiation efficiencies of antenna 1 and antenna 2 in Figure 7.1b, respectively, $S_{21} - \langle S_{21} \rangle$ is used to remove the unstirred part of the S-parameters, and $\langle S_{11} \rangle$ is the S-parameter of the antenna in free space. If the efficiency of antennas is unknown, the following estimated values can be used, which may not be accurate [2, 3]:

$$\eta_r = \begin{cases} 0.75, \text{ for log-periodic antenna} \\ 0.9, \text{ for horn antenna} \end{cases} \tag{7.5}$$

If the antennas are well matched and the unstirred part is very small, (7.4) can be simplified as:

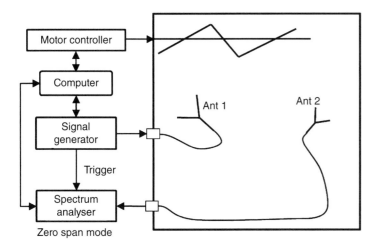

Figure 7.2 Q factor measurement in the TD.

$$Q = \frac{16\pi^2 V}{\lambda^3} \frac{\langle |S_{21}|^2 \rangle}{\eta_{1\text{rad}}\eta_{2\text{rad}}} \tag{7.6}$$

which is easy to use in the measurement.

It should be noted that, if the efficiencies of antennas are unknown and approximated values are used, the measured result may not be very accurate. However, if the Q factor is measured in the TD, this problem can be resolved. The TD measurement setup is shown in Figure 7.2. The spectrum analyser is set to zero span mode, or a power meter can be used to capture the impulse response. The schematic in Figure 7.2 is very similar to that in Figure 7.1a, but the chamber is excited with a modulated sine wave from the signal generator using antenna 1, and the 'on' and 'off' time of the signal is much larger than the chamber decay constant. The spectrum analyser should be triggered properly so that the response can be stably captured.

The received power decays exponentially with the speed of $P_0 \exp(-t/\tau_{\text{RC}})$, thus from the trace of the received power in the TD, the decay constant can be extracted [3]. The use of an oscilloscope is also feasible [4, 5], but normally a spectrum analyser can provide a larger dynamic range than an oscilloscope.

Instead of measuring the response of the RC in the TD directly, the TD response can be obtained from the inverse Fourier transform of the FD S-parameters obtained using a VNA. When doing the measurement in the FD, the number of samples needed in the frequency range of interest is important, as it affects the accuracy and measurement time. The maximum sample number is limited by the VNA. Normally, a high enough sampling rate (small sampling interval) in the FD is used to make sure the TD signal is obtained accurately (until the noise level is observed). It was found that the sampling interval was related to the coherence bandwidth or average mode bandwidth (Δf) of the RC in (5.28), and this suggested that at least 20 samples in Δf are required to calculate Δf from an autocorrelation curve [6]. However, it was shown that if we only want to obtain the chamber decay constant τ_{RC}, it is not necessary to have so many samples. Moreover, the coherence bandwidth Δf can be calculated from τ_{RC}. It was found that the sampling rate can be lower than the Nyquist criterion for the τ_{RC} extraction [7]. This makes the measurement faster over a certain frequency band or a wider band measurement can be performed for a given maximum sample number. The measurement setup is the same as Figure 7.1b. The theory will be given first below, followed by the measurement.

The Nyquist criterion in the FD is well known: the sampling frequency should be at least twice the bandwidth (upper frequency) of the signal; otherwise, an FD aliasing is observed (sampling in the TD to obtain the FD spectrum via Fourier transform). Conversely, we can get a TD signal from the FD signal, which is mathematically a dual problem (sampling in FD to obtain the TD signal via inverse Fourier transform). Assume the

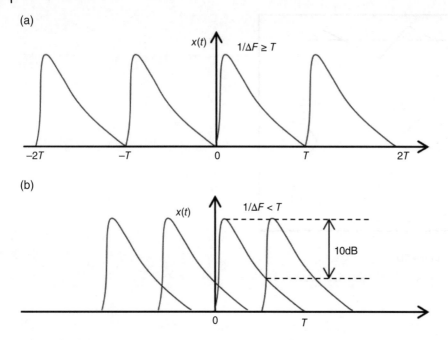

(a)

(b)

Figure 7.3 A schematic view of recovered TD signals: (a) TD signal with no aliasing and (b) TD signal with aliasing.

spectrum is sampled every ΔF (sampling interval) and T is the duration of the signal. If there is no aliasing, Nyquist criterion requires $1/\Delta F > T$ (in [6], it is written as $1/\Delta F > 2\tau$, where τ is the half period; to avoid confusion with the chamber decay constant, we use T instead of τ here). The recovered TD signal is shown in Figure 7.3. As can be seen, the Nyquist criterion is satisfied in Figure 7.3a but not in Figure 7.3b.

Since the response signal decays exponentially, as can be seen in Figure 7.3b, even though part of the signal is aliased, the decay part can still be used to extract τ_{RC} if the overlap is not too much. This offers an opportunity to use a larger sampling interval. Assume a 10 dB power decay can be observed (Figure 7.3b), where $t = \tau_{RC} \ln 10$. To obtain the TD signal in this time duration we need a sampling interval

$$\Delta F = \frac{1}{\tau_{RC} \ln 10} \tag{7.7}$$

The superposition from the tail part of the previous period is very small and can be ignored. We have $Q = f/\Delta f = \omega \tau_{RC}$ and the sampling interval can be written as

$$\Delta F = 2\pi \Delta f / \ln 10 \approx 2.73 \Delta f \tag{7.8}$$

When deriving (7.8), the early build-up time of the chamber is ignored; this will not affect the result because the build-up time is much smaller than the chamber decay time (normally the Q factor of the chamber is high) and 10 dB is a large enough threshold. As can be seen, if a 10 dB threshold is satisfied, $\Delta F \le 2.73 \Delta f$ and the sampling interval can be even larger than the coherence bandwidth. Compared with ΔF used to extract Δf from the autocorrelation curve $\Delta F \le \Delta f/20$ [6], the sampling interval in (7.8) is increased greatly (required frequency sample number is reduced). Δf can be obtained from $\Delta f = 1/(2\pi\tau_{RC})$.

The measurement was conducted in three scenarios: RC with no radio absorbing material (0 RAM), RC with 1 RAM, and RC with 3 RAMs (shown in Figure 7.4). The size of the RC at the University of Liverpool (UoL) is 3.6 m × 4 m × 5.8 m. To validate the theory, S-parameters in the range of 2.0–2.1 GHz were collected with different sample numbers (sampling intervals) and a maximum sample point of 10 001 was used (the sample point was increased until the noise level could be well observed, as can be seen in Figure 7.5). Two horn antennas were used (Rohde & Schwarz® HF 906 and SATIMO® SH 2000). One hundred stirrer positions were used with a step size of 3.5° and *S*-parameters were recorded when stirrers were steady (mode tuned method). A 10-order bandpass elliptic filter was used over the frequency band of interest to reduce the ripples from the rectangular filter. More details on the use of the bandpass filters can be found in Appendix D.

The least-squares method is used to extract τ_{RC}. To avoid the early-time response and the late-time noise level, only the exponential decay part is used for the least-squares fit. The least-squares fit at a single stirrer position is shown in Figure 7.5a, and the averaged TD signal is given in

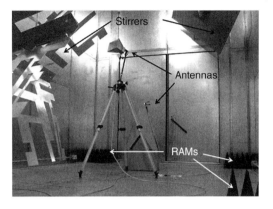

Figure 7.4 Chamber decay constant measurement.

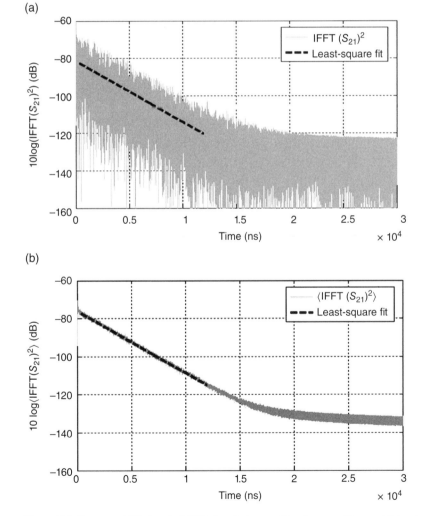

Figure 7.5 Recovered TD signals: (a) TD signal at a specific stirrer position and (b) averaged TD signal from all stirrer positions.

Figure 7.5b. $\langle \cdot \rangle$ means averaging the TD signal over all stirrer positions (power delay profile (PDP)). After the slope (k) is obtained, τ_{RC} can be yielded using

$$\tau_{RC} = -\frac{10}{k \ln 10} \tag{7.9}$$

We have 100 τ_{RC} values extracted from each stirrer position and only one τ_{RC} value extracted from the averaged TD signal. The mean value of these 100 samples is denoted as mean(τ), while the τ_{RC} from the averaged TD signal is denoted as average(τ). The results are shown in Figure 7.6. The coefficients of variance (CVs) of these 100 τ_{RC} samples with different sample numbers are given in Figure 7.7. As expected, for sampling points larger than 301 points, τ_{RC} is not sensitive to the stirrer position and we have small CVs. The loaded RAM increases the Δf of the chamber, which also reduces the CV. It is noted that average(τ) and mean(τ) are very accurate when only 301 sample points are used in this frequency range. From Figures 7.6 and 7.7, the measurement uncertainty of mean(τ) can be estimated as $1485 \pm 27\% \times 1485/\sqrt{100} = 1485 \pm 40$ ns for an empty RC (assume each measured sample is independent of the others and follows normal distribution). If we check the averaged power decay in Figure 7.8, it can be seen that, because of aliasing, the signal does not drop to the noise level but begins to increase at the end of the recovered time span. We still have around 10 dB of attenuated signal not affected by the aliasing, which allows the least-squares fit to be applied. Two hundred and one samples may not be enough, since the recovered duration is not long, which leads to errors for some stirrer positions and increases the CV value. For the chamber loaded with RAMs, because the load increases Δf, a decay of more than 10 dB can also be observed.

In this measurement, only 301 frequency samples are enough for an unloaded RC in the frequency range of 2.0–2.1 GHz. The sampling interval ΔF is 333.3 kHz, while the coherence bandwidth Δf is 118.7 kHz. $\Delta F/\Delta f \approx$ 2.8, which is very close to the theoretical value in (7.8). Note that the coherence bandwidth is the same as the average mode bandwidth [8]. In practice, by using Hill's equation, we can estimate ΔF (VNA sampling interval) by using

$$\Delta F \leq \frac{c_0^3 \eta_{1tot} \eta_{2tot}}{f^2 V \langle |S_{21}|^2 \rangle 8\pi \ln 10} \tag{7.10}$$

where η_{1tot} and η_{2tot} are the total efficiencies of the antennas and f is the frequency. If a wide frequency band is used, because Δf is smaller at higher frequency, f should be chosen as the highest frequency of interest. c_0 is the

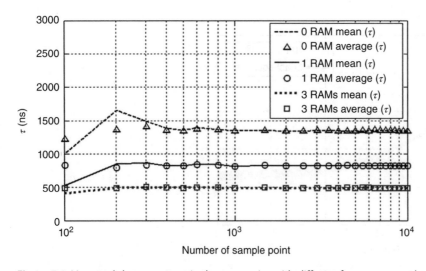

Figure 7.6 Measured decay constant in three scenarios with different frequency sample numbers.

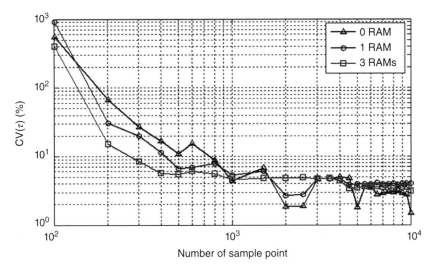

Figure 7.7 Measured CV in three scenarios with different sample numbers.

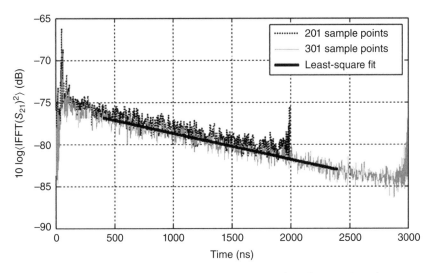

Figure 7.8 Recovered averaged TD signal and least-squares fit with 201 and 301 frequency samples in the RC (0 RAM). The peaks are the early time response dominated by the unstirred part.

speed of light in free space and V is the volume of the RC. $\langle |S_{21}|^2 \rangle$ is the averaged power transmission coefficient. Since a uniform sampling interval is used in (7.10), when the measurement frequency range of interest is very wide, the Nyquist criterion is not satisfied at higher frequencies but could still be satisfied at lower frequencies. A non-uniform sampling interval in the FD can reduce the measured sample number further but is normally not support by the measurement equipment (VNA).

It has been shown that it is not necessary to satisfy the Nyquist criterion to extract the decay constant or coherence bandwidth of an RC. The sampling interval does not have to be very small and can be even larger than the coherence bandwidth of the RC. Because the TD signal decays exponentially, aliasing only affects part of the signal and as long as a significant decay (e.g. 10 dB) is observed, τ_{RC} can be extracted accurately.

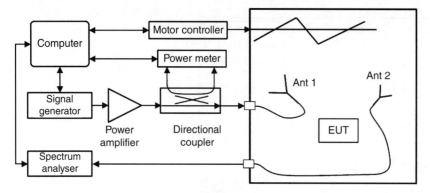

Figure 7.9 A typical radiated immunity test setup in the RC.

7.3 Radiated Immunity Test

The radiated immunity test is a classical application of the RC. Because of the high quality factor, it is easy to generate high electric field (E-field) in an RC. The radiated immunity test is to make sure the equipment under test (EUT) operates satisfactorily when subject to a strong radiated electromagnetic field. A schematic plot of the radiated immunity test is shown in Figure 7.9. Antenna 1 is used as the Tx antenna and antenna 2 in the working volume is used as the receiving antenna or a field probe. The EUT should be located at least a quarter wavelength from the chamber floor for tabletop operation and 10 cm above the floor for floor standing equipment. The support equipment should be non-metallic and low-absorbing.

A computer is used to control the measurement and the input power for each stirrer position is recorded automatically; a directional coupler can be used to monitor the input power and reflected power. Because the input power is proportional to the square of the E-field density, the required input power P_{input} to meet the test E-field E_{test} can be obtained as [2]

$$P_{\text{input}} = \left[\frac{E_{\text{test}}}{\langle E \rangle_{24\text{or}9} \times \sqrt{CLF(f)}} \right]^2 \tag{7.11}$$

where $CLF(f)$ is the chamber loading factor at frequency f and $\langle E \rangle_{24\text{or}9}$ is the average of the normalised E-field (average value or peak value) obtained from the empty chamber validation. E_{test} can be the total E-field magnitude or the rectangular component, which depends on the use of $\langle E \rangle_{24\text{or}9}$.

If the Q factor can be measured with the EUT loaded, one can also use (5.98) to obtain the required input power:

$$P_{\text{input}} = \frac{\omega \varepsilon \, V E_{\text{test}}^2}{Q} \tag{7.12}$$

The Q factor can be measured in the TD or in the FD by using antenna 1 and antenna 2. Note that $E_{\text{test}}^2 = \langle |\mathbf{E}|^2 \rangle$ in (7.12) is the average of the square of the total E-field magnitude. If the rectangular component is used, from (5.60) and (5.63) we have

$$\langle |E_x| \rangle = \langle |E_y| \rangle = \langle |E_z| \rangle = \frac{8}{15} \langle |\mathbf{E}| \rangle \tag{7.13}$$

and also from (5.47)

$$\langle |E_x|^2 \rangle = \langle |E_y|^2 \rangle = \langle |E_z|^2 \rangle = \frac{1}{3} \langle |\mathbf{E}|^2 \rangle \tag{7.14}$$

The IEC standard [2] specifies the minimum number of frequencies in the test, which should be at least 100 points per decade:

$$f_{n+1} = f_n \times 10^{\frac{1}{N-1}} \tag{7.15}$$

where $n = 1$ to N, with $N \geq 100$, and f_n is the nth test frequency. The dwell time at each frequency shall be at least 0.5 seconds to make sure the EUT is exposed to the field level for enough time. Swept frequency excitation may cause the chamber field to become non-stationary [2, 9, 10], thus the sweep rate should be specified and slow enough. The rotation of stirrers in conjunction with swept frequency testing is discouraged.

7.4 Radiated Emission Measurement

The radiated emission measurement could be the first application of the RC in the EMC area [11, 12]. The measurement setup is the same as Figure 7.9, except that the radiated power is the quantity to be measured. The substitution method is used: first measure the received power with a reference source, and then replace/turn off the reference and turn on the EUT, thus the total radiated power (TRP) of the EUT can be obtained from the received power. It should be noted that the radiated emission measurement in the RC is different from the measurement in the anechoic chamber (AC). In the RC, the TRP of the EUT is measured, but in the AC, the E-field strength at a certain distance is measured. If the TRP is measured in the AC, a 3D spherical scan is required and the received power density needs to be integrated for all angles and polarisations.

The TRP from the EUT can be determined using either average or maximum received power [2]:

$$P_{\text{Radiated}} = \frac{P_{\text{AveRec}} \eta_{\text{Tx}}}{CVF} \tag{7.16}$$

$$P_{\text{Radiated}} = \frac{P_{\text{MaxRec}} \eta_{\text{Tx}}}{CLF \times IL} \tag{7.17}$$

where P_{Radiated} is the TRP from the EUT, CVF is the chamber validation factor, CLF is the chamber loading factor, IL is the chamber insertion loss, P_{AveRec} is the received power averaged over all stirrer positions, P_{MaxRec} is the maximum power received over all stirrer positions, and η_{Tx} is the antenna efficiency of the Tx antenna used in the calibration process. (7.16) is based on the average received power and (7.17) is based on the maximum received power. The advantage of using the average received power is a lower uncertainty; the disadvantage is that the dynamic range is smaller than the maximum received power [2]. If the average received power is used, the measurement system should have sensitivity of 20 dB lower than the measured P_{MaxRec} to get an accurate average measurement, otherwise the noise level could dominate the averaged value.

The TRP measurement can also be understood from Hill's equation:

$$\langle P_{\text{r}} \rangle = \frac{\lambda^3 Q}{16\pi^2 V} P_{\text{t}} \tag{7.18}$$

First replace the P_{t} with a reference source P_{Ref}, then measure the average received power $\langle P_{\text{rRef}} \rangle$:

$$\langle P_{\text{rRef}} \rangle = \frac{\lambda^3 Q}{16\pi^2 V} P_{\text{Ref}} \tag{7.19}$$

Next measure the average received power from the EUT $\langle P_{\text{rEUT}} \rangle$:

$$\langle P_{\text{rEUT}} \rangle = \frac{\lambda^3 Q}{16\pi^2 V} P_{\text{Radiated}} \tag{7.20}$$

thus the TRP of the EUT can be obtained from (7.19) and (7.20):

$$P_{\text{Radiated}} = \frac{P_{\text{Ref}}}{\langle P_{\text{rRef}} \rangle} \langle P_{\text{rEUT}} \rangle \tag{7.21}$$

An alternative method is to measure the Q factor of the RC when the RC is loaded with the EUT, then apply (7.18) directly to obtain the TRP of the EUT [3]. The Q factor measurement was introduced in Section 7.2. The field strength generated by the EUT at a distance of R metres can be estimated by using the equation [2]:

$$E_{\text{Radiated}} = \sqrt{\frac{D \times P_{\text{Radiated}} \times \eta_0}{4\pi R^2}} \tag{7.22}$$

where E_{Radiated} is the estimated field strength generated by the EUT (in V m^{-1}), P_{Radiated} is the radiated power (in W), R is the distance from the EUT (in m) in the far-field, η_0 is the intrinsic impedance of free space and $\eta_0 \approx 377\,\Omega$, and D is the maximum directivity of the EUT. In practice, the directivity of the EUT is normally unknown and a directivity value of $D = 1.7$ is often used [2]. More generally, the expected maximum directivity $\langle D_{\text{max}} \rangle$ for an unintentional radiator can be approximated based on the radius of the smallest surrounding sphere [2, 13]:

$$\langle D_{\text{max}} \rangle = \begin{cases} 1.55 & \text{for } ka \leq 1 \\ 0.5 \left[0.577 + \ln\left(4(ka)^2 + 8ka\right) + \dfrac{1}{8(ka)^2 + 16ka} \right] & \text{for } ka > 1 \end{cases} \tag{7.23}$$

Figure 7.10 gives the plot of (7.23) with different ka.

An example is given to demonstrate this measurement [14]. The EUT used is from Invisible Systems (http://www.invisible-systems.com) and is a sensor/transceiver which collects and transmits the data (temperature, electric meters, etc.) wirelessly. The carrier frequency of the sensor is 869.5 MHz, which is much larger than the lowest usable frequency (LUF) of the RC. The measurement system and the sensor are shown in Figure 7.11a,b. The EUT and antennas are supported by empty cartons (low loss), and are positioned to reduce line-of-sight (LoS) paths.

The measurement procedure is given as follows:

1) The EUT is turned on and the signal generator (SG) is replaced with a 50 Ω load. The site attenuation (SA) reading for each stirrer position is then recorded.
2) Turn off the EUT, connect antenna 1 to the SG (reference source), and record the spectrum analyser reading for each stirrer position.

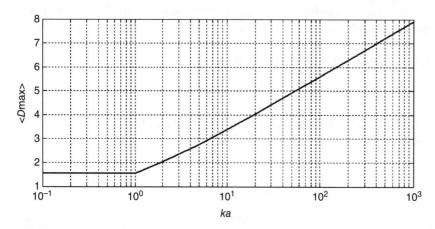

Figure 7.10 Expected maximum directivity of an unintentional radiator.

(a)

(b)

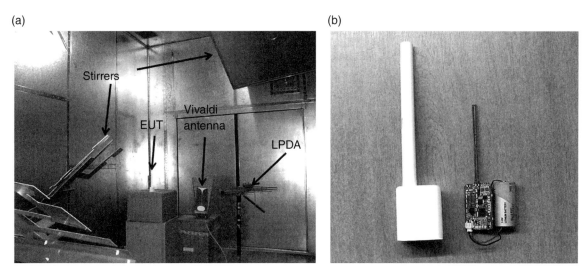

Figure 7.11 (a) The measurement system in an RC and (b) the plastic enclosure and the sensor (EUT).

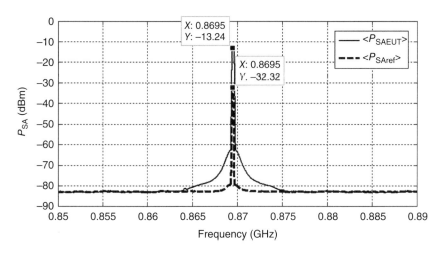

Figure 7.12 Measured averaged SA readings.

The radiated power of the EUT (P_{Radiated}) can be obtained from

$$P_{\text{Radiated}} = \frac{\langle P_{\text{SAEUT}} \rangle}{\langle P_{\text{SARef}} \rangle} P_{\text{Rad1}} \tag{7.24}$$

$$P_{\text{Rad1}} = P_{\text{Ref}} L_{\text{cable1}} \left(1 - |S_{11}|^2\right) \eta_{\text{1rad}} \tag{7.25}$$

where $\langle P_{\text{SAEUT}} \rangle$ is the averaged received power (the reading from the spectrum analyser in Figure 7.9) from the EUT for all stirrer positions, $\langle P_{\text{SARef}} \rangle$ is the averaged received power from the reference source for all stirrer positions, P_{Rad1} is the radiated power of antenna 1 when connected to the signal generator, P_{Ref} is the output power from the signal generator, L_{cable1} is the loss of the cable between antenna 1 and the signal generator, S_{11} is the reflection coefficient of antenna 1, and η_{1rad} is the radiation efficiency of antenna 1. From (7.24) and (7.25) we need to know the loss of the cable and the performance of antenna 1 (S_{11} and η_{1rad}). The loss of the cable and S_{11} of the antenna can be measured by using a VNA, and the radiation efficiency of an antenna can also be measured in the RC and is detailed in Section 7.6.

(a)

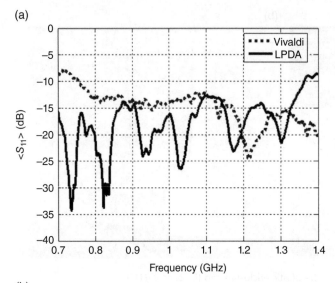

(b)

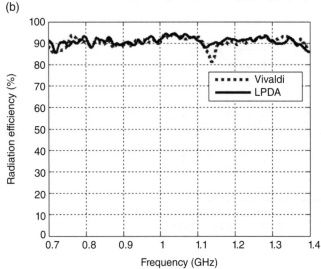

Figure 7.13 (a) Reflection coefficients of two antennas and (b) radiation efficiency of two antennas.

In the measurement, a log-periodic dipole antenna (LPDA, Rohde & Schwarz HL223) was used as antenna 2 in Figure 7.9 and a homemade wideband Vivaldi antenna was used as antenna 1. To protect the spectrum analyser, a 10 dB attenuator was connected to antenna 2 to attenuate the received signal, the reference level of signal generator was set as 0 dBm (P_{Ref} = 1 mW at 869.5 MHz), and 360 stirrer locations with a step size of $1°$ were used. The averaged readings from the spectrum analyser of the EUT and reference source are shown in Figure 7.12. The peak values are −13.24 dBm and −32.32 dBm, respectively. The reflection coefficients and radiation efficiencies of antennas were measured and are shown in Figure 7.13. It is interesting to note that the measured radiation efficiency for our LPDA is about 90%, not the recommended 75% as shown in (7.5). From (7.24) and (7.25), the radiated power of the EUT can be obtained as 17.4 dBm.

7.5 Free-Space Antenna *S*-Parameter Measurement

The *S*-parameter (or input impedance) of an antenna can be measured in an RC [15]. We have shown in Chapter 6 that the measured *S*-parameter of an antenna includes two parts: the unstirred part and the stirred part. If the stirrer efficiency η_s is high, the unstirred part is dominated by the free-space response, and the average of the stirred part is zero. Thus, the free-space S-parameter of the antenna under test (AUT) can be measured:

$$\langle S_{11} \rangle \to S_{\text{FreeSpace}} \quad \text{when } \eta_s \to 100\% \tag{7.26}$$

where $\langle S_{11} \rangle$ is the averaged complex S-parameter over all stirrer positions. The measurement setup is illustrated in Figure 7.14, where the dashed lines represent the reflections from the RC and the reference plane of port 1 is calibrated to the input of the AUT. The computer controls the rotation of the stirrers and records

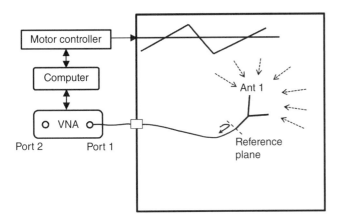

Figure 7.14 Free-space antenna S-parameter measurement in an RC.

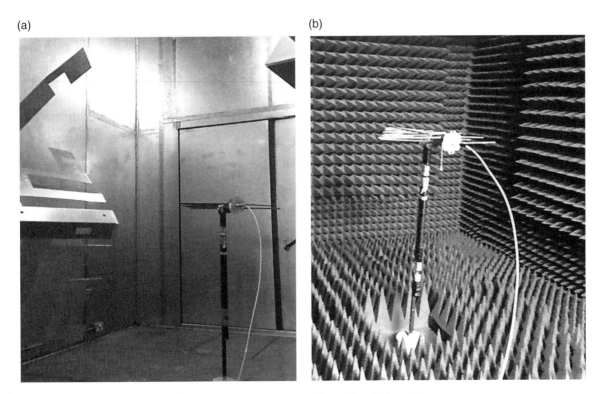

Figure 7.15 Free-space antenna S-parameter measurements in (a) an RC and (b) an AC.

data from the VNA for each stirrer position. After the complex S-parameters for all stirrer positions are measured, the free-space S-parameter of antenna can be obtained by using (7.26).

The measurement setups in the RC and AC are shown in Figure 7.15. An LPDA was used as the AUT, 100 stirrer positions were used to average the complex S-parameters measured in the RC, and the frequency sampling interval was 1 point/MHz. A comparison of measured results is illustrated in Figure 7.16. A frequency stir with 10 MHz bandwidth has also been used to average the complex S-parameters. As can be seen, a very good agreement has been obtained for both magnitude and phase.

It should be noted that the bandwidth of the frequency stir should be carefully chosen. In this measurement, a bandwidth of 10 MHz was used, which actually means that we assume the S-parameters do not have big variations in this bandwidth. If the frequency stir bandwidth is too wide, the frequency dependency of the S-parameters could be smoothed out. If the frequency bandwidth is too small, a smooth curve may not be obtained.

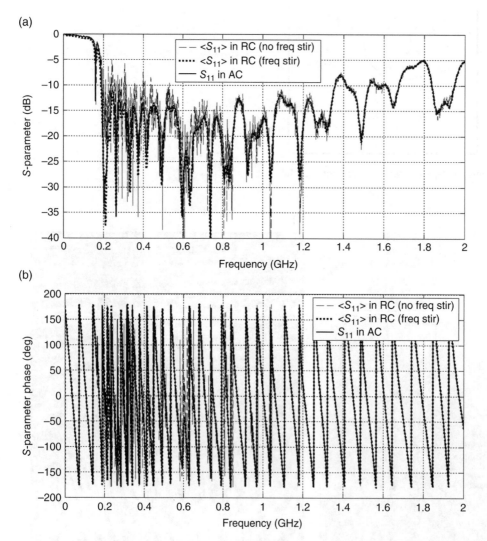

Figure 7.16 A comparison of measured free-space antenna S-parameters in an RC and an AC: (a) magnitude in dB and (b) phase in degree.

7.6 Antenna Radiation Efficiency Measurement

The antenna radiation efficiency is defined as the ratio of the total power radiated by an antenna P_{rad} to the total power accepted by the antenna P_a at its input terminal:

$$\eta_{rad} = \frac{P_{rad}}{P_a} \tag{7.27}$$

The total efficiency (or simply antenna efficiency) is defined as the ratio of the total power radiated by the antenna to the total power supplied to the antenna, which includes the mismatch of the antenna and is defined as

$$\eta_{tot} = \eta_{rad}\left(1 - |S_{11}|^2\right) \tag{7.28}$$

where S_{11} is the free-space reflection coefficient of the antenna.

There are many methods to measure the radiation efficiency of an antenna, such as the pattern integration [16], Wheeler cap [17], sliding wall cavity [18], ultra-wideband (UWB) Wheeler cap [19, 20], calorimetric [21], and RC methods. A review of antenna efficiency measurement has been given in [20]. In this section, we focus on the RC method. There are at least four methods to measure the radiation efficiency of antennas in an RC: the reference antenna method, the one-antenna method, the two-antenna method, and the three-antenna method. Different measurement methods have different preconditions. The antenna efficiency measurement in the RC has its own advantages compared with other methods: robust (no need for a high precision mechanical system), broadband, good accuracy, and no need for reference antenna (for non-reference antenna methods).

7.6.1 Reference Antenna Method

The reference antenna method has been given in IEC 61000-4-21 [2]. As shown in Figure 7.17, the computer controls the operation of the stirrers and collects the S-parameters from the VNA for each stirrer position. Suppose antenna 2 is the AUT, and the Ref Ant is the reference antenna with known radiation efficiency. The measurement procedure is given as follows.

1) Calibrate the VNA (the reference plane is calibrated at the end of the cable).
2) Place all the antennas inside the RC to ensure that the chamber Q factor does not change during the whole measurement and the main beams of each antenna are not directed to each other to reduce the unstirred part.
3) Connect antenna 1 to the VNA port 1 and antenna 2 to port 2. The reference antenna is loaded with a 50 Ω termination. Collect the complex S-parameters S_{A22} and S_{A21} for each stirrer position (S_{A^*} means the S-parameters with AUT).

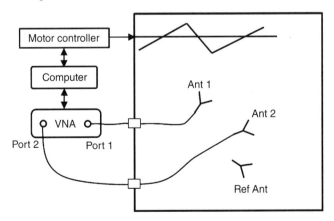

Figure 7.17 Radiation efficiency measurement setup using a reference antenna.

4) Connect the reference antenna to Port 2 of the VNA and load the AUT with a 50 Ω termination. Collect the complex S-parameters S_{R22} and S_{R21} for each stirrer position (S_{R*} means the S-parameters with the reference antenna).

If the RC has good field uniformity, based on Hill's equation, the chamber transfer function from antenna 1 to the AUT and antenna 1 to the reference antenna are the same:

$$\frac{\langle|S_{A21,s}|^2\rangle}{(1-|\langle S_{A22}\rangle|^2)\eta_A} = T_{A21} = T_{R21} = \frac{\langle|S_{R21,s}|^2\rangle}{(1-|\langle S_{R22}\rangle|^2)\eta_R} \tag{7.29}$$

where $S_{*,s}$ is the stirred part of the S-parameters:

$$S_{*,s} = S_* - \langle S_* \rangle \tag{7.30}$$

$\langle\cdot\rangle$ means the average value of the S-parameter using any stirring method (e.g. mechanical stir, frequency stir, polarisation stir, source stir, etc.). $S_{*,22}$ can also be obtained from AC, since $\langle S_{*,22}\rangle \approx S_{*,22\text{Freespace}}$ in (7.26). The radiation efficiency of AUT can be obtained as

$$\eta_A = \frac{\langle|S_{A21,s}|^2\rangle(1-|\langle S_{R22}\rangle|^2)}{\langle|S_{R21,s}|^2\rangle(1-|\langle S_{A22}\rangle|^2)}\eta_R \tag{7.31}$$

The total efficiency (including the mismatch of the antenna) is

$$\eta_{Atot} = \eta_A\left(1-|\langle S_{A22}\rangle|^2\right) = \frac{\langle|S_{A21,s}|^2\rangle}{\langle|S_{R21,s}|^2\rangle}\eta_{Rtot} \tag{7.32}$$

As can be seen, this method is direct and easy to understand. However, the problem is that we need an antenna with known radiation efficiency; this could be a problem in practice. If the accuracy requirement is not high, a well-matched horn antenna can be used as the reference antenna and an estimation value of 95% (±5%) can be used as a reference value [2].

7.6.2 Non-reference Antenna Method

Holloway et al. developed new methods to measure the radiation efficiency of antennas without involving a reference antenna [22]. The methods are based on the Q factor measured from the FD and the TD, and the enhanced backscattering effect is used. The measurement setup is shown in Figure 7.18, no reference antenna is involved.

In the FD, the Q factor can be obtained from Hill's equation [22]:

$$Q_{FDCor} = \frac{C_{RC}\langle|S_{21,s}|^2\rangle}{\eta_{1tot}\eta_{2tot}} \tag{7.33}$$

where C_{RC} is a constant:

$$C_{RC} = \frac{16\pi^2 V}{\lambda^3} \tag{7.34}$$

V is the volume of the chamber and λ is the free-space wavelength. Q_{FDCor} means that the Q factor has been corrected considering the total efficiency of the antennas. $S_{21,s}$ is the stirred part of the S-parameters defined in (7.30).

Meanwhile, the Q factor can be obtained from the TD [22]

$$Q_{TD} = \omega\tau_{RC} \tag{7.35}$$

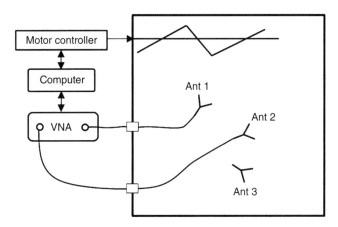

Figure 7.18 Radiation efficiency measurement without a reference antenna.

where τ_{RC} is the chamber decay constant which can be extracted from the TD response of the chamber; this was introduced in Section 7.2. τ_{RC} describes how fast the power profile decays in the TD ($\exp(-t/\tau_{RC})$). Because the Q factor obtained from the TD and the FD should be the same, thus $Q_{FDCor} = Q_{TD}$, we have:

$$\eta_{1tot}\eta_{2tot} = \frac{C_{RC}\langle|S_{21,s}|^2\rangle}{\omega\tau_{RC}} \tag{7.36}$$

Note that for each antenna, considering the enhanced backscattering effect in Section 5.4.4, we have the enhanced backscatter constant:

$$e_{b1} = \frac{\langle|S_{11,s}|^2\rangle}{\langle|S_{21,s}|^2\rangle} \tag{7.37}$$

Suppose antenna 1 and antenna 2 are exactly the same, we have $\eta_{1tot} = \eta_{2tot}$ and from (7.36) and (7.37) we have a system of equations:

$$\eta_{1tot}\eta_{2tot} = \frac{C_{RC}\langle|S_{21,s}|^2\rangle}{\omega\tau_{RC}}$$
$$e_{b1} = \frac{\langle|S_{11,s}|^2\rangle}{\langle|S_{21,s}|^2\rangle}$$
$$\eta_{1tot} = \eta_{2tot} \tag{7.38}$$

The total efficiency of antenna 1 can be obtained:

$$\eta_{1tot} = \sqrt{\frac{C_{RC}\langle|S_{11,s}|^2\rangle/e_{b1}}{\omega\tau_{RC}}} \tag{7.39}$$

If the RC is well stirred, we have $e_{b1} = 2$, thus

$$\eta_{1tot} = \sqrt{\frac{C_{RC}\langle|S_{11,s}|^2\rangle/2}{\omega\tau_{RC}}} \tag{7.40}$$

Also the radiation efficiency of antenna 1 can be expressed as:

$$\eta_{1rad} = \frac{\eta_{1tot}}{1-|\langle S_{11}\rangle|^2} = \frac{1}{1-|\langle S_{11}\rangle|^2}\sqrt{\frac{C_{RC}\langle|S_{11,s}|^2\rangle/2}{\omega\tau_{RC}}} \tag{7.41}$$

As can be seen, antenna 2 is actually not required, although we have assumed that antenna 1 and antenna 2 are exactly the same. Only antenna 1 is enough to complete the measurement and (7.41) is the formula for the one-antenna method [22].

Suppose the condition of $e_{b1} = 2$ is relaxed but let $e_{b1} = e_{b2}$. We therefore have a system of equations:

$$e_{b1} = \frac{\langle |S_{11,s}|^2 \rangle}{\langle |S_{21,s}|^2 \rangle}$$

$$e_{b2} = \frac{\langle |S_{22,s}|^2 \rangle}{\langle |S_{21,s}|^2 \rangle}$$

$$e_{b1} = e_{b2} \tag{7.42}$$

From (7.42), e_{b1} and e_{b2} can be solved and are denoted as e_b:

$$e_{b1} = e_{b2} = \frac{\sqrt{\langle |S_{11,s}|^2 \rangle \langle |S_{22,s}|^2 \rangle}}{\langle |S_{21,s}|^2 \rangle} = e_b \tag{7.43}$$

From (7.39), and applied to antenna 1 and antenna 2, we have

$$\eta_{1tot} = \sqrt{\frac{C_{RC}\langle |S_{11,s}|^2 \rangle / e_{b1}}{\omega \tau_{RC}}} = \sqrt{\frac{C_{RC}\langle |S_{11,s}|^2 \rangle / e_b}{\omega \tau_{RC}}}$$

$$\eta_{2tot} = \sqrt{\frac{C_{RC}\langle |S_{22,s}|^2 \rangle / e_{b2}}{\omega \tau_{RC}}} = \sqrt{\frac{C_{RC}\langle |S_{22,s}|^2 \rangle / e_b}{\omega \tau_{RC}}} \tag{7.44}$$

(7.44) is the formulae for the two-antenna method [22] and the radiation efficiency can be obtained in a similar way to (7.41).

It should be noted that when the AUT is very lossy (the radiation efficiency is very small), $S_{ii,s}$ could be outside the dynamic range of the VNA. In this case, we may need a power amplifier to increase the transmitting power. When the power amplifier is involved, however, the reverse signal from the output to the input is not allowed. This makes $S_{ii,s}$ hard to measure when it is very small. Thus we need an alternative method which can use the information of $S_{21,s}$ (like the gain measurement in an AC), and when the signal amplitude is small we can use a power amplifier to increase the dynamic range. If we have $e_{b1} = e_{b2} = 2$, from (7.43) we have

$$\sqrt{\langle |S_{11,s}|^2 \rangle} = \frac{2\langle |S_{21,s}|^2 \rangle}{\sqrt{\langle |S_{22,s}|^2 \rangle}} \tag{7.45}$$

Substituting (7.45) into (7.44), $S_{11,s}$ can be eliminated:

$$\eta_{1tot} = \langle |S_{21,s}|^2 \rangle \sqrt{\frac{2C_{RC}}{\omega \tau_{RC}\langle |S_{22,s}|^2 \rangle}} \tag{7.46}$$

which is a modified two-antenna method [23].

Suppose we have three antennas and no restrictions are applied on e_{b1}, e_{b2}, and e_{b3}. From (7.36), we have a system of equations

$$\eta_{1tot}\eta_{2tot} = \frac{C_{RC}\langle |S_{21,s}|^2 \rangle}{\omega \tau_{RC}} \tag{7.47}$$

$$\eta_{2tot}\eta_{3tot} = \frac{C_{RC}\langle |S_{32,s}|^2 \rangle}{\omega \tau_{RC}} \tag{7.48}$$

$$\eta_{1tot}\eta_{3tot} = \frac{C_{RC}\langle|S_{31,s}|^2\rangle}{\omega\tau_{RC}} \tag{7.49}$$

The total efficiency of antennas can be solved:

$$\eta_{1tot} = \sqrt{\frac{C_{RC}\langle|S_{21,s}|^2\rangle\langle|S_{31,s}|^2\rangle}{\omega\tau_{RC}\langle|S_{32,s}|^2\rangle}} \tag{7.50}$$

$$\eta_{2tot} = \sqrt{\frac{C_{RC}\langle|S_{21,s}|^2\rangle\langle|S_{32,s}|^2\rangle}{\omega\tau_{RC}\langle|S_{31,s}|^2\rangle}} \tag{7.51}$$

$$\eta_{3tot} = \sqrt{\frac{C_{RC}\langle|S_{31,s}|^2\rangle\langle|S_{32,s}|^2\rangle}{\omega\tau_{RC}\langle|S_{21,s}|^2\rangle}} \tag{7.52}$$

which are the formulae for the three-antenna method [22]. The radiation efficiency of antennas can be obtained in a similar way to (7.41). If the AUT is an array, the array elements can be excited simultaneously using a power divider; in this case the measured efficiency is the 'all-excited' array efficiency. The insertion loss of the power divider can be removed by using [24, 25]:

$$IL = 1 - \sum_{i=2}^{N}|S_{i,1}|^2 \tag{7.53}$$

where port 1 is the input port, and port 2 to port N are the output ports of the power divider and are connected to each element of the antenna array.

To demonstrate the non-reference antenna method, the measurements have been performed in two different chambers: the RC at the National Physical Laboratory (NPL) in the UK and the RC at the UoL, which have dimensions of 6.55 m × 5.85 m × 3.5 m and 3.6 m × 4 m × 5.8 m, respectively. The measurement setups are shown in Figure 7.19. The former has only one vertically installed paddle stirrer but the latter has two paddle

(a) 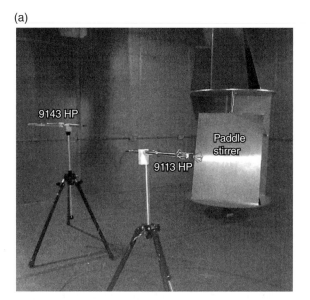　　　(b)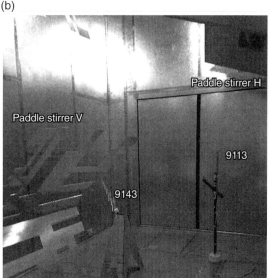

Figure 7.19 Radiation efficiency measurement setups in (a) NPL and (b) UoL.

Table 7.1 Summary of parameters used in the measurements.

Parameter	NPL	UoL
RC dimensions (m)	$6.55 \times 5.85 \times 3.5$	$5.8 \times 4 \times 3.6$
No. of stirrers	1 (vertical)	2 (vertical and horizontal)
VNA	R&S ZVB8	Keysight N9917A
No. of stirrer steps per revolution	360	
Frequency span (MHz)	200–1200	
No. of frequency sampling points	10 001	
Frequency step (kHz)	100	
IF BW (Hz)	100	
Antenna 1	Schwarzbeck log-periodic 9 143 (operating frequency range 250 MHz–7 GHz)	
Antenna 2	Schwarzbeck biconical 9 113 (operating frequency range 500 MHz–3 GHz)	
Frequency stirring BW (MHz)	20	
BPF BW (MHz)	100	
Time resolution (ns)	10	

stirrers (one is vertical and the other is horizontal). The total number of paddle steps per revolution at both NPL and UoL was set to be 360, which is equivalent to 1° per step.

The AUTs are the Schwarzbeck log periodic dipole array antenna 9143 (nominal operating frequency range 300 MHz–7 GHz, usable frequency 250 MHz–8 GHz) and the Schwarzbeck 9113 biconical antenna (operating frequency 500 MHz–3 GHz). The VNAs used at NPL and UoL were the R&S ZVB 8 and Keysight N9917A, respectively. Both VNAs were calibrated between 200 MHz and 1200 MHz with 10 001 points. When calculating the impulse responses of the chambers using inverse fast Fourier transform (IFFT), a rectangular window with 100 MHz bandwidth was used. This leads to the total time of 2000 ns and a time resolution of 10 ns in the TD. All these parameters are summarised in Table 7.1.

The measured chamber decay constants τ_{RC} are illustrated in Figure 7.20. Note that a bandpass filter with a rectangular window of 100 MHz was used, thus at the very beginning and the end of the frequency band the decay constants are not accurate. To avoid this issue, a wider measurement bandwidth can be used. In this measurement, the decay constants in the frequency range of 300–1100 MHz are reliable. Correspondingly, the Q factors are plotted in Figure 7.21. The Q factors of the RC at NPL are mostly higher than those at UoL because (i) the volume of the RC at NPL is larger than that at UoL and (ii) possibly the ohmic loss and loading (e.g. antenna support and stirrers) in the RC at NPL is less than that at the UoL. We also looked into the enhanced backscatter constant of the two RCs in the frequency range between 200 MHz and 1200 MHz. The enhanced backscatter constant is used to assess how well an RC is stirred at a particular position within the RC and can be calculated using (7.43). Figure 7.22 shows e_b of both RCs. It is noticed that the fields in the RC at UoL (at measured positions) show a better stirred profile than those in the RC at NPL for frequencies above 400 MHz as the e_b are closer to 2. As discussed in [22], $e_b = 2$ means that the one-antenna and the two-antenna methods should give similar antenna efficiency.

We then compared the electrical properties of the AUTs. Figure 7.23 compares the unstirred or 'free-space' reflection coefficients at the feed ports of antennas and the averaged receiving power between the two antennas in the two RCs. The unstirred S-parameters are defined by averaging the complex S-parameters over all paddle steps. It can be seen that although two RCs have different mechanical and electrical properties, the measured properties of the antennas agree fairly well. Figure 7.23 also shows the average received power measured in the

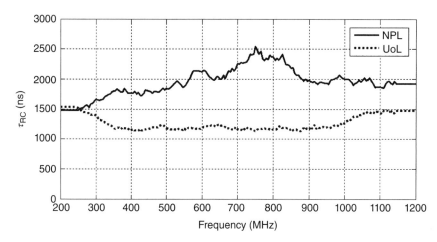

Figure 7.20 Decay constants of the RCs at NPL and UoL between 200 MHz and 1200 MHz.

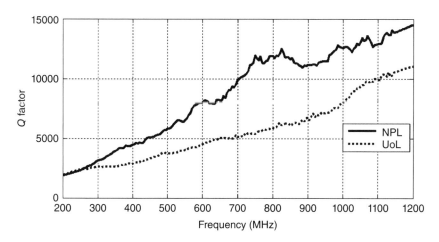

Figure 7.21 *Q* factors of the RCs at NPL and UoL, extracted from the TD response.

two chambers. The close agreement between the two tests indicates that the power losses (insertion loss or chamber transfer function) between the two chambers are similar.

Since two antennas were used in the test, both the one- and two-antenna methods can be used to measure the radiation efficiency. Figure 7.24 shows the radiation efficiency of the AUTs obtained by using the one- and two-antenna methods. For antenna 9143 (in Figure 7.24a), UoL has a very good agreement between the results calculated by using both one- and two-antenna methods for frequencies above 300 MHz but measurement at NPL provides a relatively larger difference in the efficiency. For both AUTs, the one-antenna method at NPL gives slightly higher efficiency than the two-antenna method, but the results from the two methods agree very well at UoL. This is because the enhanced backscatter constants of the RC at UoL at the test locations are closer to 2 than those at NPL, as already shown in Figure 7.22.

We have measured the antenna efficiency in different chambers using the non-reference antenna methods. It has been found that although the two RCs differ in dimensions and paddle stirrer configuration and hence in *Q* factors, decay constants, and enhanced backscatter constants, the discrepancy in the efficiency of the two antennas measured between the two chambers is less than 10% within their operational frequency bands. This should come as no surprise, as the non-reference antenna methods should give the same results in an ideal RC.

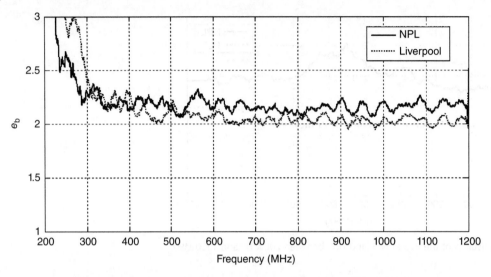

Figure 7.22 The measured enhanced backscatter constants of the RCs at NPL and UoL.

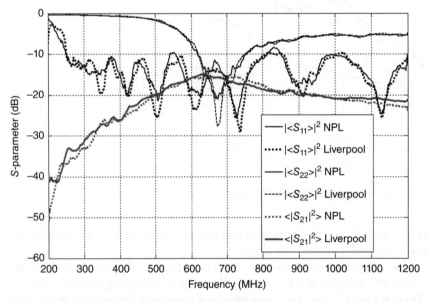

Figure 7.23 The unstirred reflection coefficients at the feed ports of antennas and the averaged receiving power. S_{11} and S_{22} refer to antenna 1 and antenna 2 in Table 7.1, respectively.

In particular, the radiation efficiency of the antennas measured using the two-antenna method between the two RCs agrees better for the omnidirectional antenna (<5%) than for the directional antenna (<10%). In addition, it is also obvious that the effect of non-perfect stirring in the RCs (corresponding to the enhanced backscatter constants being different from 2) is eliminated by the two-antenna method. Further investigation in [26] has indicated that polarisation mismatch could lead to up to 8% and 10% discrepancies for the omni-directional antenna and directional log-periodic antenna, respectively. These discrepancies can be removed by performing additional measurements with polarisation stirring [26].

(a)

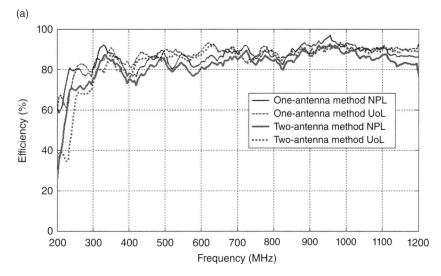

(b)

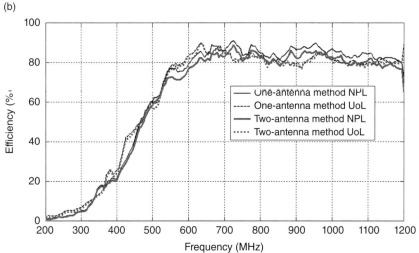

Figure 7.24 Measured radiation efficiency of antennas at NPL and UoL: (a) measured radiation efficiency of antenna 1 using the one- and two-antenna methods, and (b) measured radiation efficiency of antenna 2 using the one- and two-antenna methods.

To validate the modified two-antenna method, an antenna with known low radiation efficiency is needed. It is not easy to find a very lossy antenna with known radiation efficiency, so we could use an attenuator connected to an antenna and consider the attenuator as part of the antenna. We know how much radiation efficiency is reduced by introducing the attenuator, but in the measurement no special treatment is required, and the attenuator and the antenna become an integrated AUT. This offers an opportunity to verify the modified two-antenna method. To proceed, the measurement scenario is shown in Figure 7.25a and the size of the RC is 3.6 m × 4 m × 5.8 m. After the VNA was calibrated, antenna 1 (Rohde & Schwarz HF 906) and antenna 2 (SATIMO SH 2000) were connected to port 1 and port 2, respectively, with no LoS, and 360 stirrer locations with a step size of 1° for the stirrers and 10 001 points were set for the range of 2.8–4.2 GHz, which means that the IFFT of the collected S-parameters has a TD response of 7140 ns in length with a time step resolution of 0.24 ns.

(a)   (b)

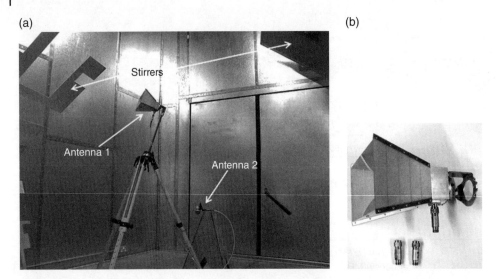

Figure 7.25 (a) The radiation efficiency measurement setup in an RC (modified two-antenna method) and (b) AUT with different attenuators.

The measured S-parameters of antenna 1 with different attenuators are shown in Figure 7.26. It can be seen that when the attenuation is large, the effect of RC could become too small to be detected in the dynamic range of the VNA. The corresponding $\ln[\text{IFFT}(S)^2]$ are shown in Figure 7.27, which can be regarded as the measured power signal in the TD. It can be found that the reflected signal is so small that it quickly drops below the noise level and cannot be detected in the limited dynamic range.

Because the power decay in the RC follows $P_0 e^{-t/\tau_{RC}}$, τ_{RC} is the measured decay time, which can be obtained from the slope (k) of $\ln[\text{IFFT}(S)^2]$, which is $\tau_{RC} = -1/k$. When the attenuation of antenna 1 is large we cannot extract τ_{RC} from $\ln[\text{IFFT}(S_{11})^2]$ so we use τ_{RC} obtained from $\ln[\text{IFFT}(S_{22})^2]$, since antenna 2 is a highly efficient antenna and the TD response has a better signal-to-noise ratio (SNR) over the time period of interest so the chamber decay time τ_{RC} becomes easy to extract. Since τ_{RC} is frequency dependent, an elliptic band-pass filter (BPF) with 200 MHz bandwidth was used to obtain τ_{RC} for different centre frequencies (Appendix D). The averaged τ_{RC} obtained from 360 stirrer locations is shown in Figure 7.28.

Three methods have been used to process the collected S-parameters: the modified two-antenna method in (7.46), the one-antenna method in (7.41), and the two-antenna method in (7.44). The frequency stir with 100 sample points was also used (14 MHz bandwidth) to provide an averaged result for each frequency – this makes $360 \times 100 = 36\,000$ samples at each frequency. The measured results are given in Figure 7.29. As expected, when the antenna radiation efficiency is high, the results from the three methods are in a good agreement, and even when the radiation efficiency is close to -10 dB, the results are still very similar. However, when the radiation efficiency decreases to -20 or -30 dB the one-antenna and two-antenna methods give inaccurate predictions but the modified two-antenna method still performs well. This problem was not caused by τ_{RC} but by $S_{11,s}$ because τ_{RC} was extracted from S_{22}. As discussed before, when antenna 1 is very lossy, the magnitude of the reflected signal can be very small (outside the dynamic range of the VNA after calibration and poor SNR). This is easy to understand because when a 30 dB attenuator is connected and the chamber loss is around 20 dB, the received signal will be around $30 + 20 + 30 = 80$ dB smaller than the transmitted signal, which is most likely comparable with noise. Also note that the error from the one-antenna method is larger than that from the two-antenna method. This was also found in [22] and is understandable from a mathematical point of view, since the precondition $e_{b1} = e_{b2} = 2$ for the one-antenna method is stricter than $e_{b1} = e_{b2}$ for the two-antenna method, thus the two-antenna method has a smaller deviation.

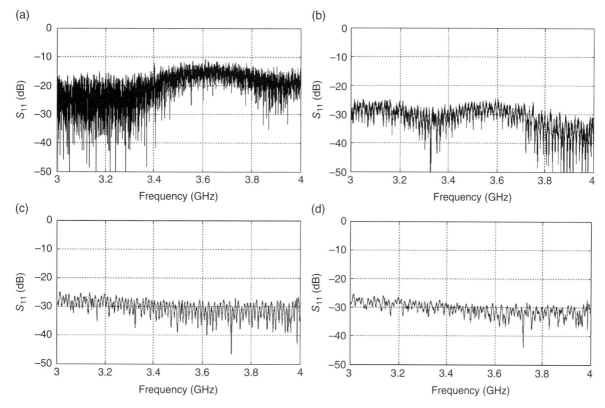

Figure 7.26 Typical measured *S*-parameters with different attenuators: (a) 0 dB, (b) 10 dB, (c) 20 dB, and (d) 30 dB.

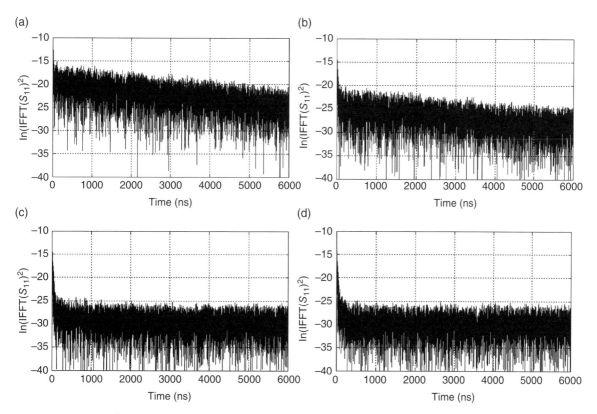

Figure 7.27 ln[IFFT(*S*)²] of the typical measured *S*-parameters with different attenuators: (a) 0 dB, (b) 10 dB, (c) 20 dB, and (d) 30 dB.

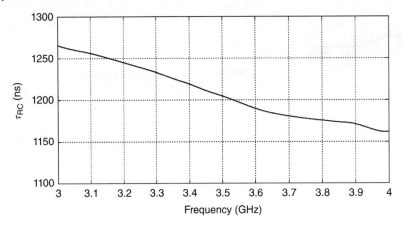

Figure 7.28 τ_{RC} extracted from $\ln[\text{IFFT}(S_{22})^2]$.

The modified two-antenna method does not need a reference antenna with known efficiency and the loss of the AUT can be arbitrary. This offers an opportunity to measure the antenna with very low radiation efficiency (e.g. implantable antenna with radiation efficiency around 0.3% [27, 28]) in the RC. The modified two-antenna method can also be generalised to non-reciprocal antennas (active antennas). If the transmitting (Tx) efficiency and receiving (Rx) efficiency of the AUT are different (non-reciprocal), the total Tx efficiency (at certain Tx power) and total Rx efficiency (at certain Rx signal strength) will be

$$\eta_{\text{tot1Tx}} = \langle |S_{21,s}|^2 \rangle \sqrt{\frac{2C_{RC}}{\omega\tau_{RC}\langle |S_{22,s}|^2 \rangle}} \tag{7.54}$$

$$\eta_{\text{tot1Rx}} = \langle |S_{12,s}|^2 \rangle \sqrt{\frac{2C_{RC}}{\omega\tau_{RC}\langle |S_{22,s}|^2 \rangle}} \tag{7.55}$$

Some issues should be noted when applying this method:

1) The RC must be well stirred. The enhanced backscatter effect is invalid when the RC is not well stirred and the e_{b1} and e_{b2} will not be 2, this is the fundamental assumption of the modified two-antenna method.
2) The assumption of 'antenna 2 being a highly efficient antenna' is to make sure that $S_{22,s}$ is in the dynamic range of the VNA. The measured η_1 does not depend on the radiation efficiency of antenna 2. Antenna 2 is used to extract the chamber transfer function and τ_{RC}, as long as η_2 can be measured accurately using the one-antenna method, the radiation efficiency of antenna 2 can be arbitrary. From Figure 7.29b it can be seen that there is 1 dB error at 4 GHz for the one-antenna method. In this scenario, as long as η_2 is larger than 10%, the error caused by η_2 will be smaller than 1 dB.
3) As can be seen in (7.46), τ_{RC} obtained from antenna 2 affects the results directly, thus τ_{RC} should be carefully measured and antenna 2 must be an efficient antenna. It should be noted that the chamber decay time is an intrinsic value related to the RC (including load); however, in practice, when it is measured using very lossy antennas, the TD response is affected by the antenna. The attenuators do not change τ_{RC}, but affect the TD response (extracted from $\text{IFFT}(S_{11})$), which makes τ_{RC} hard to extract (because of the noise level/dynamic range). τ_{RC} is determined by the chamber loss and all the loading effect, including the antennas.
4) If the radiation efficiency is so small that the $S_{21,s}$ in (7.46) is outside the dynamic range of the VNA, an amplifier can be used and the original $S_{21,s}$ can be obtained by adding the offset. τ_{RC} can also be extracted from $\text{IFFT}(S_{21})$.

(a)

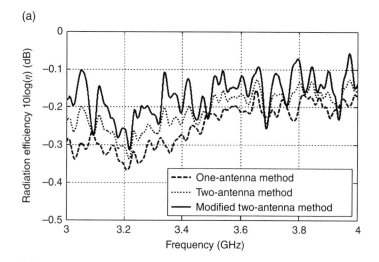

(b)

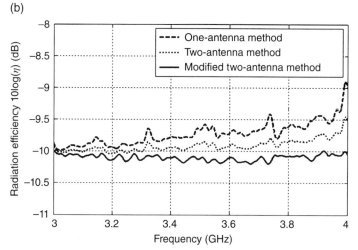

(c)

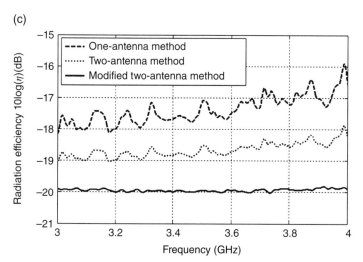

Figure 7.29 The measured radiation efficiency of antenna 1 with different attenuators: (a) 0 dB, (b) 10 dB, (c) 20 dB, and (d) 30 dB.

(d)

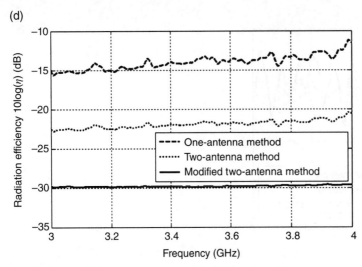

Figure 7.29 (Continued)

7.7 MIMO Antenna and Channel Emulation

Because the statistical behaviour of the field inside the RC can be controlled, the RC has been used to emulate the multipath propagation channel of mobile communications as indoor and outdoor environments [23]. In recent years, the RC has begun to emerge as a tool for wireless communication tests. Applications include the radiated power of mobile phones [29], diversity gain (DG) measurement [30–35], multiple-input multiple-output (MIMO) system measurement [30], Rician or Rayleigh channel emulation [36, 37], total isotropic sensitivity (TIS) measurement [38–45], bit error rate (BER) measurement [46–49], and Doppler spread and fading [50–65].

7.7.1 Diversity Gain Measurement

To mitigate the fading in the multipath environment, more than one antenna is used to transmit and receive the radio signal so as to increase the channel capacity and improve the SNR and BER in a MIMO system. The received signals from different channels are combined with a specific strategy such as switching, selection combining, or maximum combing schemes. The DG is an important figure of merit of the MIMO system performance, which normally gives a direct understanding of how the MIMO system outperforms the single channel system.

The definition of DG may have different forms depending on the reference used. The 'apparent' DG is the measured data relative to the branch with the strongest power level at a given level of the cumulative distribution function (CDF) of the received signal while the 'effective' DG uses the ideal reference antenna at a given CDF level and the 'actual' DG uses a single antenna as the reference at a given CDF level [31, 32]. The effective DG corrects the impedance mismatch with the transmission line and the dissipative loss, and is widely used to characterise the performance of the MIMO antenna.

A typical two-branch MIMO antenna measurement system is shown in Figure 7.30. The MIMO antenna array is considered as the AUT. In addition to the AUT, two extra antennas are needed: one is used as the transmit (Tx) antenna, the other is used as a reference antenna which must have known radiation efficiency. The measured channel samples are normalised using the radiation efficiency of the reference antenna and the

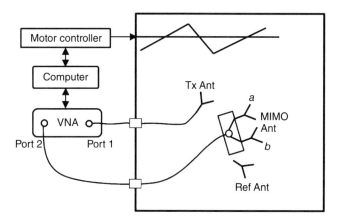

Figure 7.30 MIMO antenna measurement system in an RC.

chamber transfer function. The CDF of the normalised samples can be obtained and the DG value can be read directly from the CDF figure [30–32].

In Figure 7.30, the computer controls the operation of the stirrers and collects the S-parameters from the VNA for each stirrer position. The VNA is not the only instrument which can complete the measurement; Universal Software Radio Peripheral (USRP) has also been used to perform relevant measurement [41]. We use a VNA to demonstrate the measurement procedure:

Step 1: Calibrate the VNA, including the cables used to connect the antennas.
Step 2: Place all the antennas inside the chamber to keep the chamber Q factor constant.
Step 3: Connect the Tx antenna to the VNA port 1 and the reference antenna to port 2, load the AUT with a 50 Ω termination, and collect the full S-parameters (S_{11Tx}, S_{21Ref}, S_{22Ref}) for each stirrer position.
Step 4: Connect the AUT branch a to port 2, load the reference antenna and AUT branch b with a 50 Ω termination, collect S-parameters S_{21a}, keep the stirrer position stationary, connect the port 2 to the AUT branch b and load branch a with a 50 Ω termination, and collect S-parameters.
Step 5: Repeat Step (4) for each stirrer position.

It should be noted that in Step (4) and Step (5) the stirrer position must be the same to make sure that the environment is unchanged. An alternative way of doing this is to use an radio frequency (RF) switch to select the channel from port a or b or use a VNA with more than two ports [66]. Otherwise the received signals from different branches are decorrelated and independent, which should give an overestimated DG. After all the S-parameters are collected, the normalised channel samples for branches a and b are [31]

$$h_{21a} = |S_{21a}| \sqrt{\frac{\eta_{Ref}\left(1 - |\langle S_{11Tx}\rangle|^2\right)}{T_{Ref}}} \tag{7.56}$$

$$h_{21b} = |S_{21b}| \sqrt{\frac{\eta_{Ref}\left(1 - |\langle S_{11Tx}\rangle|^2\right)}{T_{Ref}}} \tag{7.57}$$

where η_{Ref} is the radiation efficiency of the reference antenna, T_{Ref} is the chamber transfer function (the radiation efficiency of the Tx and the reference antenna are not corrected) measured by using the reference antenna, and the definition is [31]

$$T_{Ref} = \frac{\left\langle |S_{21Ref,s}|^2\right\rangle}{\left(1 - |\langle S_{11Tx}\rangle|^2\right)\left(1 - |\langle S_{22Ref}\rangle|^2\right)} \tag{7.58}$$

$S_{*,s}$ means the stirred part of the S-parameter, which can be obtained by using $S_{*,s} = S_* - \langle S_* \rangle$. If necessary, S_{21a} and S_{21b} can be replaced by $S_{21a,s}$ and $S_{21b,s}$ to emulate a Rayleigh channel but not a Rician channel [67].

If we choose the maximum value of these two branches, the combined samples are

$$h_{comb} = \max(h_{21a}, h_{21b}) \tag{7.59}$$

Other strategies are also applicable, and the only difference is the post processing of S-parameters in (7.59). From the CDF plot of h_{21a}, h_{21b}, and h_{comb}, the DG value can be obtained.

This method can be simplified by eliminating the reference antenna using the one- or two-antenna method in Section 7.6. If the antenna of branch a is used as the reference antenna and the two-antenna method is used, from (7.44) the radiation efficiency of the reference antenna (the antenna of branch a) can be expressed as:

$$\eta_{Ref} = \eta_a = \frac{1}{1 - |\langle S_{22a} \rangle|^2} \sqrt{\frac{C_{RC} \langle |S_{22a,s}|^2 \rangle}{\omega e_b \tau_{RC}}} \tag{7.60}$$

Thus (7.56) and (7.57) can be written as [67]

$$h_{21a} = |S_{21a}| \sqrt{\frac{\left(1 - |\langle S_{11Tx} \rangle|^2\right)^2}{\langle |S_{21a,s}|^2 \rangle}} \sqrt{\frac{C_{RC} \langle |S_{22a,s}|^2 \rangle}{\omega e_b \tau_{RC}}} \tag{7.61}$$

$$h_{21b} = |S_{21b}| \sqrt{\frac{\left(1 - |\langle S_{11Tx} \rangle|^2\right)^2}{\langle |S_{21a,s}|^2 \rangle}} \sqrt{\frac{C_{RC} \langle |S_{22a,s}|^2 \rangle}{\omega e_b \tau_{RC}}} \tag{7.62}$$

To demonstrate the DG measurement, a typical measurement setup is shown in Figure 7.31a. Empty cartons are used to support the AUT for the simplicity reason. The size of the RC is 3.6 m × 4 m × 5.8 m. The frequency setting of the VNA is 1–4 GHz with 10 001 sample points. The reason for having chosen so many points is that we need to calculate the chamber decay time τ_{RC} from the IFFT of the S-parameter. Under this setting, the resolution time step of IFFT is 0.25 ns and the total time is 2500 ns. The AUT is shown in Figure 7.31b, the Tx antenna is a double-ridged waveguide horn antenna (Rohde & Schwarz HF906). The reference antenna is a homemade planar inverted-L antenna (PILA) and the detailed dimensions are given in [68]. The mechanical stirring sample number is 360 with 1° step size.

The radiation efficiency of the PILA is obtained by using the one- and two-antenna methods, including mechanical stir. Frequency stir with the nearest 50 frequency samples is also used, which means 360 × 50 = 18 000 samples for each frequency. The measured radiation efficiency is shown in Figure 7.32a. As can be seen, a good agreement is obtained between the one- and two-antenna methods. We use the values from the two-antenna method to calculate the normalised samples since it has a smaller uncertainty [22]. The chamber transfer function T_{Ref} has also been calculated using (7.58) and the result is given in Figure 7.32b.

All the samples between 2 and 3 GHz (the working frequency of the AUT) are used to plot the CDF. This means 360 × 3335 = 1 200 600 samples for each branch. The results are given in Figure 7.33. Because the antennas for the two branches are the same, the CDFs from the two branches are the same. The apparent DG and the effective DG at 1% can be obtained from Figure 7.33 as −11.05 + 21.24 = 10.19 dB and − 11.05 + 20.33 = 9.28 dB, respectively.

As discussed before, if we treat branch a as the reference antenna, the reference antenna PILA is now not needed. The chamber decay time τ_{RC} can be extracted from the S-parameters, as shown in Figure 7.34. Since τ_{RC} can be extracted from any of the S-parameters [22], the average value is used. They are very similar and we use the average value to calculate η_{Ref} in (7.60). The enhanced backscatter constant in (7.43) is extracted from the S-parameters and shown in Figure 7.35. It is very close to 2, which means the RC is well stirred [22]. Thus, the radiation efficiency of the branch a antenna can be obtained using the one- and two-antenna method, which is shown in Figure 7.36a. In addition, Figure 7.36b gives the chamber transfer function obtained from the antenna of branch a, which agrees well with that obtained using the reference antenna in Figure 7.32b.

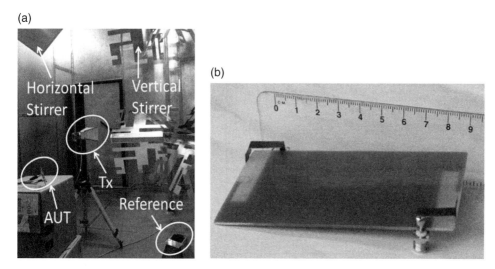

Figure 7.31 Measurement setup in an RC using the traditional method: (a) the system view and (b) the AUT.

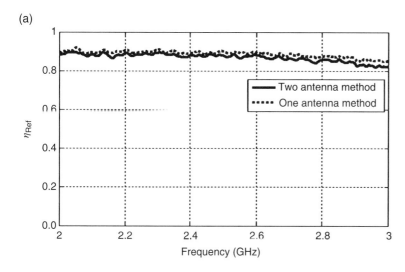

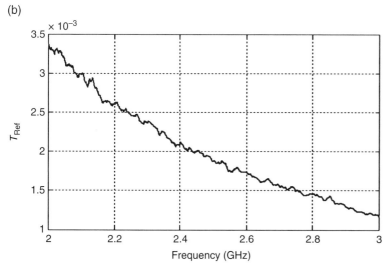

Figure 7.32 (a) Radiation efficiency of the reference antenna (PILA) using the one- and two-antenna methods, and (b) the chamber transfer function obtained from the reference antenna (PILA).

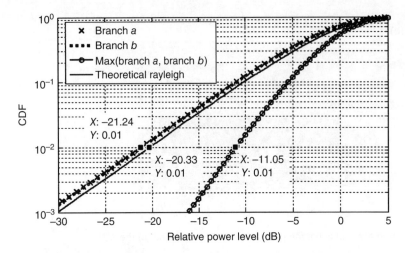

Figure 7.33 CDF plot of both branches and the combined signal. The theoretical Rayleigh distribution is also given.

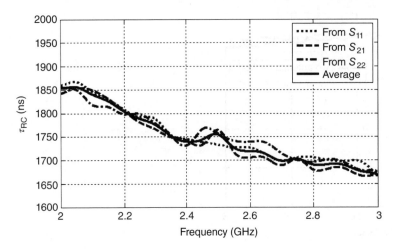

Figure 7.34 Chamber decay time extracted from *S*-parameters.

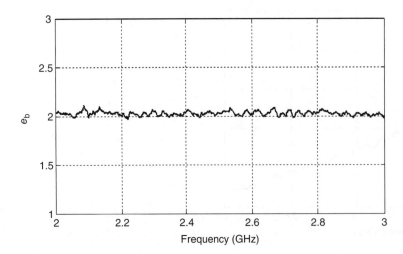

Figure 7.35 Enhanced backscatter constant extracted from *S*-parameters.

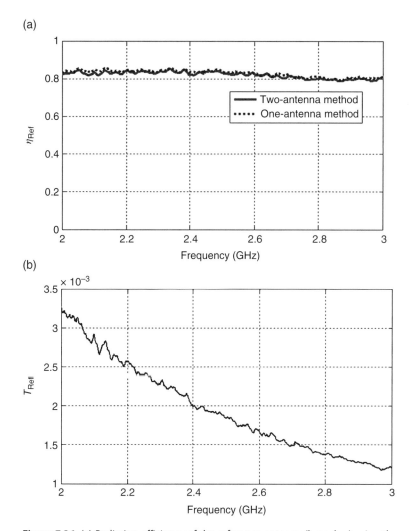

Figure 7.36 (a) Radiation efficiency of the reference antenna (branch a) using the one- and two-antenna methods, and (b) the chamber transfer function obtained from branch a.

Finally, we use the same sample number as in Figure 7.33 to generate the CDF, which is plotted in Figure 7.37. The apparent DG and the effective DG at 1% are 10.19 and 9.15 dB, respectively, which are very close to the results in Figure 7.33. Because the DG is frequency dependent, we have also divided the working frequency range 2–3 GHz into 10 bands (100 MHz each) and verified the DG values using both methods (the reference antenna method and the non-reference antenna method). The results are also in very good agreement [69]. Obviously, how the frequency band is allocated will not affect the results.

The theoretical Rayleigh curve can be derived. Suppose X is the magnitude of the received signal and follows the Rayleigh distribution

$$PDF(X) = \frac{x}{\sigma^2}\exp\left(-\frac{x^2}{2\sigma^2}\right), x \geq 0 \tag{7.63}$$

the CDF of X is

$$CDF(X) = 1 - \exp\left(-\frac{x^2}{2\sigma^2}\right) \tag{7.64}$$

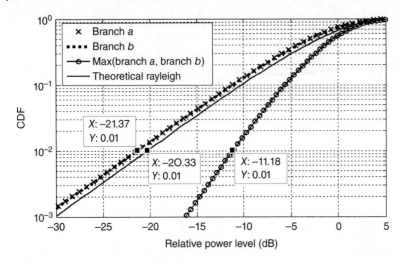

Figure 7.37 CDF plot of both branches and the combined signal using a method without reference antenna.

X^2 is the power of the received signal and follows exponential distribution

$$PDF(X^2) = \frac{1}{2\sigma^2} \exp\left(-\frac{x}{2\sigma^2}\right), x = X^2 \tag{7.65}$$

The mean value of X^2 is $2\sigma^2$, and if we normalise the signal to the mean power value, $\sigma = \sqrt{1/2}$, (7.64) becomes

$$CDF(X) = 1 - \exp\left(-x^2\right) \tag{7.66}$$

which is the theoretical Rayleigh CDF. The Matlab® code is given in Section A.9.

Besides the CDF measurements, the DG can be measured from the correlation between measured samples from two branches of the MIMO antenna array. The complex correlation coefficient can be calculated from the free-space far-field pattern [70]:

$$\rho = \frac{\iint_{4\pi} \mathbf{G}_1(\theta,\varphi) \cdot \mathbf{G}_2^*(\theta,\varphi) d\Omega}{\sqrt{\iint_{4\pi} \mathbf{G}_1(\theta,\varphi) \cdot \mathbf{G}_1^*(\theta,\varphi) d\Omega \iint_{4\pi} \mathbf{G}_2(\theta,\varphi) \cdot \mathbf{G}_2^*(\theta,\varphi) d\Omega}} \tag{7.67}$$

The complex correlation coefficient from the received voltages is defined by [70, 71]:

$$\rho = \frac{\sum V_1 V_2^*}{\sqrt{\sum V_1 V_1^* \sum V_2 V_2^*}} \tag{7.68}$$

where V_1 and V_2 are sequences of the received voltage samples under the same environment (same stirrer position). If the phases are not available, the envelope correlation can be evaluated from [70]:

$$\rho_e = \frac{\sum |V_1 V_2^*|}{\sqrt{\sum |V_1|^2 \sum |V_2|^2}} \tag{7.69}$$

Measurements have confirmed that the power correlation, magnitude squares of complex correlation, and the envelope correlation are equal to each other [71] and the analytical derivation is given in [72]. The complex correlation coefficient can also be obtained from the S-parameters measured at the antenna ports [70, 73, 74]:

$$\rho = \frac{-\left(S_{11}^* S_{12} + S_{21}^* S_{22}\right)}{\sqrt{\left[1 - |S_{11}|^2 - |S_{21}|^2\right]\left[1 - |S_{12}|^2 - |S_{22}|^2\right]}} \tag{7.70}$$

However, the following conditions need to be satisfied: (i) the antenna is lossless (the radiation efficiency is high), (ii) the antenna system is positioned in a uniform multipath environment, and (iii) the load termination for the antenna not connected to equipment is 50 Ω [73].

For a MIMO array with two elements, the effective DG (DG_{eff}) can be obtained from the apparent DG (DG_{app}) [70]:

$$DG_{eff} = \eta_{rad} DG_{app} \tag{7.71}$$

where η_{rad} is the radiation efficiency of the element antenna, and the radiation efficiencies of the two elements are assumed to be the same. The relation between the apparent DG and the correlation can be approximated by [70]

$$DG_{app} = 10.5\sqrt{1 - |\rho|^2} \tag{7.72}$$

When the correlation is very close to unity, a more accurate expression is [70]

$$DG_{app} = 10.5\sqrt{1 - |0.99\rho|^2} \tag{7.73}$$

which has a less than 0.1 dB error of the apparent DG at 1% CDF. If the radiation efficiencies of two elements are different, an empirical formula has been obtained in [70, 75]:

$$DG_{app} = \sqrt{\left(1 + \frac{\eta_{min}}{\eta_{max}}\right)^2 + 105\frac{\eta_{min}}{\eta_{max}}\left(1 - |\rho|^2\right)} \tag{7.74}$$

where η_{min} and η_{max} correspond to the minimum and maximum radiation efficiencies of two ports.

For an (N, N) MIMO antenna array, the envelope correlation can be generalised to [76]:

$$\frac{\left|\iint_{4\pi} \mathbf{G}_i(\theta,\varphi) \bullet \mathbf{G}_j^*(\theta,\varphi) d\Omega\right|^2}{\iint_{4\pi} \mathbf{G}_i(\theta,\varphi) \bullet \mathbf{G}_i^*(\theta,\varphi) d\Omega \iint_{4\pi} \mathbf{G}_j(\theta,\varphi) \bullet \mathbf{G}_j^*(\theta,\varphi) d\Omega} - \frac{\left|\sum_{n=1}^{N} S_{i,n}^* S_{n,j}\right|^2}{\prod_{k=(i,j)}\left[1 - \sum_{n=1}^{N} S_{i,n}^* S_{n,k}\right]} \tag{7.75}$$

and the effect of power losses has been discussed in [76]. It can be seen in Section 7.8 that the correlation coefficient can be used to invert the axial ratio (AR) and the radiation pattern of an antenna.

7.7.2 Total Isotropic Sensitivity Measurement

The TIS P_{TIS} is the power estimated at the input port of the antenna of the device under test (DUT) when it reports a specified maximum acceptable BER. The measurement setup is very similar to the radiation efficiency measurement, but with different measurement instrument. P_{TIS} is given by [40]

$$P_{TIS} = \frac{\eta_M\left(1 - |\Gamma_M|^2\right)}{|1 - \Gamma_M\Gamma_{RX}|^2}\left(\frac{1}{M}\sum_{r=1}^{M} G_{ref,r}\right)\frac{1}{N}\left(\sum_{n=1}^{N}\frac{1}{P_{BSE}(n)}\right)^{-1} \tag{7.76}$$

where Γ_{RX} is the reflection coefficient of the receiver assembly and Γ_M is the free-space reflection coefficient of the measurement antenna. η_M is the radiation efficiency of the measurement antenna, which can be measured by using the non-reference antenna method. The harmonic mean is used to calculate the average value of the base station emulator (BSEs) output power because it can minimise the effects of outliers. Other averaging arrangements can also be used [40]. $P_{BSE}(n)$ corresponds to the power emitted from the BSE when the DUT reports the threshold BER, $G_{ref,r}$ is the power transfer function of one DUT position, the average over M means the position stirring is used, and also a single value of G_{ref} can be used to save time:

$$G_{ref} = \frac{\langle |S_{21}|^2 \rangle}{\eta_1 \eta_2 \left(1 - |S_{11}|^2\right)\left(1 - |S_{22}|^2\right)} \tag{7.77}$$

7.7.3 Channel Capacity Measurement

Channel capacity is a measure of how many bits per second can be transmitted through a radio channel per Hz, also referred to as the spectral efficiency [31]. The measurement procedure is very similar to the DG measurement. From (7.56) or (7.57), suppose a MIMO system has M transmit antennas and N receive antennas, and the normalised measured samples are [31]

$$h_{ij} = |S_{ij}| \sqrt{\frac{\eta_{Ref}\left(1 - |\langle S_{ii} \rangle|^2\right)}{T_{Ref}}}, j = 1, 2, \ldots, M, i = 1, 2, \ldots, N \tag{7.78}$$

which means the signal is transmitted from antenna j and received by antenna i. Thus for each stirrer position, we have a channel matrix $\mathbf{H}_{N \times M}$:

$$\mathbf{H}_{N \times M} = \begin{bmatrix} h_{11} & \ldots & h_{1M} \\ \ldots & \ldots & \ldots \\ h_{N1} & \ldots & h_{NM} \end{bmatrix} \tag{7.79}$$

and the instantaneous maximum channel capacity of the MIMO system can be obtained as [31, 32, 70, 77]:

$$C_{N \times M} = \log_2\left[\det\left(\mathbf{I}_N + \frac{SNR}{M} \mathbf{H}_{N \times M} \mathbf{H}_{N \times M}^H\right)\right] \tag{7.80}$$

where $\mathbf{I}_N$ is the identity matrix, SNR is the signal-to-noise ratio, and $[]^H$ means the conjugate (Hermitian) transpose of the matrix. The mean channel capacity can be obtained by averaging $C_{N \times M}$ over all stirrer positions [31, 32, 70, 77]:

$$\langle C_{N \times M} \rangle = \left\langle \log_2\left[\det\left(\mathbf{I}_N + \frac{SNR}{M} \mathbf{H}_{N \times M} \mathbf{H}_{N \times M}^H\right)\right] \right\rangle \tag{7.81}$$

The non-reference antenna method in the DG measurement can also be generalised to the channel capacity measurement.

7.7.4 Doppler Effect

The Doppler effect is a phenomenon in which the frequency of a source signal is shifted when the transmitter is moving towards or away from the receiver, or vice versa [58]. It can be observed when there is relative difference in velocity between the transmitter and receiver. In a multipath environment, most received signals go through multiple paths after being reflected by reflectors/scatterers around the receiver, thus the movement of the reflector/scatterer could result in a Doppler shift. A multipath environment can be generated in an RC and the Doppler effect can be emulated by moving stirrers [50, 59]. The Doppler effect can also be used to characterise the effectiveness of the stirrers. If the stirrers can scatter the field more effectively, the Doppler effect can be more easily observed, that is the Doppler spread should be larger [50].

The characteristic of the channel in an RC can be described from the time-variant transfer function $H(f, t)$. Then its time autocorrelation function is [56, 57]

$$R_H(f, \partial t) = H(f, \partial t) \otimes H^*(f, -\partial t) = \mathrm{E}[H(f, t)H^*(f, t + \partial t)] \tag{7.82}$$

where $\mathrm{E}[\cdot]$ represents the mathematical expectation. The Doppler spectrum is [56, 57]

$$D(f, \rho) = \int_{-\infty}^{\infty} R_H(f, \partial t) \exp(-j2\pi\rho\partial t) d(\partial t) = H(f, \rho)H^*(f, \rho) = |H(f, \rho)|^2 \tag{7.83}$$

where $H(f, \rho)$ is the Fourier transform of $H(f, t)$ with respect to time t. The channel transfer function $H(f, t)$ is equal to the S-parameter $S_{21}(f, t)$ measured with a VNA. The Doppler spread can be obtained by selecting the root mean square (RMS) Doppler bandwidth. The RMS Doppler bandwidth at a certain frequency f_0 is given by [56, 57]

$$\rho_{\mathrm{RMS}} = \sqrt{\frac{\int \rho^2 D(f_0, \rho) d\rho}{\int D(f_0, \rho) d\rho}} \tag{7.84}$$

The Doppler effect measurements were performed in the RC at UoL, which has a size of 3.6 m × 4 m × 5.8 m. It has two mode stir paddles: the vertical one is mounted in a corner while the horizontal one is set close to the ceiling. Two double-ridged waveguide horn antennas were used as antenna 1 (SATIMO SH 2000) and antenna 2 (Rohde & Schwarz HF 906). The measurement setup is shown in Figure 7.38. Antenna 1 and antenna 2 were kept far enough apart and oriented towards the vertical stirrer to make the electromagnetic wave fully interact with the stirrer and to avoid LoS between these two antennas. During the measurement, the stirrers were rotated continuously with a speed of 6° per second (1 minute per round). 10 001 points were sampled by the VNA at discrete frequencies of 2, 2.2, 2.4, 2.6, 2.8, and 3 GHz, respectively. The intermediate frequency (IF) bandwidth was set to be 100 Hz, thus the total sweeping time was 242 seconds for each single frequency. The typical Rayleigh fading can be observed when the stirrers are rotating continuously, as can be seen from Figure 7.39. At each frequency, three sets of measurement were done. In the first measurement, the vertical stirrer was rotating and the horizontal stirrer was kept stationary. In the second measurement, the horizontal stirrer was rotating while the vertical stirrer was stationary. In the last measurement, both the vertical and the horizontal stirrers were rotating.

The Doppler spectrum at 2 GHz of the three sets of measurements is shown in Figure 7.40. As can be seen, when two stirrers are rotating, the Doppler spectrum is wider than that from only one rotating stirrer. The Doppler spreads are calculated and compared with the standard deviation (SD) of the E-field, as shown in Figure 7.41. When two stirrers are rotating, the standard deviations are smaller than that from only one rotating stirrer, which indicates that the stirring effectiveness of two stirrers is better. Thus, the Doppler spread is positively correlated to the stirring effectiveness of the stirrers, and the Doppler spread could be an effective parameter to characterise the effectiveness of the stirrers. Also the Doppler spread measurement is very quick as it takes about 1 minute for one revolution, which is much shorter than the standard deviation measurement.

It should be noted that in this measurement the Doppler spectrum was measured when the stirrer was rotating continuously [50]. In this case the VNA sampling frequency should be high enough (or the

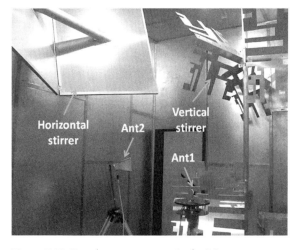

Figure 7.38 Doppler measurement in the RC.

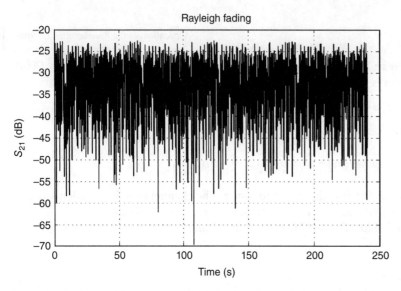

Figure 7.39 Measured *S*-parameters at 2 GHz when the vertical stirrer is rotating continuously.

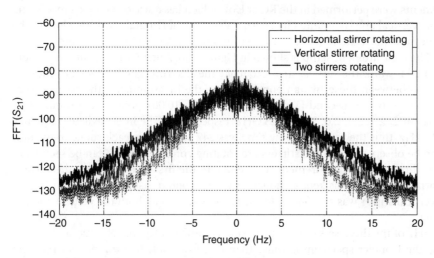

Figure 7.40 The Doppler spectrum under three different stirring scenarios.

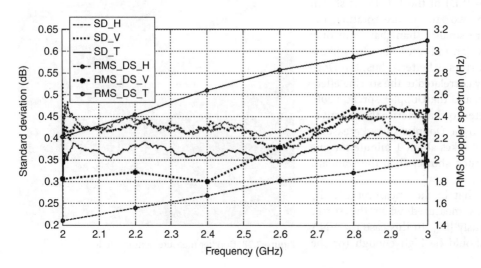

Figure 7.41 Comparison of RMS Doppler spread and standard deviation of the E-field. H, horizontal stirrer; V, vertical stirrer; T, two stirrers; SD, standard deviation; RMS_DS, RMS Doppler spectrum.

rotation speed is slow enough) so that the measured *S*-parameter is correlated with the previous one. The Doppler spectrum can also be measured when the stirrer is rotated step-wise [25, 57] and the rotation degree of each step should be small enough for the same reason; this is very different from the RC operations in most applications and should be treated carefully, especially when the stirrer is very large.

7.8 Antenna Radiation Pattern Measurement

7.8.1 Theory

Because of multipath effects, it is not easy to measure the radiation pattern of an antenna in an RC. It has been shown that by using a deconvolution technique [78], the radiation pattern of the AUT can be reconstructed in a non-anechoic environment. It is also possible to measure the pattern directly in an RC when the LoS response dominates the unstirred part. In this case, the pattern can be extracted by using the *K*-factor [79–81] or Doppler shift [53, 82].

A typical measurement setup is shown in Figure 7.42a, where two antennas are connected to the measurement instrument. A high directivity transmitting (Tx) or receiving (Rx) antenna is fixed and directed towards the AUT. The non-line-of-sight (NLoS) part can either be averaged out by rotating the stirrer [80, 81] or be

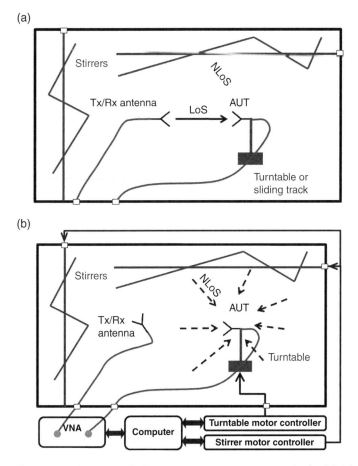

Figure 7.42 Antenna radiation pattern measurement setup in the RC: (a) existing methods and (b) proposed method.

filtered by using the Doppler shift (moving the AUT along a sliding track) [53, 82]. Time gating can also be used but the LoS and NLoS signals need to be distinguished in the TD. By rotating the turntable, the angle dependency of the radiation pattern can be measured. That is, the information hidden behind the LoS/unstirred response is extracted while removing the NLoS/stirred response. It is also possible to extract the radiation pattern by using the time-reversal technique in the RC [83–85], but the time-reversal behaviour of the RC needs to be carefully characterised and calibrated.

We have proposed a new method which can extract the 3D pattern of the AUT from the NLoS/stirred response while the LoS/unstirred response is not needed [86]. This is very different from previous methods. The measurement setup is shown in Figure 7.42b where the Tx/Rx antenna is not directed to the AUT. Since the RC is inherently a rich multipath environment, using the stirred part is actually easier than using the unstirred part. In Figure 7.42a, the LoS/unstirred part needs to be extracted by rotating the stirrers or moving the AUT (along the sliding track) for each AUT angle, this could be very time consuming. In the proposed method, the radiation pattern of the AUT is decomposed into spherical harmonics with unknown coefficients, and the pattern measurement becomes a generalised mode matching problem. By measuring the self-correlation coefficient of the radiation pattern in the RC, the coefficients of the spherical wave modes/harmonics can be inverted and thus the radiation pattern can be reconstructed.

In the K-factor method [79–81], by using (5.200), if the distance, the RC volume, and the Q factor are the same for the case with an AUT and a reference antenna, we have

$$K(\theta,\varphi) \propto D(\theta,\varphi) \tag{7.85}$$

By using the substitution method the directivity of the AUT is estimated from its relative directivity value with respect to a reference antenna and the directivity of the AUT can be obtained as [80, 81]

$$D_{\mathrm{AUT}}(\theta,\varphi) = \frac{D_{\mathrm{Ref}}}{K_{\mathrm{Ref}}} K_{\mathrm{AUT}}(\theta,\varphi) \tag{7.86}$$

where D_{Ref} is the directivity of the reference antenna, K_{Ref} is the measured K-factor of the reference antenna, and K_{AUT} is the measured K-factor of the AUT.

In the proposed method (spherical mode matching method) [86], the radiation pattern measurement is an inverse problem of the correlation coefficients measurement problem. We first introduce the forward problem and then move on to the inverse problem.

It is well known that the far-field radiation pattern of an antenna can be decomposed into a superposition of spherical wave harmonics [87–91]. Two types of decomposition have been used: vector spherical harmonics (VSHs) with scalar coefficients and scalar spherical harmonics (SSHs) with vector coefficients. It has been shown that SSHs with vector coefficients can avoid the singularity problem of the antenna field on the poles [89–91], therefore we use SSHs with vector coefficients.

The far-field of an antenna can be expressed as [90]:

$$\mathbf{E}(\theta,\varphi) = \lim_{L \to \infty} \sum_{l=0}^{L} \sum_{m=-l}^{l} \boldsymbol{a}_{lm} Y_l^m(\theta,\varphi) \tag{7.87}$$

where θ and φ are the polar angle and azimuthal angle in the spherical coordinate system respectively, and $Y_l^m(\theta,\varphi)$ is the SSH with level l and mode m defined by

$$Y_l^m(\theta,\varphi) = \sqrt{\frac{(2l+1)}{4\pi} \frac{(l-m)!}{(l+m)!}} P_l^m(\cos\theta) e^{jm\varphi} \tag{7.88}$$

$P_l^m(\cos\theta)$ is the associated Legendre functions of the first kind [90]. The vector coefficients $\boldsymbol{a}_{lm}$ can be obtained by using the orthogonality of the SSHs [89] with each component

$$a_{x,lm} = \int_0^{2\pi} \int_0^{\pi} E_x(\theta,\varphi) Y_l^{m*}(\theta,\varphi) \sin\theta d\theta d\varphi$$

$$a_{y,lm} = \int_0^{2\pi} \int_0^{\pi} E_y(\theta,\varphi) Y_l^{m*}(\theta,\varphi) \sin\theta d\theta d\varphi$$

$$a_{z,lm} = \int_0^{2\pi} \int_0^{\pi} E_z(\theta,\varphi) Y_l^{m*}(\theta,\varphi) \sin\theta d\theta d\varphi \qquad (7.89)$$

where $\boldsymbol{a}_{lm} = [a_{x,lm}\ a_{y,lm}\ a_{z,lm}]$, $\mathbf{E}(\theta,\varphi) = [E_x(\theta,\varphi)\ E_y(\theta,\varphi)\ E_z(\theta,\varphi)]$, and $*$ means the complex conjugate.

If the far-field of the radiation pattern is known, using (7.89), the far-field can be decomposed into SSHs in (7.87) with a truncation of level L [90]. Suppose we have two antennas, where the radiation patterns are $\mathbf{E}_1(\theta,\varphi)$ and $\mathbf{E}_2(\theta,\varphi)$, respectively, and the correlation coefficient ρ_E between these two antennas is [70, 92]

$$\rho_E = \frac{\left| \int\int \mathbf{E}_1(\theta,\varphi) \cdot \mathbf{E}_2^*(\theta,\varphi) d\Omega \right|}{\sqrt{\int\int |\mathbf{E}_1(\theta,\varphi)|^2 d\Omega \int\int |\mathbf{E}_2(\theta,\varphi)|^2 d\Omega}} \qquad (7.90)$$

where $\int\int [] d\Omega = \int_0^{2\pi} \int_0^{\pi} [] \sin\theta d\theta d\varphi$ means the integral over a unit spherical surface. Suppose that $\mathbf{E}_1(\theta,\varphi)$ and $\mathbf{E}_2(\theta,\varphi)$ are decomposed using the SSHs with order L:

$$\mathbf{E}_1(\theta,\varphi) = \lim_{L \to \infty} \sum_{l=0}^{L} \sum_{m=-l}^{l} \boldsymbol{a}_{lm} Y_l^m(\theta,\varphi)$$

$$\mathbf{E}_2(\theta,\varphi) = \lim_{L \to \infty} \sum_{l=0}^{L} \sum_{m=-l}^{l} \boldsymbol{b}_{lm} Y_l^m(\theta,\varphi) \qquad (7.91)$$

Since the SSHs are orthogonal to each other on a unit spherical surface, substituting (7.91) into (7.90) gives the correlation coefficient

$$\rho_E = \frac{\left| \sum_{l=0}^{L} \sum_{m=-l}^{l} \boldsymbol{a}_{lm} \cdot \boldsymbol{b}_{lm}^* \right|}{\sqrt{\sum_{l=0}^{L} \sum_{m=-l}^{l} \|\boldsymbol{a}_{lm}\|^2} \sqrt{\sum_{l=0}^{L} \sum_{m=-l}^{l} \|\boldsymbol{b}_{lm}\|^2}} \qquad (7.92)$$

Since $a_{x,lm}$, $a_{y,lm}$, and $a_{z,lm}$ are complex numbers, we use $\|\boldsymbol{a}_{lm}\|^2$ to represent the square of the magnitude of each coefficient, which is $\|\boldsymbol{a}_{lm}\|^2 = |a_{x,lm}|^2 + |a_{y,lm}|^2 + |a_{z,lm}|^2$. Next we introduce the concept of the self-correlation coefficient.

Suppose $\mathbf{E}_2(\theta,\varphi)$ can be transformed from $\mathbf{E}_1(\theta,\varphi)$ by using rotations. That is, $\mathbf{E}_2(\theta,\varphi)$ is a transformed version of $\mathbf{E}_1(\theta,\varphi)$, so they have the same shape but different reference coordinate systems. It should be noted that, to obtain a rotated version of $\mathbf{E}_1(\theta,\varphi)$, we only need to apply a rotation matrix to the coefficients at each level of l [89, 93], which is

$$\begin{aligned}
&[b_{xl,-l}...b_{xl,l}\ b_{yl,-l}...b_{yl,l}\ b_{zl,-l}...b_{zl,l}] = \\
&[a_{xl,-l}...a_{xl,l}\ a_{yl,-l}...a_{yl,l}\ a_{zl,-l}...a_{zl,l}] \mathbf{M}_l(\alpha,\beta,\gamma)
\end{aligned} \qquad (7.93)$$

where $\mathbf{M}_l(\alpha,\beta,\gamma)$ is the rotation matrix at level l. (α,β,γ) means the radiation pattern is rotated around the x-axis first with angle α, then rotated around y-axis with angle β, and finally rotated around the z-axis with angle γ, all with the right-hand rule, which means

$$\mathbf{M}_l(\alpha,\beta,\gamma) = \mathbf{M}_l(\alpha,0,0)\mathbf{M}_l(0,\beta,0)\mathbf{M}_l(0,0,\gamma) \qquad (7.94)$$

Since the TRP values of $\mathbf{E}_1(\theta, \varphi)$ and $\mathbf{E}_2(\theta, \varphi)$ are the same, we have

$$\sum_{l=0}^{L} \sum_{m=-l}^{l} \|\boldsymbol{a}_{lm}\|^2 = \sum_{l=0}^{L} \sum_{m=-l}^{l} \|\boldsymbol{b}_{lm}\|^2 \tag{7.95}$$

Equation (7.92) becomes

$$\rho_{SC}(\alpha,\beta,\gamma) = \frac{\left| \sum_{l=0}^{L} \sum_{m=-l}^{l} \boldsymbol{a}_{lm} \bullet T_{\alpha\beta\gamma}^*(\boldsymbol{a}_{lm}) \right|}{\sum_{l=0}^{L} \sum_{m=-l}^{l} \|\boldsymbol{a}_{lm}\|^2} \tag{7.96}$$

which is the definition of the 'self-correlation coefficient' (ρ_{SC}), where we use $T_{\alpha\beta\gamma}(\boldsymbol{a}_{lm})$ to represent the transformed version of $\boldsymbol{a}_{lm}$ by using (7.93). Since $T_{\alpha\beta\gamma}$ depends on the rotation angles, ρ_{SC} is also a function of rotation angles.

As can be seen, the forward problem is well defined, and once the radiation pattern of the antenna $\mathbf{E}_1(\theta, \varphi)$ is known, the self-correlation coefficient can be calculated using (7.93) and (7.96). The calculation procedure for the rotation matrix $\mathbf{M}_l(\alpha, \beta, \gamma)$ can be found in [89, 93].

If the angle dependency of ρ_{SC} in (7.96) is known, by solving (7.96) for $\boldsymbol{a}_{lm}$ the 3D radiation pattern of the AUT can be reconstructed. As can be seen, the inverse problem is actually a generalised mode matching problem, that is, the coefficients of the spherical wave modes need to be solved to match the angle dependency of ρ_{SC}.

Luckily, this self-correlation coefficient can be measured in an RC using the setup shown in Figure 7.42b. The measurement procedure is very similar to the DG measurement, the only difference being that in the DG measurement $\mathbf{E}_1(\theta, \varphi)$ and $\mathbf{E}_2(\theta, \varphi)$ in (7.90) are the radiation patterns of the two MIMO antenna branches; in this measurement, $\mathbf{E}_2(\theta, \varphi)$ is a rotated version of $\mathbf{E}_1(\theta, \varphi)$. We just need to rotate the AUT for a set of angles (α, β, γ) instead of switching the receiving/transmitting port in the DG measurement.

The correlation coefficient can be obtained by measuring the S-parameters in the RC [71, 92, 94]:

$$\rho_S(\alpha,\beta,\gamma) = corr(S_{31}, S_{32}) = \frac{\left| \sum_{k=1}^{N} [S_{31}(k) - \langle S_{31} \rangle][S_{32}(k) - \langle S_{32} \rangle]^* \right|}{\sqrt{\sum_{k=1}^{N} |S_{31}(k) - \langle S_{31} \rangle|^2} \sqrt{\sum_{k=1}^{N} |S_{32}(k) - \langle S_{32} \rangle|^2}} \tag{7.97}$$

where N represents the total number of samples collected and $\langle \cdot \rangle$ means the average value of N samples. In the DG measurement, S_{31} and S_{32} represent the measured S-parameters between the Tx/Rx antenna (in Figure 7.42b) and two MIMO branches, respectively. In this measurement, S_{31} and S_{32} represent the measured S-parameters between the Tx/Rx antenna and the AUT at two different positions, respectively. Since position 2 is the same antenna rotated from position 1, and it has been shown that the correlation of measured S-parameters equals the correlation of the radiation patterns [71, 92], thus we have

$$\rho_S(\alpha,\beta,\gamma) = \rho_{SC}(\alpha,\beta,\gamma) \tag{7.98}$$

To solve $\boldsymbol{a}_{lm}$, two more conditions are needed: the radiated power should be normalised to 1 (otherwise $\boldsymbol{a}_{lm}$ can be scaled arbitrarily) and the far-field condition should be satisfied. This is because we use the SSHs rather than VSHs, while in the VSHs, the far-field condition can be satisfied automatically (but the singularity problem needs to be treated separately [90]). Mathematically, these two conditions are

$$\sum_{l=0}^{L} \sum_{m=-l}^{l} \|\boldsymbol{a}_{lm}\|^2 = 1 \tag{7.99}$$

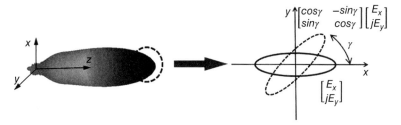

Figure 7.43 Self-correlation of a directional antenna. The integral result is dominated by the field in the main beam direction.

$$\hat{\mathbf{r}}(\theta,\varphi) \cdot \sum_{l=0}^{L} \sum_{m=-l}^{l} \mathbf{a}_{lm} Y_l^m(\theta,\varphi) = 0 \tag{7.100}$$

where $\hat{\mathbf{r}}(\theta,\varphi)$ is the unit vector in the radial direction.

As can be seen, the inverse problem is a generalised mode matching problem defined by the nonlinear equation system (7.98)–(7.100), and these can be solved numerically by using the Levenberg–Marquardt algorithm [95] in Matlab.

It is interesting to note that, if only the AR is of interest, the problem can be simplified. Assume that the antenna is a directional antenna. The integral in (7.90) can therefore be approximately evaluated using the field in the main beam, as shown in Figure 7.43.

Suppose that the main beam is aligned with the z-axis and rotated around it by an angle γ. The integral (7.90) can be approximated as

$$
\begin{aligned}
\rho_{SC}(\gamma) &= \frac{\left| \int\int \mathbf{E}_1(\theta,\varphi) \cdot \mathbf{E}_2^*(\theta,\varphi) d\Omega \right|}{\sqrt{\int\int |\mathbf{E}_1(\theta,\varphi)|^2 d\Omega \int\int |\mathbf{E}_2(\theta,\varphi)|^2 d\Omega}} \\[2mm]
&\approx \frac{\left| \begin{bmatrix} \cos\gamma & -\sin\gamma \\ \sin\gamma & \cos\gamma \end{bmatrix} \begin{bmatrix} E_x \\ jE_y \end{bmatrix} \cdot \begin{bmatrix} E_x \\ jE_y \end{bmatrix}^* \right|}{E_x^2 + E_y^2} \\[2mm]
&= \frac{\left| \left(E_x^2 + E_y^2 \right) \cos\gamma - 2jE_xE_y \sin\gamma \right|}{E_x^2 + E_y^2} \\[2mm]
&= \frac{\left(E_x^2 + E_y^2 \right)^2 \cos^2\gamma + \left(2E_xE_y \right)^2 \sin^2\gamma}{E_x^2 + E_y^2}
\end{aligned}
\tag{7.101}
$$

where $\mathbf{E}_1(\theta,\varphi)$ is the original pattern and $\mathbf{E}_2(\theta,\varphi)$ is the rotated pattern with angle γ. It can be found from (7.101) that, when $\gamma = 90°$, the $\rho_{SC}(\gamma)$ has the minimum value

$$\rho_{SC}(\gamma)_{min} \approx \frac{2E_xE_y}{E_x^2 + E_y^2} \tag{7.102}$$

Note that the AR is defined as E_x/E_y (when $E_x \geq E_y$), thus the minimum of $\rho_{SC}(\gamma)$ can be related to the AR of the antenna

$$\rho_{SC}(\gamma)_{min} \approx \frac{2E_xE_y}{E_x^2 + E_y^2} = \frac{2\mathrm{AR}}{1 + \mathrm{AR}^2} \tag{7.103}$$

and the AR can be obtained as

$$AR \approx \frac{1 + \sqrt{1 - \rho_{SC}^2(\gamma)_{min}}}{\rho_{SC}(\gamma)_{min}} \tag{7.104}$$

When $E_y > E_x$, the expression is the same. This offers an opportunity to measure the AR in the RC for directional antennas.

7.8.2 Simulations and Measurements

Simulations and measurements were conducted to demonstrate the proposed method. Before conducting the measurements, numerical simulations were carried out. In the numerical simulation, once the geometrical structure of the AUT is defined, the radiation pattern of the AUT can be obtained by using the full wave simulation software (e.g. CST Microwave Studio). Thus the simulated radiation pattern can be used as the reference, and we can compare the reconstructed pattern (obtained from the self-correlation coefficients) with the pattern obtained by using CST to verify the proposed method. Then measurements in an RC and an AC were conducted and the results were compared. We use the measured results obtained in the AC as the reference to confirm the effectiveness of the proposed method.

A typical rectangular horn antenna shown in Figure 7.44 is used as the AUT in the numerical simulation. The definitions of the rotation angles are also given (right-hand rule). The simulated radiation pattern at 5 GHz is obtained by using CST Microwave Studio and given in Figure 7.45 with maximum E-field magnitude normalised to $1\,V\,m^{-1}$.

The radiation pattern can be decomposed into SSHs with coefficients a_{lm} (the forward problem) by using (7.89). The magnitude and phase of each component of a_{lm} are shown in Figure 7.46. We use L up to 15 in Figure 7.46, and as can be seen, when $l > 8$ the magnitude of the SSHs is already very small (<-27 dB). Then (7.96) is used to obtain the self-correlation coefficients of the radiation pattern. Since the AUT is rotated only around the x-, y-, and z-axes, we use $\rho_{SC}(\alpha), \rho_{SC}(\beta)$, and $\rho_{SC}(\gamma)$ instead of $\rho_{SC}(\alpha, 0, 0), \rho_{SC}(0, \beta, 0)$, and $\rho_{SC}(0, 0, \gamma)$, respectively.

After the self-correlation coefficients of the radiation pattern are obtained, the radiation pattern of the AUT can be reconstructed (the inverse problem). Mathematically, by combining (7.98)–(7.100), the inverse problem can be expressed as

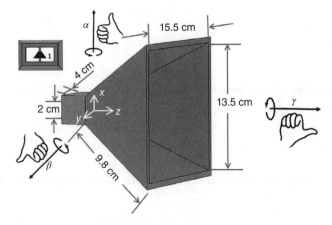

Figure 7.44 The dimensions of the horn antenna and the definition of the rotation angles α, β, γ, and axes x, y, and z. The antenna is excited by using a lumped port located at the centre of the waveguide.

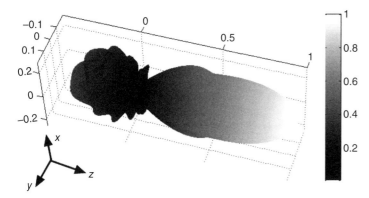

Figure 7.45 The simulated E-field magnitude pattern (linear scale) at 5 GHz. The maximum value is normalised to $1\,\mathrm{V\,m^{-1}}$.

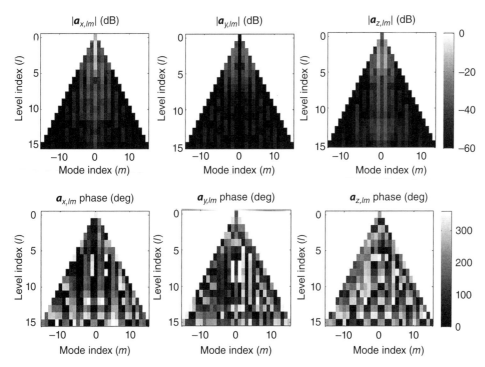

Figure 7.46 The decomposed magnitude (dB scale, $20\log_{10}|\cdot|$) and phase (degree) of the SSHs of the radiation pattern in Figure 7.45. $\boldsymbol{a}_{lm}$ is normalised to make sure $\sum_{l=0}^{N}\sum_{m=-l}^{l}\|\boldsymbol{a}_{lm}\|^2 = 1$.

$$
\begin{cases}
\dfrac{\left|\displaystyle\sum_{l=0}^{L}\sum_{m=-l}^{l} \boldsymbol{a}_{lm}\bullet T_{\alpha}^{*}(\boldsymbol{a}_{lm})\right|}{\displaystyle\sum_{l=0}^{L}\sum_{m=-l}^{l} \|\boldsymbol{a}_{lm}\|^{2}} = \rho_{SC}(\alpha),\alpha=1,2,\ldots,360^{\circ} \\[4ex]
\dfrac{\left|\displaystyle\sum_{l=0}^{L}\sum_{m=-l}^{l} \boldsymbol{a}_{lm}\bullet T_{\beta}^{*}(\boldsymbol{a}_{lm})\right|}{\displaystyle\sum_{l=0}^{L}\sum_{m=-l}^{l} \|\boldsymbol{a}_{lm}\|^{2}} = \rho_{SC}(\beta),\beta=1,2,\ldots,360^{\circ} \\[4ex]
\dfrac{\left|\displaystyle\sum_{l=0}^{L}\sum_{m=-l}^{l} \boldsymbol{a}_{lm}\bullet T_{\gamma}^{*}(\boldsymbol{a}_{lm})\right|}{\displaystyle\sum_{l=0}^{L}\sum_{m=-l}^{l} \|\boldsymbol{a}_{lm}\|^{2}} = \rho_{SC}(\gamma),\gamma=1,2,\ldots,360^{\circ} \\[4ex]
\displaystyle\sum_{l=0}^{L}\sum_{m=-l}^{l} \|\boldsymbol{a}_{lm}\|^{2} = 1 \\[3ex]
\hat{\mathbf{r}}(\theta,\varphi)\bullet\displaystyle\sum_{l=0}^{L}\sum_{m=-l}^{l} \boldsymbol{a}_{lm}Y_{l}^{m}(\theta,\varphi)=0,(\theta,\varphi)\in \mathbf{S}
\end{cases}
\tag{7.105}
$$

where $L=8$ is chosen, $\rho_{SC}(\alpha),\rho_{SC}(\beta)$, and $\rho_{SC}(\gamma)$ are sampled at 1°/step, $\mathbf{S}$ is the point set chosen on the sphere, and 5°/step is used for both θ and φ angles. The nonlinear system of Eq. (7.105) can be solved/optimised using the Levenberg–Marquardt algorithm [95] in Matlab. After optimisation, the reconstructed $\boldsymbol{a}_{lm}$ are obtained and shown in Figure 7.47. As can be seen, compared with Figure 7.46, they have a very similar magnitude but different phases, which is not an issue since the phase is a relative value. The reconstructed ρ_{SC} calculated from the reconstructed $\boldsymbol{a}_{lm}$ are shown in Figure 7.48, and comparisons between the reconstructed ρ_{SC} and ρ_{SC} of the original pattern (in Figure 7.45) are given. As can be seen, a very good agreement has been obtained except at some angles.

The 3D radiation pattern can be obtained quickly using

$$
\mathbf{E}(\theta,\varphi)\approx\sum_{l=0}^{8}\sum_{m=-l}^{l} \boldsymbol{a}_{lm}Y_{l}^{m}(\theta,\varphi)
\tag{7.106}
$$

and is shown in Figure 7.49. Compared with the original pattern in Figure 7.45, a very similar pattern is reconstructed. Comparisons of co-polarisation (CP) and cross-polarisation (XP) components in the YOZ plane and the relative error in all angles are given in Figure 7.50. As can be seen, a good agreement is obtained in the main lobe of the pattern. However, the back lobe, side lobe, and the cross-polarisation component are not exactly the same; we discuss the possible error sources later.

The AR measurement result in an RC can also be verified through simulations. We use a conical horn antenna with two perpendicular excited ports to generate a directional pattern with different AR values. To improve the radiation pattern, a layer of absorbing material is used to suppress the side lobe and back lobe. The model is shown in Figure 7.51. By tuning the phase difference between port 1 and port 2, waves with different AR can be generated.

Patterns with different AR values are generated and decomposed into SSHs with $\boldsymbol{a}_{lm}$, thus $\rho_{SC}(\gamma)$ can be obtained by using (7.96) with different rotation angles. Results are given in Figure 7.52 with minimum value markers. Finally, (7.104) is used to extract the AR values from $\rho_{SC}(\gamma)_{\min}$ and compare them with the AR values obtained from CST (shown in Figure 7.53). As can be seen, a very good agreement is obtained.

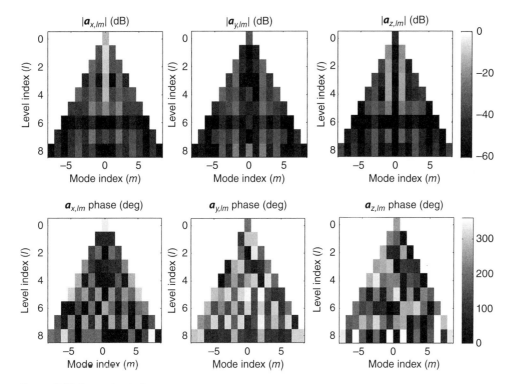

Figure 7.47 Reconstructed a_{lm}.

Measurements were conducted in the AC and RC at UoL. The radiation pattern measurement setup in the AC is shown in Figure 7.54 and the SATIMO SH 2000 horn antenna was used as the AUT. In the AR measurement, we used a homemade wide band log-periodic cross dipole as the AUT (Figure 7.55). Measurement setup in the RC is also shown in Figure 7.56; foams were used to hold the AUT for three different rotation axes. The measurement scenarios and procedures were the same for different AUTs and different rotation axes.

In the measurement of pattern reconstruction, the procedure was the same as in the simulation, the only difference was that $\rho_{SC}(\alpha)$, $\rho_{SC}(\beta)$, and $\rho_{SC}(\gamma)$ were measured in the RC rather than simulated. The self-correlation coefficients were measured at 4 GHz with 50 points of frequency stir (in 10 MHz bandwidth), the turntable was rotated with 1°/step for 360° and the stirrers were rotated with three stirrer positions. Therefore, we had frequency stir [96], source stir [15, 97, 98], and mechanical stir for each rotation angle, and $N = 50 \times 360 \times 3 = 54000$ sample points in (7.97) for each angle of self-correlation coefficient calculation. It should be noted that the turntable and the stirrer were not rotated simultaneously. When the turntable was rotated for a full revolution, the stirrer position was fixed at one position to make sure that the environment was the same (shown in Figure 7.57), and we assumed that the size of the antenna was not so large that rotating the antenna would not perturb the field in the RC greatly, otherwise the results would be decorrelated and always give small correlation coefficients. After all the S-parameters were collected, (7.97) was used to obtain the self-correlation coefficient for different angles of α, β, and γ.

The measured ρ_{SC} are shown in Figure 7.58, and the reconstructed ρ_{SC} from a_{lm} are also given. As expected they agree well with the measured results. The reconstructed 3D pattern from a_{lm} using the summation of SSHs is also shown in Figure 7.59. To validate the results, the patterns in the XOY plane and the XOZ plane were also measured in the AC, and the results are compared in Figure 7.60. As can be seen, very good agreements are obtained for the main beam; however, the error becomes large when the magnitude of the pattern becomes small (side lobes, back lobes, and cross-polarisation). The maximum error for the co-polarisation

(a)

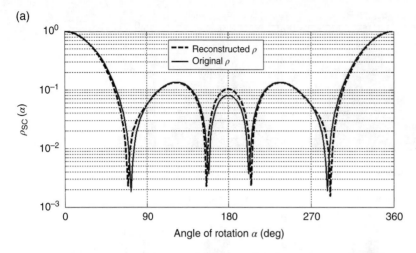

(b)

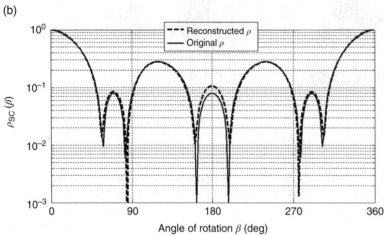

(c)

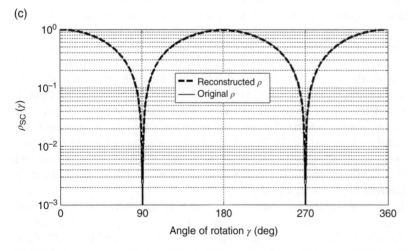

Figure 7.48 Reconstructed ρ_{SC} and the original ρ_{SC}. (a), (b), and (c) are the self-correlation rotated around the *x*-, *y*-, and *z*-axes, respectively.

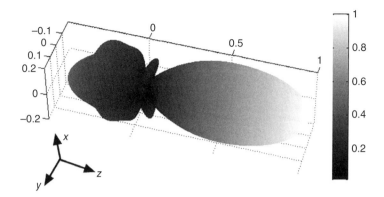

Figure 7.49 Reconstructed E-field magnitude pattern (linear scale) at 5 GHz. The maximum value is normalised to $1\,\mathrm{V\,m^{-1}}$.

(a)

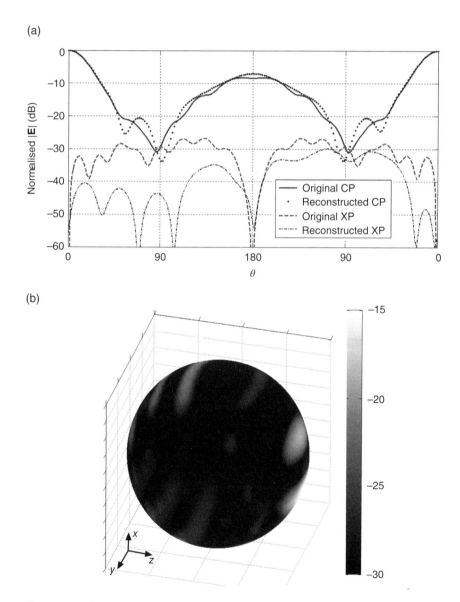

(b)

Figure 7.50 (a) Co-polarisation (CP) and cross-polarisation (XP) of the original pattern and the reconstructed pattern in the YOZ plane, normalised to the peak value in dB. (b) $20\log_{10}\|\mathbf{E}| - |\mathbf{E}'\|$, relative error in all angles in dB, where $\mathbf{E}$ and $\mathbf{E}'$ are the original pattern and the reconstructed pattern, respectively ($\mathbf{E}$ and $\mathbf{E}'$ are normalised to the peak value of $1\,\mathrm{V\,m^{-1}}$).

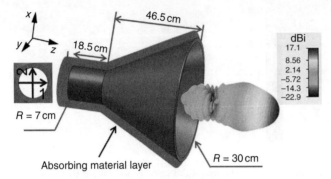

Figure 7.51 The conical antenna and a typical radiation pattern at 1.5 GHz. Two perpendicular lumped ports are used to synthesise waves with different AR values.

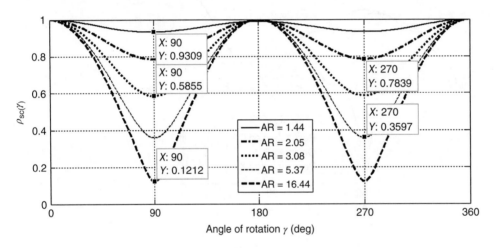

Figure 7.52 $\rho_{SC}(\gamma)$ with different AR values. The AR values in the legend are obtained from simulation in CST.

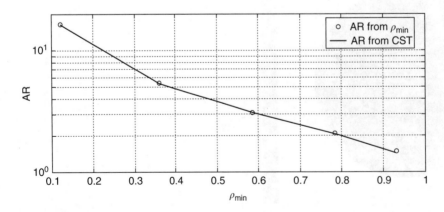

Figure 7.53 AR obtained from $\rho_{SC}(\gamma)_{min}$ using (7.104) and AR obtained from CST.

Figure 7.54 Radiation pattern measurement in the AC and the definition of the x-, y-, and z-axes.

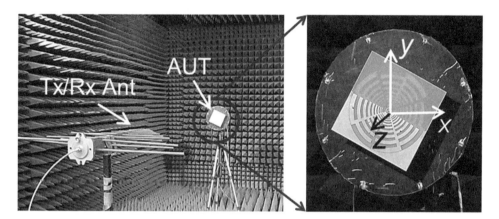

Figure 7.55 AR measurement in the AC.

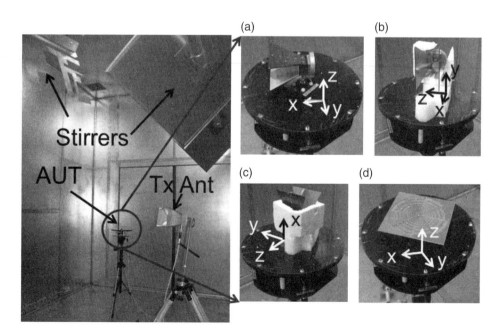

Figure 7.56 Measurement setup in the RC. The AUT rotated around the z-, y-, and x-axes are shown in (a), (b), and (c), respectively. AR measurement is shown in (d).

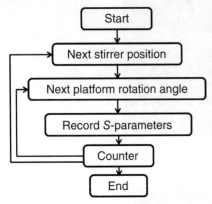

Figure 7.57 Self-correlation measurement flowchart in the RC.

component in the XOY plane occurs at $\theta = 90°$, $\varphi = 252°$, where the measured value is -23.8 dB and the reconstructed value is -15.0 dB. In the XOZ plane, the maximum error for the co-polarisation component occurs at $\theta = 113°$, $\varphi = 180°$, where the measured value is -34.4 dB and the reconstructed value is -26.1 dB. This phenomenon is very similar to that in the numerical simulation and will be discussed in the next section.

In the AR measurement in Figure 7.55, the AUT (the measured S_{11} is given in Figure 7.61) was rotated on the turntable with $1°$/step in the frequency range from 200 MHz to 5 GHz, S-parameters were collected for each degree between the AUT and the Tx antenna, and the maximum and minimum values in a revolution are shown in Figure 7.62. The AR values can be obtained using [99]

(a)

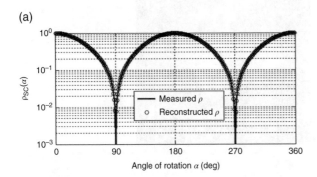

(b)

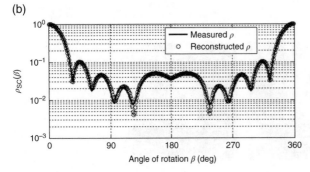

(c)

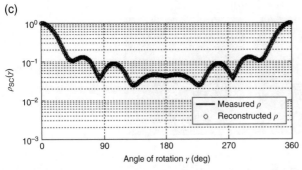

Figure 7.58 Measured and reconstructed ρ_{SC} of the AUT (SATIMO SH 2000). (a), (b), and (c) are the self-correlation rotated around the x-, y-, and z-axes, respectively. The definitions of the three axes are shown in Figure 7.54.

Figure 7.59 Reconstructed E-field magnitude pattern (linear scale) of the AUT (SATIMO SH 2000). The maximum value is normalised to $1 \, V \, m^{-1}$.

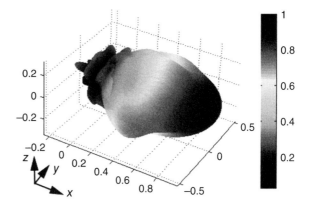

(a)

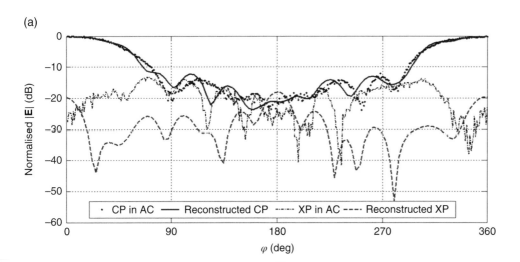

(b)

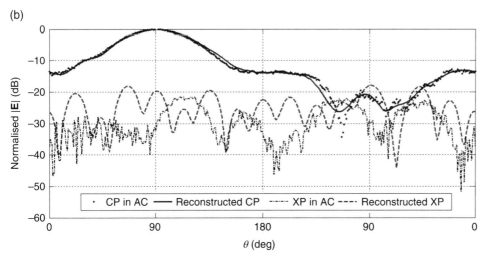

Figure 7.60 Measured and reconstructed radiation pattern in the (a) XOY and (b) XOZ planes. The peak value is normalised to 0 dB. CP, co-polarisation component; XP, cross-polarisation component.

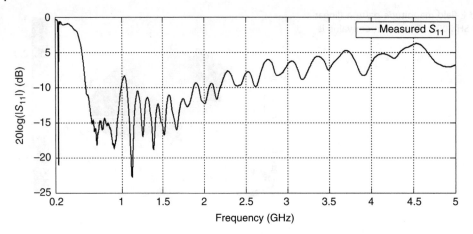

Figure 7.61 Measured S_{11} of the AUT (wide band log-periodic cross dipole).

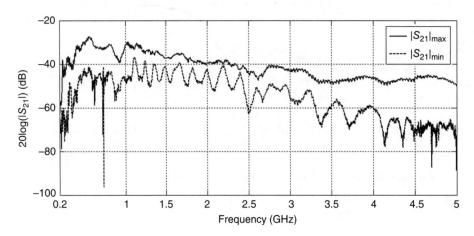

Figure 7.62 Measured maximum and minimum S_{21} in a revolution in the AC.

$$AR = \frac{|S_{21}|_{max}}{|S_{21}|_{min}} \tag{7.107}$$

Similarly, the AR measurement was conducted in the RC using (7.104). Measured $\rho_{SC}(\gamma)$ in the RC are shown in Figure 7.63 at each frequency and the minimum values were extracted to calculate the AR. Finally, the obtained AR in the AC and RC are shown in Figure 7.64. As can be seen, a very good agreement is obtained when the AUT is close to the circular polarisation (1–3.5 GHz).

7.8.3 Discussion and Error Analysis

It should be noted that, in both simulations and measurements, the reconstructed patterns are similar to the original pattern but not exactly the same (Figures 7.49a and 7.60), possible error sources and error analysis are as follows:

1) The SSHs are truncated at level L. To analyse the error caused by L, different L are used to decompose the far-field in Figure 7.45. By comparing the far-field calculated from $\boldsymbol{a}_{lm}$ in (7.87) with the original pattern, the

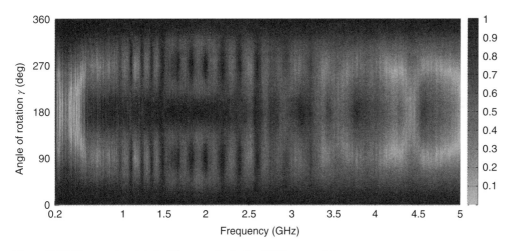

Figure 7.63 Measured $\rho_{SC}(\gamma)$ at all frequencies in the RC (linear scale).

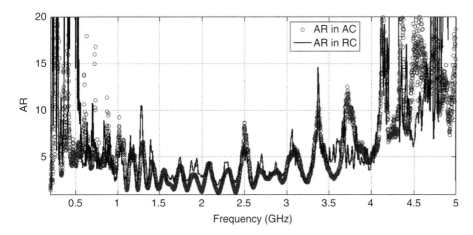

Figure 7.64 Measured $\rho_{SC}(\gamma)$ at all frequencies in the RC (linear scale). The directivity in 1–3.5 GHz is about 5–6.5 dBi.

truncation effect can be observed in Figure 7.65, where both average error and maximum error are given. As can be seen, for the antenna pattern in Figure 7.45, when $L = 8$, the average pattern error is quite small.

2) The inverse problem is a complex nonlinear problem and could have multiple solutions. ρ_{SC} for all α, β, γ angles in Figures 7.46 and 7.47 can be calculated using (7.96) and is shown in Figure 7.66. In the measurements, because of the limitation of the facility, only three cut planes were measured, which correspond to the values on the three axes in Figure 7.66 ($\alpha = 1°\sim360°$, $\beta = 0$, $\gamma = 0$; $\alpha = 0$, $\beta = 1°\sim360°$, $\gamma = 0$; $\alpha = 0$, $\beta = 0$, $\gamma = 1°\sim360°$). As can be seen, in some regions these two ρ_{SC} are different, which means that it could be possible to have two sets of ρ_{SC} with the same value on three axes but have differences in some regions. To quantify the relation between the pattern error and the error in the self-correlation coefficients, ρ_{SC} was perturbed with random values 2000 times. The pattern error was calculated in each case. Results are given in Figure 7.67, which shows a direct statistical understanding between the error in the self-correlation coefficients and the error in the pattern. It can be seen that when the reconstructed self-correlation coefficients are accurate, statistically a more accurate pattern is likely. It is also possible to sample $\rho_{SC}(\alpha, \beta, \gamma)$ in 3D to reconstruct the radiation pattern. To simulate this procedure, the original

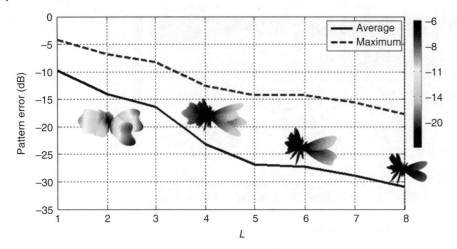

Figure 7.65 Average and maximum pattern error with different level *L*. The average error is defined as 20 log(mean‖**E**| − |**E**′‖) and the maximum error is defined as 20 log(max‖**E**| − |**E**′‖), where **E** and **E**′ are the original pattern and the pattern approximated by using SSHs, respectively (**E** and **E**′ are normalised to the peak value of 1 V m⁻¹). *mean* and *max* are the average and maximum values over all angles, respectively. The error patterns (20 log ‖**E**| − |**E**′‖) are also given when *L* = 2, 4, 6, 8.

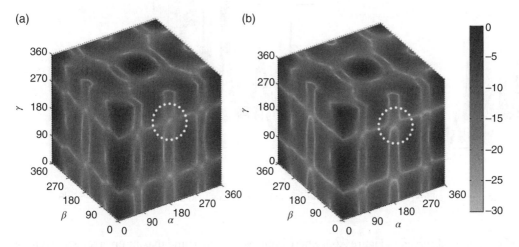

Figure 7.66 Calculated $\rho_{SC}(\alpha, \beta, \gamma)$ from (a) the original $\boldsymbol{a}_{lm}$ in Figure 7.46 and (b) the reconstructed $\boldsymbol{a}_{lm}$ in Figure 7.47. A different region is marked with a dotted circle (the grayscale represents the value of 10 log $\rho_{SC}(\alpha, \beta, \gamma)$, dB scale).

$\rho_{SC}(\alpha, \beta, \gamma)$ in Figure 7.66a was sampled in 3D with different degrees per step for all α, β, and γ. Using the resampled ρ_{SC} to reconstruct the pattern, the pattern errors with different step size can be obtained and are shown in Figure 7.68. As expected, more 3D samples can improve the accuracy of the reconstructed pattern, but more time is needed in the optimisation.

3) The inverse problem is a multi-goal optimisation problem. The reconstruction accuracy is limited by the Levenberg–Marquardt algorithm; the optimised $\boldsymbol{a}_{lm}$ could be at a local minimum rather than the global minimum. Normally, the main beam with co-polarisation has the highest magnitude thus has higher weight than the other components (side lobes, back lobes, cross-polarisations, etc.).

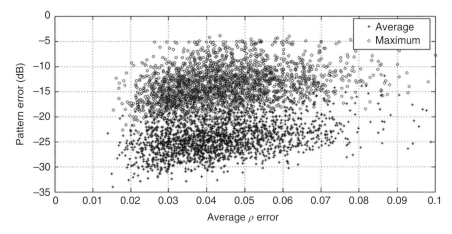

Figure 7.67 The average self-correlation coefficient error (averaged over all selected sample angles) and the pattern error in dB: the average error is defined as $20\log(\text{mean}\||\mathbf{E}| - |\mathbf{E}'|\|)$ and the maximum error is defined as $20\log(\text{max}\||\mathbf{E}| - |\mathbf{E}'|\|)$, where $\mathbf{E}$ and $\mathbf{E}'$ are the original pattern and the reconstructed pattern, respectively ($\mathbf{E}$ and $\mathbf{E}'$ are normalised to the peak value of $1\ \text{V m}^{-1}$). *mean* and *max* are the average and maximum values over all angles, respectively.

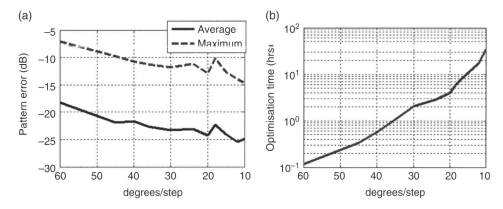

Figure 7.68 (a) Typical reconstructed pattern error and the step size of 3D (α, β, γ), e.g. $10°$/step means all α, β, and γ are sampled every $10°$ in the range of $0°$ to $360°$, thus there are 37 samples in each angle dimension and 50 653 samples in total. The definitions of *average* and *maximum* are the same as in Figure 7.67. (b) Optimisation time in hours for different degrees/step, the results are based on the same initial values and convergence tolerance as in the Levenberg–Marquardt algorithm.

4) To investigate the measurement convergence of ρ_S, different sample numbers are used to repeat the calculation. The average ρ_S error with different sample numbers is shown in Figure 7.69. As can be seen, when the sample number is large the measured ρ_S converges with small uncertainties.

Although the reconstructed pattern in the RC is not as accurate as that measured in an AC, an RC is more cost-effective than an AC. There are also other advantages. For example, the measurement can be conducted in 2D while the reconstructed radiation pattern is always in 3D. The method is based on the NLoS/stirred part of the measured S-parameters. The AUT and the Tx/Rx antenna do not need to be carefully aligned as they would in an AC, which makes the measurement setup more robust and insensitive to antenna positions. The measurement time is shorter than some of the methods using the LoS/unstirred part (at each angle, the stirred part needs to be cancelled out by averaging S-parameters at many stirrer positions or using the Doppler shift). The

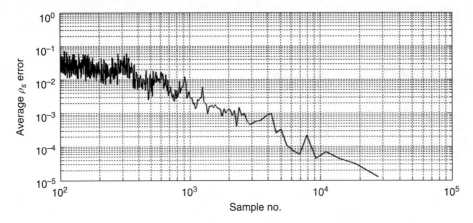

Figure 7.69 The average ρ_S error (averaged over all selected sample angles) and the sample number. ρ_S values with 54 000 samples are used as the reference.

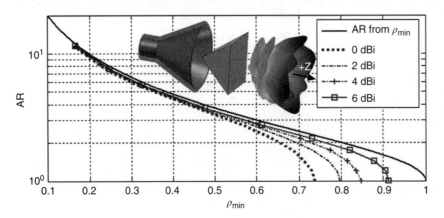

Figure 7.70 Degradation curves with different directivity. AR curves deviate from (7.104) (AR from ρ_{min}) when the directivity reduces in the +z direction.

measurement time could be even shorter than the direct measurement of the 3D radiation pattern in an AC with an acceptable loss of accuracy, since 3D sample points could be much larger than 2D sample points. Also, more time is needed in the post processing of the measurement data.

The AR of a directional antenna has also been measured approximately in an RC. This is under the assumption that the pattern integral is dominated by the main beam in (7.101). To investigate how the accuracy degrades over the directivity, we use a pyramid scatterer to block the wave in the +z direction. By tuning the size of the pyramid we can tune the directivity but not change the AR in the +z direction. Results with different directivity are shown in Figure 7.70. As can be seen, even when the directivity is 0 dBi in the +z direction, the maximum error of AR in (7.104) is 1.3. Therefore (7.104) is a very good approximation and different antenna may have different degradation curves.

It is also interesting to note that by combing the existing measurement methods in the RC, with the same measurement setups but different data post-processing techniques, nearly all antenna parameters (such as radiation efficiency, 3D pattern, S_{11}, and gain) can be obtained in one measurement.

There are also potential issues: when the directivity of the AUT is high (e.g. a parabolic antenna with a large reflector) ρ_{SC} becomes uncorrelated very quickly, and a very fine rotation step for the turntable is needed, which

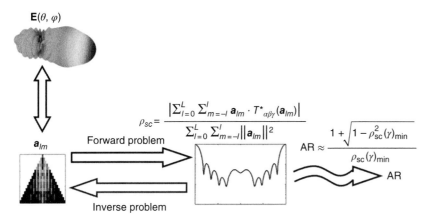

Figure 7.71 An overview of the framework in the 3D radiation pattern reconstruction.

could be time consuming for a whole revolution. Also, to approximate a high directivity pattern, more coefficients are required [100], which further increases the optimisation time when solving (7.105), and the unknown numbers increase quickly in the order of $O(L^2)$. For $L = 8$ in this measurement, the optimisation time is around 2 hours on a personal computer. How to quickly reconstruct a_{lm} of high directivity antenna could be challenging.

It has been shown that the 3D radiation pattern of the AUT can be reconstructed using the self-correlation coefficient of the antenna pattern, which can be measured in an RC. The proposed method only used the NLoS components and it can be considered as a generalised mode matching method which optimises the SSH coefficients to match the measured self-correlation coefficients. This results in a system of nonlinear equations which can be solved/optimised using the well-developed Levenberg–Marquardt algorithm. It has also been shown that the AR of an antenna can be measured in an efficient manner (rotated around one axis) in an RC. An overview of the framework of this method is shown in Figure 7.71. Simulations and measurements have been conducted to verify the proposed theory. Error sources have been analysed and quantified, and have been shown to be small for all of the cases considered in this measurement.

7.9 Material Measurements

Recently, it has been found that the RC can be used to measure the material properties, such as absorption cross section (ACS), average absorption/reflection coefficient, permittivity, and total scattering cross section (TSCS). It is interesting that these applications were not expected in the early years of the RC measurement, as most applications were in the EMC area. In this section, we review the applications of the RC in the material measurement, including recently developed methods.

7.9.1 Absorption Cross Section

The ACS of a lossy object is defined as the ratio of the power dissipated in the object and the power density of the incident wave. The averaged ACS of a lossy object, which is averaged over all angles of incidence and polarisation, can be measured in the RC. The measurement of the ACS of a lossy object is required for many applications, including the characterisation of the effect of lossy objects in multipath environments such as interiors of mass transit vehicles or aircraft loaded with cargoes or passengers [101], and biometrics electromagnetic exposure studies such as the human body's specific absorption rate (SAR) [102].

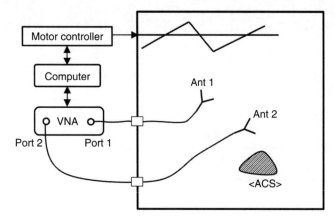

Figure 7.72 ACS measurement in the RC.

From (5.107), the averaged ACS can be written in terms of the measured loaded and unloaded chamber Q factors Q_l and Q_u:

$$\langle \text{ACS} \rangle = \frac{2\pi V}{\lambda} \left(Q_l^{-1} - Q_u^{-1} \right) \tag{7.108}$$

Thus the measurement of ACS is converted to the measurement of Q factors in a loaded and unloaded RC. We have shown in Section 7.2 that the Q factor can be measured in the FD or in the TD. A typical measurement setup is shown in Figure 7.72.

In the FD, if we define the chamber transfer function T as

$$T = \frac{\langle |S_{21,s}|^2 \rangle}{\left(1 - |\langle S_{11} \rangle|^2 \right) \left(1 - |\langle S_{22} \rangle|^2 \right) \eta_{1\text{rad}} \eta_{2\text{rad}}} \tag{7.109}$$

where $\langle \cdot \rangle$ means the averaged value using any stirring method (mechanical stir, frequency stir, source stir, etc.), $S_{21,s} = S_{21} - \langle S_{21} \rangle$ means the stirred part of the S-parameters, and $\eta_{1\text{rad}}$ and $\eta_{2\text{rad}}$ are the radiation efficiencies of antenna 1 and antenna 2 used in the measurement. By using Hill's equation, the averaged ACS in (7.108) can be determined from the net power transfer function with and without the lossy objects [1]:

$$\langle \text{ACS} \rangle = \frac{\lambda^2}{8\pi} \left(T_l^{-1} - T_u^{-1} \right) \tag{7.110}$$

where T_l means the net transfer function of the RC loaded with the object under test (OUT) and T_u means the RC without the OUT. The radiation efficiency of the two antennas should be known in advance, and can be measured by using non-reference antenna methods, for example. For a well-stirred RC, the enhanced backscatter constant is

$$e_b = \frac{\sqrt{\langle |S_{11,s}|^2 \rangle \langle |S_{22,s}|^2 \rangle}}{\langle |S_{21,s}|^2 \rangle} = 2 \tag{7.111}$$

Assuming two identical antennas are used in the measurement (another antenna does not necessary exist), by using (7.111), (7.109) can be written as [103]

$$T = \frac{\langle |S_{11,s}|^2 \rangle}{2 \left(1 - |\langle S_{11} \rangle|^2 \right)^2 \eta_{1\text{rad}}^2} \tag{7.112}$$

Thus, only one antenna is required to complete this measurement, and we only need to know the radiation efficiency of one antenna and reflection coefficient S_{11}.

The TD method is realised by performing the measurement in the FD and then transforming the results to the TD. In the TD, the loaded and unloaded chamber Q can be determined from the chamber time/decay constant τ_{RC}. By using $Q = \omega\tau_{RC}$, (7.108) can be written as

$$\langle ACS \rangle = \frac{V}{c}\left(\frac{1}{\langle\tau_l\rangle} - \frac{1}{\langle\tau_u\rangle}\right) \tag{7.113}$$

where c is the speed of light in free space, τ_l is the loaded chamber decay constant, and τ_u is the unloaded chamber decay constant. This technique requires knowledge of the chamber decay time. To obtain τ_{RC}, we first need to obtain the PDP of the RC from the inverse Fourier transform of S_{11}. Because the TD power in the RC decays exponentially, τ_{RC} can be obtained from the slope of ln(power) in the TD, as shown in Section 7.2. Compared with the FD method, the TD method is simpler and more accurate because we do not require the knowledge of the antenna efficiency and the systematic error caused by antenna efficiency estimation can be avoided.

Measurements were conducted in the RC at UoL, which has a size of 3.6 m × 4 m × 5.8 m. Two double-ridged waveguide horn antennas were used as antenna 1 (SATIMO SH 2000) and antenna 2 (Rohde & Schwarz HF 906). Antenna 1 was mounted on a turntable platform to introduce source stir positions and connected to port 1 of a VNA via a cable running through the bulkhead of the chamber, and antenna 2 was connected to port 2 of the VNA via another cable through the bulkhead of the chamber. During the measurement, the turntable platform was rotated stepwise to three source stir positions (20° for each step). At each source stir position the two paddles were moved simultaneously and stepwise to 100 positions (3.6° for each step). At each mode stir position and for each source stir position a full frequency sweep was performed by the VNA and the S-parameters were collected. Thus, for each frequency, we have 300 stirrer positions (three source stir positions and 100 mode stir positions for each source stir position). A piece of RF absorber was selected as the OUT. The measurement setup is shown in Figure 7.73. The measurement procedure is as follows:

1) Calibrate the VNA, including the cables, according to the standard calibration procedure.
2) Place the two antennas, the turntable platform, and the support (excluding the OUT) inside the RC.
3) Connect antenna 1 to port 1 of the VNA and antenna 2 to port 2 of the VNA, and collect the full S-parameters for each stirrer position.
4) Keep the previous measurement setup unchanged and place the OUT on the support, and repeat step (3).

In this measurement, 10 001 points were sampled in the frequency range from 3.8 to 5.2 GHz. The ACS of the OUT was calculated using two antennas in the FD (7.109), one antenna in the FD (7.111), and one antenna in

(a) 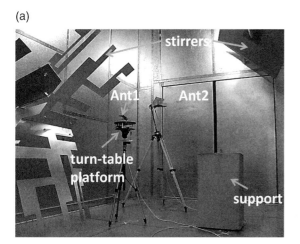 (b)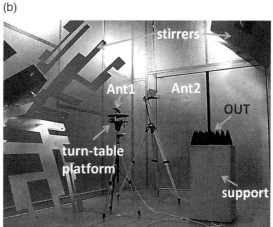

Figure 7.73 ACS measurement setup in the RC: (a) unloaded and (b) loaded with the object under test.

the TD (7.113). In the FD, the enhanced backscatter coefficients (e_b) in the loaded and unloaded RC are obtained and shown in Figure 7.74. We can see that they are very close to 2, which means the RC is well stirred [104]. The chamber transfer functions using one antenna ($T_{FD, 1}$) and two antennas ($T_{FD, 2}$) in the loaded and unloaded RC are shown in Figure 7.75. As can be seen, the chamber transfer function is reduced when the chamber is loaded because of the increase in the power loss. $T_{FD, 1}$ is very close to $T_{FD, 2}$ for both loaded and unloaded scenarios, which manifests the effectiveness of the one-antenna method in the FD.

In the TD, we use a band-pass elliptic filter of order 10 to filter the measured S_{11} with 200 MHz bandwidth, as shown in Figure 7.76a, and then the IFFT is applied to the filtered S_{11}. Since the received power decays exponentially ($\exp(-t/\tau_{RC})$) in the TD, the least-squares fit is applied to ln(power) to obtain the slope k, and τ_{RC} can be extracted by getting the negative inverse of the slope ($\tau_{RC} = -1/k$). To avoid the fit error caused by the noise level, only part of the signal is used for least-squares fit, as shown in Figure 7.76b. By sweeping the centre frequency of the filter, τ_{RC} at different centre frequencies are obtained. The measured chamber decay times using one antenna in the loaded and unloaded RC are depicted in Figure 7.77. The thin curves are the measured τ_{RC} for different stir samples and the thick dash curves are the averaged τ_{RC} for all samples. As expected, the chamber decay time is reduced when the chamber is loaded.

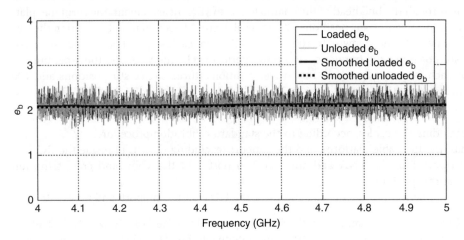

Figure 7.74 e_b in the loaded and unloaded scenarios.

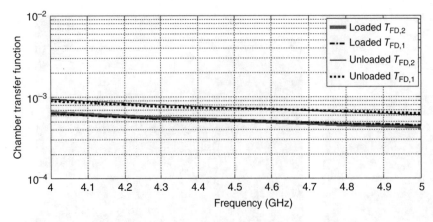

Figure 7.75 The measured chamber transfer function using one antenna ($T_{FD,1}$) and two antennas ($T_{FD,2}$) in the loaded and unloaded RC.

(a)

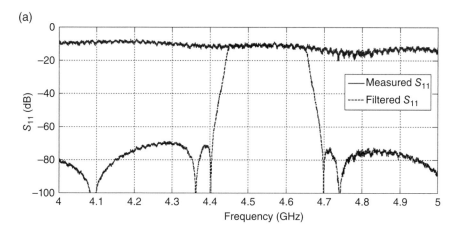

(b)

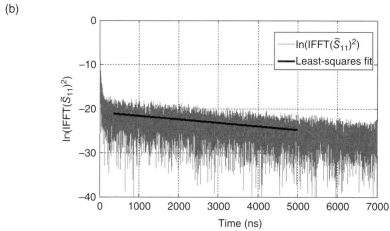

Figure 7.76 Extracting the chamber decay time from S_{11}: (a) measured S_{11} and filtered S_{11}, and (b) the TD response: $\ln([\mathrm{IFFT}(\widetilde{S}_{11})]^2)$ and the least-squares fit.

Another thing to be noted is that the decay constant does not vary much for different stir samples. The reason is that the chamber decay time is determined by the loss of diffused waves. For a given frequency, the chamber loss will vary due to the change in the paddle positions and the change in the chamber modes. However, when the number of resonant modes is massive, the diffused loss is not sensitive for different paddle positions and shows an averaged value for a huge number of standing waves. As shown in Figure 7.77, in the unloaded RC the variation between the τ_{RC} for one sample and the averaged τ_{RC} is within about ±10%, while in the loaded RC the variation is within about ±5%. It is easy to understand, when the RC is loaded, that the excited mode number is larger than the unloaded RC, which leads to more stable τ_{RC} values. This is also observed in Section 7.2, where the TD method actually offers an opportunity to extract τ_{RC} by merely a few stir samples, thus the ACS can be measured rapidly and accurately, which will be discussed later. The ACS measurement results are shown in Figure 7.78. In the FD, 200 MHz frequency stir is adopted. The efficiencies of antenna 1 and antenna 2 in 4–5 GHz are around 78% and 95%, respectively. It can be seen clearly that the measured ACSs using the three methods are all around $0.1\,\mathrm{m}^2$ and the maximum variation is within 10%.

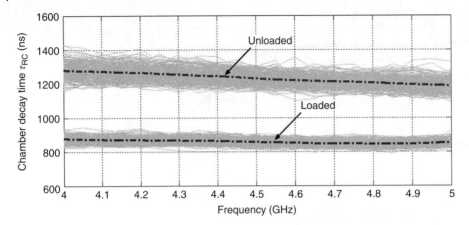

Figure 7.77 The measured chamber transfer function using one and two antennas in the loaded and unloaded RC.

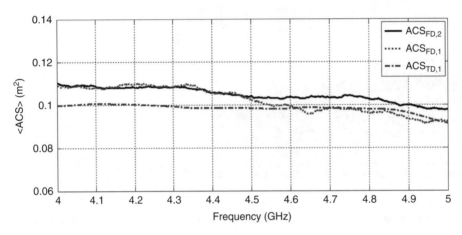

Figure 7.78 The measured average ACS of the object under test.

To study the convergence of the three methods, the root mean square error (RMSE) of the measured ACS is analysed. The RMSE in the frequency range of 4–5 GHz with different number of stirrer positions is adopted to evaluate the convergence, and the RMSE is defined as

$$
\text{RMSE}_i = \sqrt{\frac{\sum_{j=1}^{N}\left(\text{ACS}_{i,j} - \text{ACS}_{M,j}\right)^2}{N}} \quad i = 1, 2, \ldots, M
\tag{7.114}
$$

where i is the number of stirrer positions, M is the maximum number of stirrer positions, j is the frequency sampling point number, and N is the number of frequency sampling points in 4–5 GHz. In our case, $M = 300$ and $N = 7143$. The calculated results are shown in Figure 7.79. As can be seen, the convergence speeds of the one-antenna method and the two-antenna method in the FD are close but the TD method converges faster than the FD method because the chamber decay time τ_{RC} is not sensitive to the stirrer positions and only depends on the overall loss of the RC. The FD S-parameters are very sensitive to the stirrer positions and need many samples to obtain the average values. Thus, $\text{ACS}_{\text{TD, 1}}$ converges faster than $\text{ACS}_{\text{FD, 1}}$ and $\text{ACS}_{\text{FD, 2}}$. It is

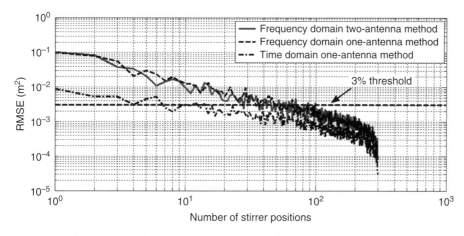

Figure 7.79 The RMSE with the increase in the number of stirrer positions for different methods.

worth mentioning that, in the TD, the RMSE is always below 10% (compared with the averaged ACS in the full frequency span, about 0.1 m^2 from Figure 7.78) and drops below 3% after 15 stirrer positions. However, in the FD, the one-antenna method and the two-antenna method have similar convergence behaviour, and the RMSEs are always above 10% before the first 10 stirrer positions and drop down slowly afterwards. They are below 3% after 100 stirrer positions.

Considering the robustness of the chamber decay time and the fast convergence property of the TD method, the measurement setup can be further simplified by using an electrically large conducting cavity, i.e. an RC is not necessary. To verify this, the paddles of the RC were set stationary, therefore no mode stir was introduced during the measurement, and thus the RC would merely act as an electrically large cavity. To extract the correct τ_{RC} of the electrically large cavity, a simple source stir was introduced by rotating the turntable platform. Based on the convergence speed of the one-antenna method in the TD, 20 source stir positions were adopted in our measurement. The turntable platform was moved stepwise to 20 source stir positions (18° for each step). A double-ridged waveguide horn antenna (SATIMO SH 2000) was mounted on the turntable platform and connected to port 1 of a VNA via a cable running through the bulkhead of the cavity. The measurement procedure was similar to that in the RC. The cavity decay time with and without the OUT was extracted from S_{11}. The measurement results are shown in Figure 7.80.

As can be seen, the results from the mode stir and source stir are in good agreement. The difference is within 4% and the whole measurement time for the source stir was about 7 minutes while the measurement time for the mode stir was more than 8 hours, which means that the ACS can be measured in the TD rapidly and accurately. The major contribution to the measurement time is the sweeping time and the time of transferring data from the VNA to the computer. The measurement time of this method is comparable with that of the rapid method proposed in [105]. It should be pointed out that the cavity should be large enough to support sufficient cavity modes to ensure enough independent samples can be obtained at the lowest frequency of the measurement, or the OUT could not fully 'submerge' into the field-uniform area and the measurement result could be wrong. This is the main consideration for the selection of the size of the conducting cavity in the measurement.

Here we have presented three methods for determining the ACS of the OUT in the FD and in the TD. The commonly used RC technique for determining the ACS of the OUT requires two antennas and the radiation efficiency of the two antennas should be known. In this section, we first presented the one-antenna method in the FD which requires only one antenna (with known efficiency) by making use of enhanced backscatter effect. Thus, the measurement setup was simplified. Then, we presented the one-antenna method in the TD which needs no knowledge of the efficiency of the antenna. The experimental setup was illustrated and measurement results were presented. It can be seen that the measured ACSs by the three methods are in good agreement.

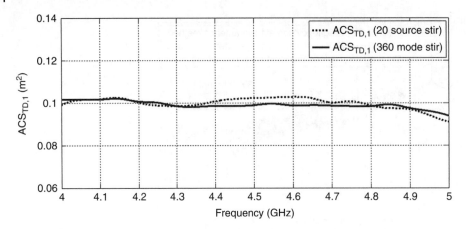

Figure 7.80 The comparison of the measured ACS in the TD with 20 source stir positions and 360 mode stir positions.

Table 7.2 Comparison of different measurement methods.

Measurement method	Number of antennas needed	Measurement time	Measurement facility needed
FD two-antenna method	2	Approx. 8 h	Reverberation chamber
FD one-antenna method	1	Approx. 8 h	Reverberation chamber
TD one-antenna method (mode stir)	1	Approx. 7 min	Reverberation chamber
TD one-antenna method (source stir)	1	Approx. 7 min	Electrically large cavity

We have also investigated the robustness of the chamber decay time and the convergence speed of the three methods and found that the TD method converges much faster than the FD methods. A rapid and accurate measurement can be achieved in the TD based on this finding by using the source stir technique, which makes it quite suitable for human absorption and exposure measurement. Furthermore, in the TD approach, the RC can be replaced by a suitable electrically large conducting cavity, which will greatly reduce the hardware requirement. The method was validated in the RC by setting the paddles stationary and the results agree well with that measured in the RC using mode stir. The comparison of these measurement methods is shown in Table 7.2. It is demonstrated that the TD method is much more efficient and its hardware requirement is much lower than the FD method.

Some points need to be emphasised. The proposed methods assume that the RC is well stirred. When the RC is not well stirred, the OUT could not fully 'submerge' in the field-uniform area. The measured chamber transfer function and chamber decay time will be inaccurate, hence the measured ACS will have considerable errors. During the measurement, the OUT should be set far away from the antennas to avoid the proximity effect [106].

The measurement procedure of the TSCS (scattering cross sections averaged over all angles of incidence and polarisations) is very similar to that of the ACS measurement and was introduced in Section 6.4.2. The duality principle between the TSCS and ACS was discussed in Section 6.5.

7.9.2 Average Absorption Coefficient

We have discussed the measurement of ACS in the RC. The ACS depends on the shape and material properties of the OUT. The ACS of a given object can be measured in the RC; however, if we want to characterise the material of the object, the size and shape of the object need to be considered. A big OUT with small losses could have the same ACS value as an OUT with a smaller size but higher losses. It is worth mentioning that if the OUT is planar, the average absorption/scattering coefficient can be obtained by comparing the ACS value to an ideal absorber with the same size [107–109].

If this method can be generalised to an arbitrarily shaped OUT measurement, it could be very useful for applications such as material characterisation, microwave absorption of a human body, predicting the ACS using exposed surface area, or quantifying the stealth ability of an aircraft averaged over all incident angles and polarisations. The average absorption coefficient (AAC) can be understood as a system-level parameter to characterise the overall absorptive ability of an arbitrarily shaped object.

By generalising the idea in [107–109], the problem becomes how to obtain the ACS value of an arbitrarily shaped perfect absorber. It has been shown in [110] that the ACS of perfect absorbers with some regular objects can be derived analytically and is found to be a quarter of its surface area. It could be very easy to mistakenly use the overall surface area to calculate the ACS of an arbitrarily shaped perfect absorber. If the OUT shape is concave, because of the shadow, using the overall surface area will give an overestimated value. From a mathematical point of view, the reason for this is that the shadow will block the incoming wave at some angles, which will limit the integral angle region [111, Eq. (31)]. This phenomenon is illustrated in Figure 7.81a. At the infinitesimal area of P_{out}, the surface can interact with all the waves coming from outside. However, only the waves from the gap can interact with the absorber at P_{in}, there are no waves coming from the directions in dashed lines. This can also be understood from a physical point of view (boundary condition). An equivalent boundary can be chosen as shown in Figure 7.81b to shorten the gap (Huygen's principle). Because not all surfaces can interact with waves coming from all angles (maximum half space), using the surface area to calculate the ACS of a perfect absorber may give an overestimated value.

In this section we deal with this problem by calculating the projected area in all angles numerically. For an electrically large perfect absorber, the ACS in each direction is the projected area. A 3D model can be represented by using a point cloud; each point can be projected into a 2D plane and then a numerical algorithm can be applied to calculate the area of the point cloud in 2D. Thus once the projected area in all directions is obtained, the averaged value can be calculated, which is the reference value in the RC.

Suppose, we have a perfect/ideal electrically large absorber which has the same shape as the OUT and we denote the average ACS value of it as $\langle\text{ACS}_\infty\rangle$, which means it is the theoretical limit (reference value). By comparing the measured ACS ($\langle\text{ACS}_{\text{RC}}\rangle$) to $\langle\text{ACS}_\infty\rangle$, the AAC $\langle\alpha\rangle$ can be defined as [112]

$$\langle\alpha\rangle = \frac{\langle\text{ACS}_{\text{RC}}\rangle}{\langle\text{ACS}_\infty\rangle} \tag{7.115}$$

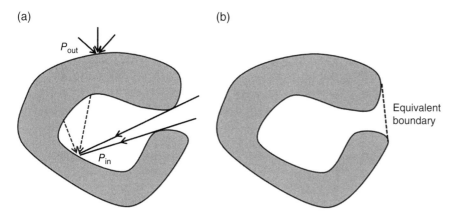

Figure 7.81 A perfect absorber of arbitrary (concave) shape: (a) mathematical point of view and (b) physical point of view.

where $\langle \cdot \rangle$ means the averaged value over all incident angles and polarisations. The value of $\langle \alpha \rangle$ is between 0 and 1. Similarly, the average power scattering coefficient $\langle \Gamma \rangle$ (including reflection and transmission) can be obtained as

$$\langle \Gamma \rangle = 1 - \langle \alpha \rangle \tag{7.116}$$

The measurement of $\langle \mathrm{ACS_{RC}} \rangle$ has been given in Section 7.9.1, and can be obtained in the FD or in the TD. The problem is the determination of $\langle \mathrm{ACS_\infty} \rangle$. For an arbitrarily shaped object, there is no analytical solution but a numerical result can be obtained. For an electrically large perfect absorber, the ACS in each direction is the projected area. As long as we can obtain the projected area in all directions ($\mathrm{ACS_\infty}(\theta, \varphi)$), the averaged value $\langle \mathrm{ACS_\infty} \rangle$ can be obtained.

For the arbitrarily shaped ideal absorber shown in Figure 7.82a, a plane wave impinges from a direction (θ, φ) given in a spherical coordinate system. The vectors $\hat{r}_p$ and $\hat{r}_q$ are two unit vectors in the constant phase plane ($\hat{r}_p \perp \hat{r}_q, \hat{r}_p \perp \hat{r}, \hat{r}_q \perp \hat{r}$). To obtain the projected area, the 3D model is discretised into a point cloud; each point can be projected into the 2D plane by applying

$$P_{2Dp} = \hat{r}_p \cdot \vec{P}_{3D}$$
$$P_{2Dq} = \hat{r}_q \cdot \vec{P}_{3D} \tag{7.117}$$

as shown in Figure 7.82b. Next, the 2D plane is discretised into small grids. By counting the grids occupied by the projected points (the shaded grids in Figure 7.83a), the projected area in 2D can be obtained, as shown in Figure 7.83b. By repeating this procedure for each direction (θ, φ), $\langle \mathrm{ACS_\infty} \rangle$ can be obtained. Because the projected areas in the directions $\mathbf{r}(\theta, \varphi)$ and $-\mathbf{r}(\theta, \varphi)$ are the same, only directions in the hemisphere need to be calculated. This procedure can also be calculated in real time by using a graphics processing unit (GPU) [113]. After $\mathrm{ACS_\infty}(\theta, \varphi)$ is obtained for each direction (θ, φ), the average ACS $\langle \mathrm{ACS_\infty} \rangle$ can be obtained as [110]

$$\langle \mathrm{ACS_\infty} \rangle = \frac{\int_0^{2\pi} \int_0^\pi \mathrm{ACS_\infty}(\theta, \varphi) \sin\theta d\theta d\varphi}{\int_0^{2\pi} \int_0^\pi \sin\theta d\theta d\varphi} = \frac{\int_0^{2\pi} \int_0^{\pi/2} \mathrm{ACS_\infty}(\theta, \varphi) \sin\theta d\theta d\varphi}{2\pi} \tag{7.118}$$

which can be calculated numerically.

To verify the accuracy of the algorithm, we use three different shapes to calculate $\langle \mathrm{ACS_\infty} \rangle$ and compare the results with analytical values. It has been found that for a sphere, a rectangular parallelepiped, and a circular cylinder, $\langle \mathrm{ACS_\infty} \rangle$ is 1/4 of the surface area [111]. Actually, for electrically large objects, as long as the shape is

(a)   (b)

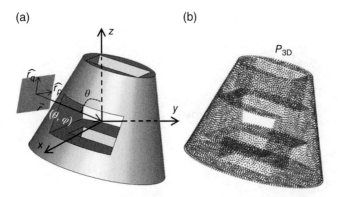

Figure 7.82 (a) An arbitrary (concave) shape 3D model illuminated by a plane wave coming from direction (θ, φ) and (b) projected points in 2D.

(a)

(b)

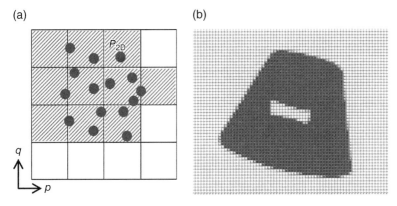

Figure 7.83 (a) Grids with 2D points and (b) projected area in 2D.

Table 7.3 Numerical verification.

Shape	$r = 10$	$h = 20$	$r = 10$ $h = 30$
Number of points	9 950	11 190	10 578
Analytical $\langle ACS_\infty \rangle$	314.2	600	628.3
Simulated $\langle ACS_\infty \rangle$	314.5	600.4	630.6
Relative error (%)	0.10	0.07	0.36

Units can be arbitrary.

convex, $\langle ACS_\infty \rangle$ is 1/4 of the surface area (the average projected area theorem) [114, 115], while for arbitrary (concave) shape objects, numerical evaluation of (7.118) is required.

Three objects are used to verify the numerical procedure, as shown in Table 7.3, since the analytical solutions have been found in [110]: a sphere with radius $r = 10$, a cube with edge length $h = 20$, and a cylinder with radius $r = 10$ and height $h = 30$ (units can be arbitrary). They are represented by point clouds. Since the surface areas can be obtained analytically, they can be used as reference values to verify the accuracy of the algorithm. As can be seen, the results are very accurate and the relative errors are smaller than 1%.

Measurements were performed in the RC at UoL for arbitrary (concave) shapes of OUTs. Three measurement scenarios are shown in Figure 7.84. Two horn antennas were used (Rohde & Schwarz HF 906 and SATIMO SH 2000) and 10 001 samples of S-parameters in the range of 9.8–10.2 GHz were collected at each stirrer position. One hundred stirrer positions were used with 3.6°/step, and a 10th-order elliptic band pass filter with 10 GHz centre frequency and 200 MHz bandwidth was used to filter the S-parameters. The PDPs can be obtained from the IFFT of the filtered S-parameters. We have shown in Section 7.9.1 that the ACS measurement in the TD is more accurate and has lower uncertainties [103] since the early time response, insertion loss of antennas, and cables will not affect the extraction of Q factors, and $\langle ACS_{RC} \rangle$ can be obtained using (7.113).

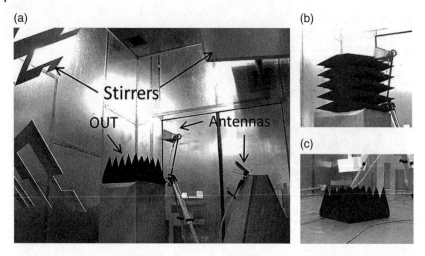

Figure 7.84 Three measurement scenarios: (a) one piece of RAM positioned in the centre of the RC, (b) two back-to-back pieces of RAM positioned in the centre of the RC, and (c) one piece of RAM positioned on the ground.

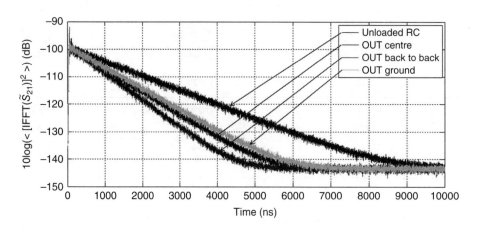

Figure 7.85 Measured PDPs of three scenarios. $\widetilde{S}_{21}$ represents the filtered S-parameter between two antennas.

The least-squares fit method [116] is used to extract the decay time constants of PDPs shown in Figure 7.85. The results are shown in Table 7.4. For the unloaded RC (including the empty carton used to support the OUT), the decay constant of the unloaded RC is $\tau_u = 860.4$ ns. Three different positions of the OUT were measured; the maximum difference of the decay time caused by different positions is 4 ns. The maximum relative error of $\langle ACS_\infty \rangle$ (0.36%) has also been included. From (7.115), the uncertainty of $\langle \alpha \rangle$ depends on the uncertainties of both $\langle ACS_{RC} \rangle$ and $\langle ACS_\infty \rangle$, giving a maximum uncertainty of 0.02 of $\langle \alpha \rangle$, which is shown in Table 7.4. Three independent measurements could be insufficient to accurately estimate the measurement uncertainty; nevertheless, the maximum differences were used which gave reliable estimated boundaries of the uncertainties.

To obtain the AACs, the OUTs are discretised into point clouds. The models are shown in Figures 7.86a and 7.87a. The simulated $ACS_\infty(\theta, \varphi)$ patterns with a resolution of $2°$/step are given in Figures 7.86b and 7.87b. By using the two-dimensional trapezoidal rule, the integral in (7.118) is calculated numerically, thus $ACS_\infty(\theta, \varphi)$ are obtained. Finally, $\langle \alpha \rangle$ in the three scenarios are obtained using (7.115). Results are summarised and given in Table 7.4.

Table 7.4 Measured ACS and average absorption coefficients @ 10 GHz.

Scenario	Figure 7.84a	Figure 7.84b	Figure 7.84c
τ_l (ns)	540.1 ± 2	453.7 ± 2	591.5 ± 2
$\langle \mathrm{ACS}_\infty \rangle$ (m^2)	0.23 ± 0.001	0.34 ± 0.001	0.17 ± 0.001
$\langle \mathrm{ACS_{RC}} \rangle$ (m^2)	0.192 ± 0.003	0.290 ± 0.003	0.147 ± 0.002
$\langle \alpha \rangle$	0.83 ± 0.02	0.85 ± 0.01	0.86 ± 0.02

(a) (b)

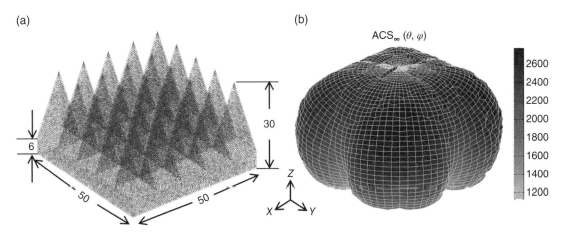

Figure 7.86 (a) Point cloud model of a piece of RAM with 62 286 points (unit: cm) and (b) simulated ACS$_\infty(\theta, \varphi)$ (unit: cm^2).

At 10 GHz, the OUTs are electrically large. As can be seen in Table 7.4, although the measured average ACSs are different, the AACs $\langle \alpha \rangle$ are very close and not sensitive (but not independent) to the shape and size of the OUT since they are made of the same materials. Note that $\langle \alpha \rangle$ in Figure 7.84b is larger than that in Figure 7.84a. This is because the OUT in Figure 7.84b has a larger corrugated area ratio (corrugated surface area/overall surface area) than in Figure 7.84a, thus has a larger $\langle \alpha \rangle$. An intuitive explanation is shown in Figure 7.88.

It should also be noted that the ACS in Figure 7.84c is half of the ACS in Figure 7.84b but they have similar $\langle \alpha \rangle$. This is because of the image of the ground plane. This effect is shown in Figure 7.89a. However, this does not affect the proposed method. The model with its image can be considered as an integrated model (Figure 7.87a), after the numerical evaluation of (7.118), and by dividing the result by 2, the final $\langle \mathrm{ACS}_\infty \rangle$ can be obtained. Similarly, if the model is placed close to the corner, as shown in Figure 7.89b, the model and its images can be considered as an integrated model while the ACS needs to be divided by 4 for a 2D problem and divided by 8 for a 3D problem.

It can be seen that when the OUT is far from the boundary of the RC, the image effect is reduced; the overlapping area between the model and its images becomes ignorable. This can also be used to obtain the limit of TSCS of stirrers in Section 6.6, since TSCS and ACS are dual quantities. In Section 6.6, an equivalent boundary is used to wrap the OUT, and the limit of TSCS (ACS) is a quarter of the equivalent boundary surface area. However, the equivalent boundary with minimum surface area may be hard to find for an arbitrarily shaped OUT (especially when the OUT is close to the boundary of the RC). The numerical procedure in this section is a general method and does not have this problem, and the equivalent boundary is not necessary. $\langle \mathrm{ACS}_\infty \rangle$ can be obtained directly using numerical methods.

(a)

(b)

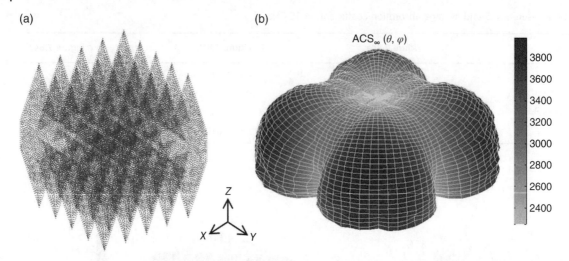

Figure 7.87 (a) Point cloud model of two back-to-back pieces of RAM with 46 505 points and (b) simulated $ACS_\infty(\theta, \varphi)$ (unit: cm^2).

Figure 7.88 A corrugated surface absorbs waves more than once, thus has a higher average absorption coefficient than a planer surface.

(a)

(b)

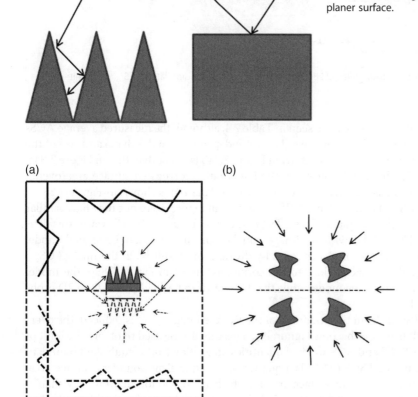

Figure 7.89 (a) The effect of the ground plane, images are represented by dashed lines. (b) A general case when the object is positioned at the corner of the RC, waves coming from arbitrary directions are represented by arrows.

In this section, a general definition of the AAC has been given. For an electrically large perfect absorber, when the shape is convex, from the average projected area theorem [114, 115], $\langle ACS_\infty \rangle$ is a quarter of the surface area. When the shape is arbitrary, the proposed numerical method can be applied to obtain $\langle ACS_\infty \rangle$. By comparing the measured average ACS to $\langle ACS_\infty \rangle$, the average absorption/scattering coefficient of arbitrarily shaped object can be obtained. A system-level average absorption/scattering coefficient could be very useful for many applications. The method can be used to compare the absorptive/scattering ability of an object with the same material but different shape, evaluate the AAC of a human body, predict the ACS using the exposed surface area [117], and even quantify the stealth ability of an aircraft. Note that the size needs to be electrically large when using this method, otherwise $ACS_\infty(\theta, \varphi)$ needs to be calculated using full wave methods [110].

7.9.3 Permittivity

The measured average ACS can be further used to extract the material properties of dielectric objects [111]. This method is limited to electrically large objects; only when the skin depth is much smaller than the sample cross dimension can the real part (ε_r) of the permittivity be obtained. When the skin depth is much larger than the cross dimension the relationship between the conductivity and permittivity ($\sigma/\sqrt{\varepsilon_r}$) can be found but not the specific values of the conductivity and permittivity. For general cases, a much more complex relationship can be obtained but it is hard to evaluate.

At a given frequency, the ACS is a single value, but the permittivity has a real part and an imaginary part. It is impossible to find a unique solution of two variables from one equation, which is why the real part and imaginary part cannot be obtained simultaneously in [111]. If we consider the frequency dependency, and assume that the permittivity and conductivity will not change much over a certain frequency range, it is possible to obtain both the real part and the imaginary part. If both permittivity and conductivity are also frequency dependent, this will limit the proposed method since we do not have enough information to solve the unknown variables (solution may not be unique). We limit the object shape to a sphere since for a spherical object the ACS is independent of the incident angle; this directly links the measured ACS value in an RC with the theoretical value because the measured ACS in an RC is the averaged value of all the incident angles. By considering the ACS value at different frequencies, the real part permittivity and conductivity that satisfy the ACS value at all the frequencies can be found, an evaluation function is defined, like the probability distribution function (PDF), the PDF-like evaluation function will make the results self-explanatory, no presumptions, and uncertainty analysis are needed after the measurement and calculation. The effect of the container and the limitations of the method are also discussed.

The measurement setup is the same as the ACS measurement shown in Figure 7.72. Suppose a plastic sphere filled with water (Figure 7.90) is used as the OUT. The measurement process is the same as the ACS measurement in the RC. The averaged ACS of OUT can be obtained from both in the FD (7.110) or in the TD (7.113).

The theoretical ACS can be calculated analytically by using Mie theory [118]:

$$ACS_{sphere} = \pi r^2 (Q_{ext} - Q_{sca}) \tag{7.119}$$

where r is the radius of the sphere, Q_{ext} is the extinction efficiency, and Q_{sca} is the scattering efficiency:

$$Q_{ext} = \frac{2}{(kr)^2} \sum_{n=1}^{\infty} (2n+1) \operatorname{Re}(a_n + b_n) \tag{7.120}$$

$$Q_{sca} = \frac{2}{(kr)^2} \sum_{n=1}^{\infty} (2n+1) \left(|a_n|^2 + |b_n|^2 \right) \tag{7.121}$$

$k = 2\pi/\lambda$ is the wave number, and a_n and b_n are the Mie coefficients, which are different for coated or uncoated spheres [118–120], and they will be different when considering the holder of the OUT.

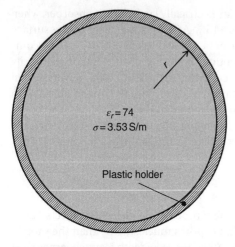

Figure 7.90 A plastic sphere filled with water.

When considering the frequency dependency, the permittivity ε_r and conductivity σ can be obtained by solving the equation system

$$\text{ACS}\big|_{f=f_i} = \pi r^2 (Q_{\text{ext}} - Q_{\text{sca}})\big|_{f=f_i}, \quad i = 1,...,N \qquad (7.122)$$

where the left-hand side of (7.122) is the measured ACS and the right-hand side is the theoretical value, which is a function of ε_r and σ. (7.122) is a transcendental equation system which can be solved graphically [121].

To illustrate how the ε_r and σ can be obtained, we assume that the OUT is the sea water in a sphere with a radius $r = 9.7$ cm, the sea water relative permittivity is $\varepsilon_r = 74$, and the conductivity is $\sigma = 3.53$ S/m, and we assume that they do not change over the frequency of interest. A plastic spherical container is used to ensure that the water is in a spherical shape. The effect of the container will be discussed later, and here we assume that it has the same material property as air. Figure 7.91 gives the ACS values calculated from the Mie series.

If we sweep the ε_r and the σ at each given frequency, the ACS value can be calculated using (7.119) and is shown in Figure 7.92. For each frequency we can find a contour line with the corresponding value in Figure 7.91. Next, we overlay all these contour lines together; they will all cross a common point, which is the permittivity and conductivity of the OUT. Figure 7.93 shows the overlay plot of the contour lines at all selected frequencies. As can be seen all the contour lines share one common point (74, 3.53), which is exactly the point corresponding to the relative permittivity and conductivity of the OUT.

It is important to note that in the ideal situation the measurement uncertainty is zero, which means the ACS value at each frequency point is exactly measured. In reality, all the measurement results have uncertainty, and the contour lines may not always cross a common point, thus there is an uncertainty issue. Since the contour lines can be extracted from measurement as shown in Figure 7.92 for each frequency, an evaluation function can be defined to illustrate how close the permittivity and conductivity test points are to the contour lines. We define the distance as

$$d_i = \min \sqrt{\left[\varepsilon_r - C(x,y)_x \right]^2 + \left[10\log \frac{\sigma}{C(x,y)_y} \right]^2} \qquad (7.123)$$

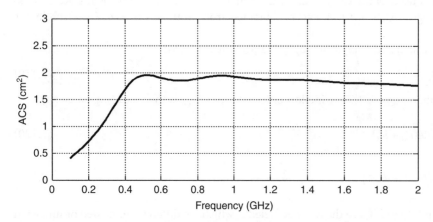

Figure 7.91 ACS value with different frequencies.

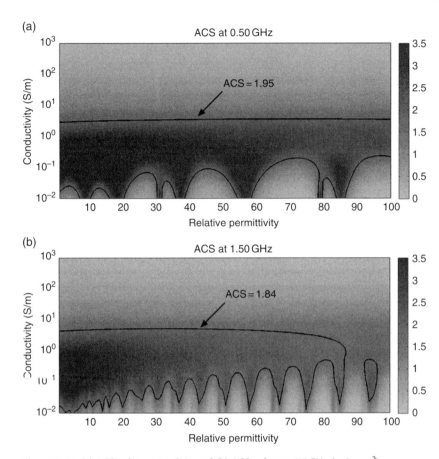

Figure 7.92 (a) ACS value at 0.5 GHz and (b) ACS value at 1.5 GHz (unit: cm^2).

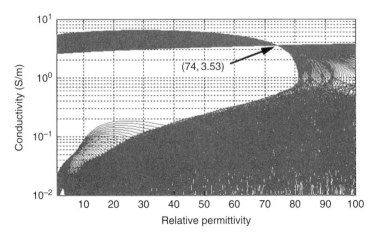

Figure 7.93 Overlay of all the contour lines.

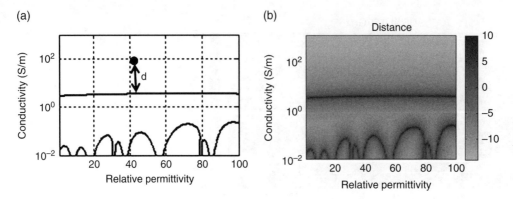

Figure 7.94 (a) Geometric view of distance definition and (b) distance plot in dB.

where $C(x, y)$ is the contour line equation, which is described numerically by the coordinates $C(x, y)_x$ and $C(x, y)_y$. Since the logarithmic axis is used for the σ value, the distance in logarithmic axis is defined and min means the nearest distance is chosen. A geometric view of (7.123) is shown in Figure 7.94a. As can be seen, for each grid point a value can be found to illustrate the distance between the grid point and the contour line. 100×100 points in the ε_r axis and the σ axis are used to calculate the distance; we use $-10 \log d_i$ to show the distance in dB scale in Figure 7.94b. The same procedure can be applied to all the contour lines for all the frequency points. The average distance can be defined as

$$d_{\mathrm{avr}} = \frac{1}{N} \sum_{i=1}^{N} d_i \tag{7.124}$$

If 100 frequency points are chosen in the range of 200 MHz–1.5 GHz to calculate the average distance using (7.124), $-10 \log d_{\mathrm{avr}}$ in dB unit is shown in Figure 7.95. We can consider d_{avr} as the evaluation function which considers all the contour lines for all the measured frequencies. It is easy to see that the point (74, 3.53) is the nearest point to all the contour lines. It can also be observed that for a given uncertainty, e.g. $-10 \log d_{\mathrm{avr}} > 5$ dB, the trust region of permittivity and conductivity can be easily found. Like the PDF which is used to describe the random variables, this evaluation function explains the measurement results by using the measurement data, which makes it self-explanatory.

Normally, the measured data is not as smooth as that shown in Figure 7.91. Suppose we have measured ACS values superimposed with random noise; the noise follows Gaussian distribution with mean value 0 and standard deviation 0.05 as shown in Figure 7.96. The same procedure can be applied as well and the result is shown in Figure 7.97. The trust region with different threshold values can also be easily obtained. Because of the noise, the result is less significant than in Figure 7.95. Another peak point at the left corner actually means that the object with smaller relative permittivity and conductivity may have similar results.

The container only affects the Mie coefficients in (7.120) and (7.121). We assume that the outer radius of the container is 10 cm, the relative permittivity of the container is 4, and the conductivity is 0. The ACS values are calculated and shown in Figure 7.98. Similarly, the reference ACS value of different permittivity and conductivity needs to be updated. Figure 7.92b becomes Figure 7.99. The other procedures are the same. As can be seen, the container does not affect the measurement method, but the permittivity and conductivity of the container need to be known in advance.

In this section a new method has been proposed to measure the permittivity and conductivity of a spherical object. Compared with the existing method [111], the frequency dependency of ACS was considered which has enabled the determination of the absolute values of *both* relative permittivity and conductivity. The effect of the

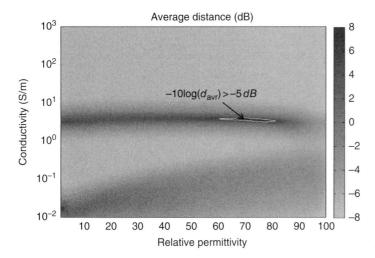

Figure 7.95 Average distance for all the frequency points and the trust region.

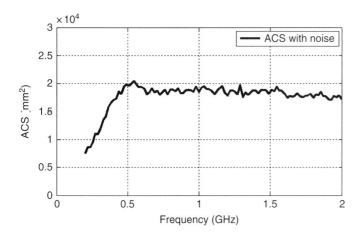

Figure 7.96 ACS value superimposed with random noise.

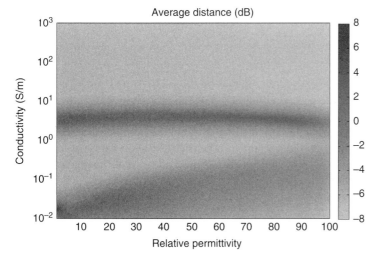

Figure 7.97 Average distance for all the frequency points (consider random noise).

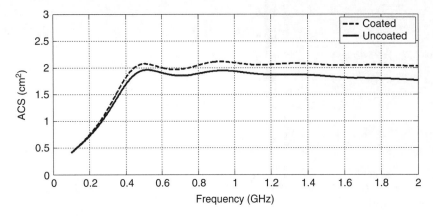

Figure 7.98 ACS comparison of coated and uncoated spheres.

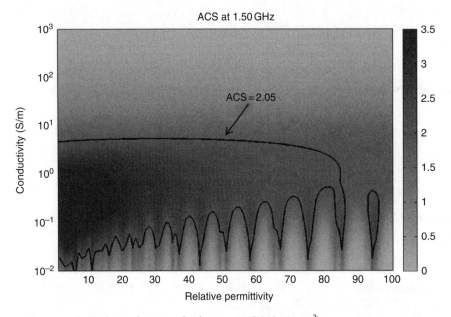

Figure 7.99 ACS value with a coated sphere at 1.5 GHz (unit: cm^2).

container has been discussed and did not affect the proposed method. For the practical measurement, noise may interfere with the results, an evaluation function has been defined to treat the noisy data, and PDF-like results for the permittivity and conductivity have been given. The distribution of the permittivity and conductivity has made the results self-explanatory. It was also important to note that when the data was interfered by noise, the trust region may become not unique, which would reduce the significance of the results. In this case, prior knowledge may be needed to help to choose the correct trust region.

In practice, when the conductivity is large, as can be seen in Figure 7.92, the ACS value will become less dependent on ε_r, which will increase the uncertainty for the measured ε_r. Special care should be taken when the ACS is not in the measurable range of the RC [122], as when the measurement uncertainty (noise) is too large the results will obviously become insignificant.

7.9.4   Material Shielding Effectiveness

According to [1, 123], compared with conventional definitions, a better way to define the shielding effectiveness (SE) of a material sample is to use the averaged transmission cross section (TCS) (the SE of a box or cavity will be discussed in Section 7.10). The SE is defined as [1, 123]

$$SE = 10\log_{10}\frac{\langle P_{\mathrm{t,ns}}\rangle/\langle S_{\mathrm{ns}}^{\mathrm{inc}}\rangle}{\langle P_{\mathrm{t,s}}\rangle/\langle S_{\mathrm{s}}^{\mathrm{inc}}\rangle} \tag{7.125}$$

where $\langle P_{\mathrm{t,s}}\rangle$ is the averaged power transmitted through the aperture with a sample, $\langle P_{\mathrm{t,ns}}\rangle$ is the averaged power transmitted through the same aperture with no sample (open aperture), $\langle S_{\mathrm{s}}^{\mathrm{inc}}\rangle$ and $\langle S_{\mathrm{ns}}^{\mathrm{inc}}\rangle$ are, respectively, the scalar power densities incident on the aperture with and without the sample. Since the average transmitted power and the averaged incident power density can be related by using the averaged cross section [1, 123]:

$$\langle P_{\mathrm{t,s}}\rangle = \langle \sigma_{\mathrm{t,s}}\rangle\langle S_{\mathrm{s}}^{\mathrm{inc}}\rangle \ \text{ and } \langle P_{\mathrm{t,ns}}\rangle = \langle \sigma_{\mathrm{t,ns}}\rangle\langle S_{\mathrm{ns}}^{\mathrm{inc}}\rangle \tag{7.126}$$

where $\langle \sigma_{\mathrm{t,s}}\rangle$ and $\langle \sigma_{\mathrm{t,ns}}\rangle$ are the averaged TCSs of the aperture (over all incidence angles and polarisations) with and without the sample, respectively. Thus (7.125) becomes [1, 123]

$$SE = 10\log_{10}\frac{\langle \sigma_{\mathrm{t,ns}}\rangle}{\langle \sigma_{\mathrm{t,s}}\rangle} \tag{7.127}$$

which is a function of the material under test.

To measure the SE of a sample, nested RCs, as shown in Figure 7.100, are normally used to measure the average TCS of the aperture with and without the sample under test. From (5.120) we have [1, 123]

$$\langle \sigma_{\mathrm{t,s}}\rangle = \frac{\langle S_{\mathrm{in,s}}\rangle}{\langle S_{\mathrm{o,s}}\rangle}\frac{4\pi V}{\lambda Q_{\mathrm{in,s}}}$$

$$\langle \sigma_{\mathrm{t,ns}}\rangle = \frac{\langle S_{\mathrm{in,ns}}\rangle}{\langle S_{\mathrm{o,ns}}\rangle}\frac{4\pi V}{\lambda Q_{\mathrm{in,ns}}} \tag{7.128}$$

where $\langle S_{\mathrm{in,s}}\rangle$ and $\langle S_{\mathrm{in,ns}}\rangle$ are the averaged scalar power densities in the inner chamber with and without the sample, $\langle S_{\mathrm{o,s}}\rangle$ and $\langle S_{\mathrm{o,ns}}\rangle$ are the averaged scalar power densities in the outer chamber with and without the sample, λ is the wavelength, and V is the volume of the inner chamber. From (5.99), the average received power by an ideal antenna (well matched, 100% radiation efficiency) is

$$\langle P_r\rangle = \frac{1}{2}\langle S\rangle\frac{\lambda^2}{4\pi} \tag{7.129}$$

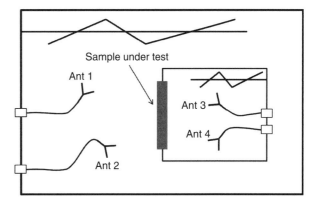

Figure 7.100 SE measurement of a material sample in a nested chamber.

thus (7.127) becomes

$$SE = 10\log_{10} \frac{\frac{\langle S_{\text{in,ns}} \rangle}{\langle S_{\text{o,ns}} \rangle} \frac{4\pi V}{\lambda Q_{\text{in,ns}}}}{\frac{\langle S_{\text{in,s}} \rangle}{\langle S_{\text{o,s}} \rangle} \frac{4\pi V}{\lambda Q_{\text{in,s}}}} = 10\log_{10} \frac{\langle P_{\text{r,in,ns}} \rangle \langle P_{\text{r,o,s}} \rangle Q_{\text{in,s}}}{\langle P_{\text{r,in,s}} \rangle \langle P_{\text{r,o,ns}} \rangle Q_{\text{in,ns}}} \tag{7.130}$$

From Hill's equation [1], the Q factors can be expressed as:

$$Q_{\text{in,s}} = \frac{16\pi^2 V}{\lambda^3} \frac{\langle P_{\text{rQ,in,s}} \rangle}{P_{\text{Tx,in,s}}}$$

$$Q_{\text{in,ns}} = \frac{16\pi^2 V}{\lambda^3} \frac{\langle P_{\text{rQ,in,ns}} \rangle}{P_{\text{Tx,in,ns}}} \tag{7.131}$$

where $\langle P_{\text{rQ,in,s}} \rangle$ is the average measured power in the inner chamber with a sample in the aperture for a transmitting antenna located in the inner chamber with an output power $P_{\text{Tx,in,s}}$, and we have similar definitions of $\langle P_{\text{rQ,in,ns}} \rangle$ and $P_{\text{Tx,in,ns}}$. The SE in (7.130) can be expressed as:

$$SE = 10\log_{10} \frac{\langle P_{\text{r,in,ns}} \rangle \langle P_{\text{r,o,s}} \rangle \langle P_{\text{rQ,in,s}} \rangle P_{\text{Tx,in,ns}}}{\langle P_{\text{r,in,s}} \rangle \langle P_{\text{r,o,ns}} \rangle \langle P_{\text{rQ,in,ns}} \rangle P_{\text{Tx,in,s}}} \tag{7.132}$$

If a VNA is used to perform the measurement in Figure 7.100, (7.132) can be expressed using measured S-parameters:

$$SE = 10\log_{10} \frac{\langle |S_{31\text{ns}}|^2 \rangle \langle |S_{21\text{ws}}|^2 \rangle \langle |S_{34\text{ws}}|^2 \rangle}{\langle |S_{21\text{ns}}|^2 \rangle \langle |S_{31\text{ws}}|^2 \rangle \langle |S_{34\text{ns}}|^2 \rangle} \tag{7.133}$$

We use 'ws' to represent 'with sample' to avoid the confusion with the 'stirred part', and 'ns' represents the measured S-parameters with no sample. If the antennas are not ideal, the input S-parameters and the radiation efficiency need to be corrected by using

$$\langle |S_{\text{ab}}|^2 \rangle \Rightarrow \frac{\langle |S_{\text{ab}}|^2 \rangle}{\left(1 - |\langle S_{\text{aa}} \rangle|^2\right)\left(1 - |\langle S_{\text{bb}} \rangle|^2\right)\eta_{\text{a}}\eta_{\text{b}}} \tag{7.134}$$

where a and b are the port indices. In this case, (7.133) still keeps the same form, if we assume the averaged input S-parameters and the radiation efficiency are the same for the scenarios with and without the sample under test. More details and measurement results are given in [1, 123].

7.10 Cavity Shielding Effectiveness Measurement

To protect an electronic device from unexpected signals or to reduce the emission of the device itself, an enclosure is normally employed to protect it. The shielding effectiveness is an important figure of merit to characterise the electromagnetic isolation performance of an enclosure. In this section, we introduce two methods to measure the SE of a cavity: the nested RC method and the one-antenna method.

Generally, the SE can be measured by following the Institute of Electrical and Electronics Engineers (IEEE) standard 299 [2] or using a nested RC [124], as shown in Figure 7.101. Suppose the radiation efficiency of antennas are high, the SE value can be obtained as

$$SE_{\text{RC}}(\text{dB}) = 10\log_{10} \frac{\langle |S_{32}|^2 \rangle / \left(1 - |\langle S_{22} \rangle|^2\right)}{\langle |S_{31}|^2 \rangle / \left(1 - |\langle S_{11} \rangle|^2\right)} \tag{7.135}$$

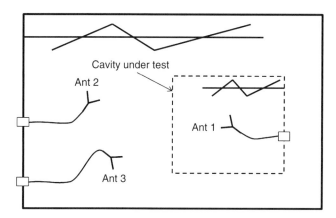

Figure 7.101 SE measurement of a cavity in a nested chamber.

where $\langle\cdot\rangle$ means the average value of S-parameters with any stirring method (mechanical stir, frequency stir, source stir, etc.).

For an electrically large well-stirred enclosure, it has been shown that the SE (defined by using the ratio of the power density outside the cavity to the power density inside the cavity) is related to the TCS and the Q factor in (5.121):

$$SE(\text{dB}) = 10\log_{10}\frac{S_i}{S_c} = 10\log_{10}\frac{4\pi V}{\langle\sigma_{TCS}\rangle\lambda Q_{\text{tot}}} \tag{7.136}$$

where S_i is the power density of the uniformly random incident plane wave (external), S_c is the power density inside the cavity (internal), V is the volume of the cavity, and λ is the wavelength of the frequency of interest. $\langle\sigma_{TCS}\rangle$ is the averaged TCS of the apertures. Q_{tot} is the total Q factor of the enclosure, which can be expressed as [1]

$$Q_{\text{tot}}^{-1} = Q_{\text{surf}}^{-1} + Q_{ACS}^{-1} + Q_{\text{ant}}^{-1} + Q_{TCS}^{-1} \tag{7.137}$$

where Q_{surf} is the contribution from the loss of the walls of the RC, Q_{ACS} is the contribution from the lossy object in the RC, Q_{ant} is the contribution from the antennas in the RC, and Q_{TCS} is the contribution from the TCS of the aperture.

If we assume that the aperture can be properly covered, the contribution from the aperture can be identified by using the covered and uncovered Q factors. The measurement scenarios are shown in Figure 7.102. Let Q_{totu} be the total Q factor for the uncovered EUT and Q_{totc} be the total Q factor for the covered EUT (assume the aperture is perfectly covered, $\langle\sigma_{TCS}\rangle = 0$). From (7.137) Q_{TCS} can be obtained as [125]

$$Q_{TCS}^{-1} = Q_{\text{totu}}^{-1} - Q_{\text{totc}}^{-1} \tag{7.138}$$

Thus (7.136) can be written as

$$SE_Q(\text{dB}) = 10\log_{10}\frac{\left(Q_{\text{totu}}^{-1} - Q_{\text{totc}}^{-1}\right)^{-1}}{Q_{\text{totu}}} \tag{7.139}$$

thus the SE is now only determined by the Q factors.

More specifically, the Q factor can be obtained from the FD and the TD. In the TD, it has been shown that $Q = \omega\tau$, ω is the angular frequency, and τ is the decay time constant of the EUT (the same as the definition of τ_{RC} used in the antenna radiation efficiency measurement). (7.139) can be further expressed as:

$$SE_Q(\text{dB}) = SE_{TD}(\text{dB}) = 10\log_{10}\frac{\tau_c}{\tau_c - \tau_u} \tag{7.140}$$

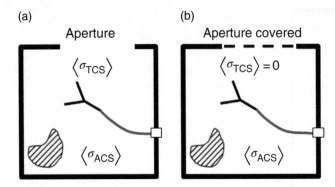

Figure 7.102 (a) Enclosure with aperture uncovered and (b) enclosure with aperture covered.

where τ_c is the decay time constant of the covered EUT and τ_u corresponds to the decay time constant of the uncovered EUT. Both τ_c and τ_u can be obtained from the direct measurement in the TD or from the IFFT of the S-parameters measured in the FD (Section 7.2). Note that τ is the average decay time, which means the waves interact with the EUT uniformly. Imagine a high gain antenna with main lobe directs to a large aperture: the antenna cannot even see the cavity. In this case, τ cannot be accurately extracted. It is interesting to examine the insight of the SE from the TD. The SE value is only related to the decay time constant and is independent of the volume of the EUT. As expected, when the EUT is well shielded, $\tau_u \rightarrow \tau_c$, $SE \rightarrow 0$; when the EUT is completely open and the antenna radiates freely, $\tau_u \rightarrow 0$, $SE \rightarrow 0$.

In the FD, from Hill's equation, we have

$$Q_{FD} = \frac{16\pi^2 V \left\langle |S_{21,s}|^2 \right\rangle}{\lambda^3 \eta_{1tot} \eta_{2tot}} \tag{7.141}$$

where $S_{*,s}$ is the stirred part of S-parameters (no LoS), and η_{1tot} and η_{2tot} are the total efficiencies (including mismatch) of transmitting and receiving antennas in the cavity. When the EUT is well stirred, the enhanced backscatter constant is

$$e_b = \frac{\sqrt{\left\langle |S_{11,s}|^2 \right\rangle \left\langle |S_{22,s}|^2 \right\rangle}}{\left\langle |S_{21,s}|^2 \right\rangle} = 2 \tag{7.142}$$

Like the one-antenna method in Section 7.6.2, assuming we have two identical antennas in the EUT (not necessary in practice), we have $\left\langle |S_{11,s}|^2 \right\rangle = \left\langle |S_{22,s}|^2 \right\rangle = 2\left\langle |S_{21,s}|^2 \right\rangle$, thus (7.139) can be expressed as

$$SE_Q(dB) = SE_{FD}(dB) = 10\log_{10} \frac{\left\langle |S_{11,sc}|^2 \right\rangle}{\left\langle |S_{11,sc}|^2 \right\rangle - \left\langle |S_{11,su}|^2 \right\rangle} \tag{7.143}$$

where $\left\langle |S_{11,sc}|^2 \right\rangle$ and $\left\langle |S_{11,su}|^2 \right\rangle$ are the stirred parts of the S-parameters with covered and uncovered EUT. It should be noted that (7.143) requires $e_b = 2$ for both covered and uncovered EUT. In practice, e_b could depend on the stirring effectiveness. When the aperture is large and the frequency is high, the cavity could not be well stirred and $e_b \neq 2$ [104]. To measure the stirred part of S-parameters, some kind of stirring mechanism should be involved. Normally, it is not easy to stir the EUT mechanically, but source stir and frequency stir can be easily applied, as shown in the measurement process.

Measurements were conducted to verify the method. The nested RC method in (7.135) was first used to conduct the measurement, and the results from a nested RC were compared with that from the one-antenna method. The measurement setup is shown in Figure 7.103. Because there was no stirrer in the EUT, a nylon wire was fixed with antenna 1 to realise a simple source stir (by pulling the nylon wire manually), as shown in Figure 7.103b.

A wideband monopole antenna was used as antenna 3, and two double-ridged waveguide horn antennas were used as antenna 2 (Rohde & Schwarz HF 906) and antenna 1 (SATIMO SH 2000). A VNA was employed, and after cables were connected to the VNA and two-port calibration was conducted, antenna 2 was first connected to port 2 of the VNA and antenna 1 was connected to a 50 Ω load. Antenna 2 was then connected to a 50 Ω load and antenna 1 was connected to port 2 of the VNA. All the S-parameters were recorded and saved for post processing. In this measurement we used 10 001 points in the range of 2.8–5.2 GHz and 100 rotation steps with a step size of 3°. The size of the RC is 3.6 m $\times$ 4 m $\times$ 5.8 m and the size of the EUT is 0.8 m $\times$ 1 m $\times$ 1.1 m, which is about 8λ $\times$ 10λ $\times$ 11λ at 3 GHz. According to Weyl's law, the mode number is around 7300, which is large enough. Two apertures of different sizes (shown in Figure 7.104) were tested, and a foam block was used to control the aperture size and ensure its stability in the measurement process. When using the one-antenna method, to reduce the effect of the RC on S_{11}, the doors of the RC were kept open and only S_{11} is recorded.

To obtain the τ in (7.140) we need to measure the S-parameters in the FD and then apply the IFFT to the measured S-parameters. 10 001 points were collected in the range of 2.8–5.2 GHz, which gave a resolution of 0.2 ns in the TD with a length of 4166 ns. Because τ is frequency dependent, a five-order elliptic BPF with 200 MHz bandwidth was used to filter the S-parameters, as shown in Figure 7.105a, then the IFFT was applied to the filtered S-parameters. Because the TD power in the cavity decays exponentially ($\exp(-t/\tau)$), the least-squares fit was applied to $\ln(power)$ to extract the slope ratio k, and $\tau = -1/k$ can be extracted. To avoid the fit error caused by the noise level, only part of the signal was used for the least-squares fit, as shown in Figure 7.105b. By sweeping the centre frequency of the filter, τ with different centre frequencies were obtained.

To measure the stirred part of the S-parameters in (7.143), in practice both source stir and frequency stir are used: for each frequency, S-parameters with 10 source potions are used and a window of 1000 sample points (200 MHz) is used to apply the frequency stir. The measured decay time constants for the covered and uncovered apertures are shown in Figure 7.106a. For each case, 10 source positions are used and the averaged values are used to calculate the SE_{TD}. The measured SEs of the two apertures using (7.135), (7.140), and (7.143) are given in Figure 7.106b,c.

The SE values are relatively small due to the aperture sizes and a good agreement between SE_{TD} and SE_{RC} is obtained. However, SE_{FD} shows peaks and troughs and has some deviations from the other two results at some frequencies, which means $e_b \neq 2$ at these frequencies (not well stirred). Limitations and more discussions of the one-antenna method of SE measurement can be found in [125].

(a)　　　　　　(b)

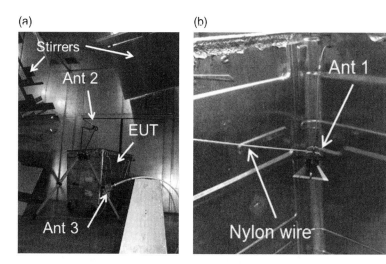

Figure 7.103 SE measurement using the nested RC: (a) measurement scenario and (b) antenna in the EUT.

(a)

(b)

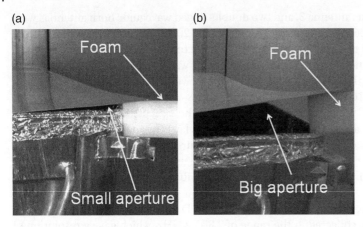

Figure 7.104 EUT with different apertures: (a) a small aperture and (b) a big aperture.

(a)

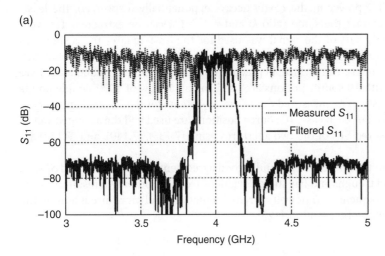

(b)

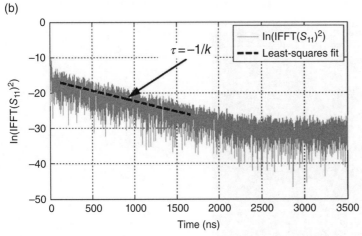

Figure 7.105 τ extraction procedure: (a) measured S_{11} and filtered S_{11} (EUT with big aperture) and (b) TD response: $\ln[\text{IFFT}(S_{11})^2]$ and least-squares fit.

(a)

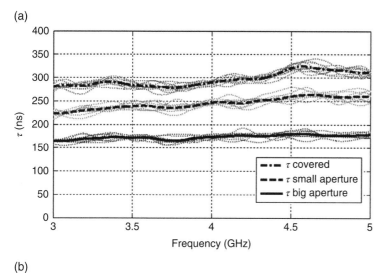

(b)

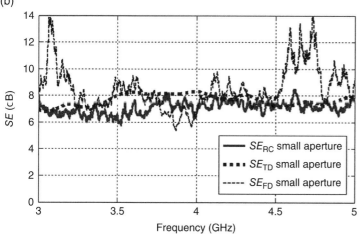

(c)

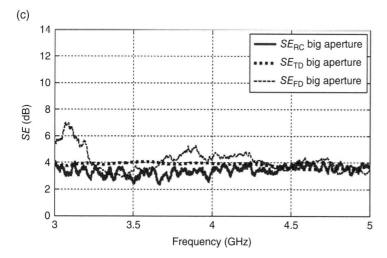

Figure 7.106 SE measurement results using three equations: (a) averaged τ (solid lines) and τ with different source positions (light dots), (b) SE values of EUT with a small aperture, and (c) SE values of EUT with a big aperture.

7.11 Volume Measurement

It is interesting to note that the method used to measure the radiation efficiency of antennas can be used to measure/sense the volume of a large cavity [126]. This could be useful for some applications, such as shipbuilding industry or environment sensing [127]. From Hill's equation, if we solve for the cavity volume V, we have

$$V = \frac{\lambda^2 c_0 \langle \tau \rangle \eta_{1tot} \eta_{2tot}}{8\pi \langle |S_{21,s}|^2 \rangle} \tag{7.144}$$

where $S_{21,s}$ is the stirred part of the transmission coefficient, $\langle \cdot \rangle$ means the average value using any stirring method (e.g. mechanical stir, frequency stir, polarisation stir, source stir, etc.), and λ is the free-space wavelength. c_0 is the speed of light in free space (3×10^8 m/s), and η_{1tot} and η_{2tot} are the total efficiencies (having taken the loss and impedance matching into account) of antenna 1 and antenna 2, respectively. Here we use $\langle \tau \rangle$ instead of $\langle \tau_{RC} \rangle$ because we obtain τ in different scenarios and use the average value. The scenario is not necessary in an RC so we omit the subscript.

If the cavity is well stirred, by using the enhanced backscatter constant we have the one-antenna approach:

$$V = \frac{\lambda^2 c_0 \langle \tau \rangle \, \eta_{itot}^2}{4\pi \langle |S_{ii,s}|^2 \rangle}, \; i = 1 \text{ or } 2 \tag{7.145}$$

The measurement scenario is the same as the antenna efficiency measurement shown in Figure 7.107. Two horn antennas are used as antenna 1 (Rohde & Schwarz HF 906) and antenna 2 (SATIMO SH2000). One hundred stirrer positions with 3.5°/step are used. At each stirrer position, 10 001 frequency points are collected in the frequency range of 2.8 GHz to 4.2 GHz. The volume of the RC is 3.6 m × 4 m × 5.8 m = 83.52 m³. The product of the total efficiency of two antennas can be measured using the two-antenna method in a cavity with known volume and is given in Figure 7.108.

To validate the method, we change the environment: load the RC with RAMs, and open the door with 45° and 90° (shown in Figure 7.109b,c respectively). In practice, a cavity may have an entrance without a conducting door, so we open the door to emulate an imperfect cavity.

The techniques in Section 7.2 are used to obtain the τ values in different scenarios, and the measured values are shown in Figure 7.110. The measured $\langle |S_{21,s}|^2 \rangle$ is given in Figure 7.111. Finally, we apply (7.144) to obtain the volume of the cavity, shown in Figure 7.112. The whole measurement procedure is also repeated by using the one-antenna format in (7.145), and results are shown in Figures 7.113 and 7.114, respectively. It should be noted that in practice we may not have stirrers to change the field distribution in the cavity under test, but the source stir and the frequency stir are also applicable. To verify this, we keep the stirrers steady. Mount antenna 2 on a rotation platform (shown in Figure 7.115) and 100 platform rotation angles with 3.5°/step are used. The whole calibration and measurement procedure is repeated, and the results are given in Figures 7.116 and 7.117. Note antenna 1 is not rotated; the one-antenna method from antenna 1 is not available.

A fast measurement method to measure the volume of a large cavity is proposed in this section. 'Large' means the cavity is large compared with the wavelength of the used electromagnetic/acoustic wave. The method is non-destructive and the system can be assembled as a portable volume probe to measure large cavities like ship, granary, and storehouse. The measurement can be real time and the shape of the cavity can be arbitrary (complex shape has better field uniformity). After the measurement system is calibrated, only τ and S-parameters need to be measured. The measured volume values in the frequency range of 3–4 GHz are summarised in Table 7.5.

It is interesting to note that, although the proposed method does not depend on the Q factor of the cavity, results from a small Q cavity tend to have a larger standard deviation. This is easy to understand because when the Q factor reduces, the environment degrades from the rich isotropic multipath environment (RIMP) environment to the free space, and the transmission coefficient becomes highly dependent on the antenna patterns, orientations, and the distance between antennas (from Hill's equation to the Friis equation). Thus, it could be

Figure 7.107 Volume measurement in the RC.

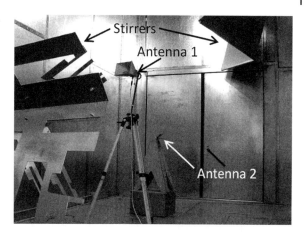

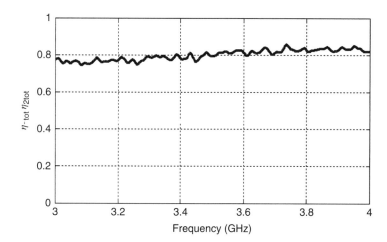

Figure 7.108 Calibrated $\eta_{1tot}\,\eta_{2tot}$ in the calibration process.

hard to realise a well-stirred low Q cavity and the measurement error could be large if the measurement sample number is small (or a narrow band antenna is used). A wideband antenna/broadband frequency sweep is better.

The relative error is calculated from the deviation from 83.52 m³. It is found that when the door is open (from 45° to 90°), the measured volume is larger than the real value. This is because when the door is open, we have an equivalent larger cavity. This means the uncovered apertures could make the measured volume larger than the real value. In this measurement the area of the door is 1.94 m² (1.6% of the surface area of the inner walls). Theoretically, the RAMs have occupied a certain volume, as we can see from the results of the source stir in Table 7.5, and the measured volumes for the RC with loaded RAMs are indeed smaller, this phenomenon is not significant for the results of the mechanical stir. This may be due to the total volume of RAMs used being too small (~0.1 m³) to be sensed accurately.

For the one-antenna method in (7.145), the condition of $e_b = 2$ should be met. If we compare the results from antenna 1 and antenna 2 in the mechanical stir, it can be found that the results from antenna 2 have a larger error than antenna 1, which is due to the error from e_b because the S-parameters are recorded in the same environment and system configuration.

There is an equation similar to Hill's equation in acoustics and we can also use sound/supersonic waves to detect the volume of a cavity (no longer limited to metallic cavities). A more general form can be written as

(a)

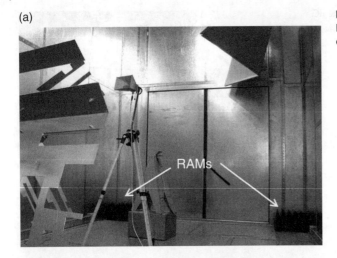

Figure 7.109 Measurement scenarios: (a) RC loaded with RAMs, (b) RC with 45° door open, and (c) RC with 90° door open.

(b)   (c)

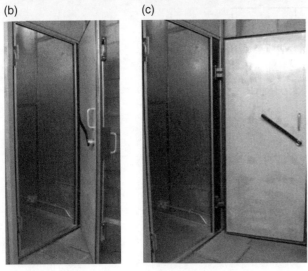

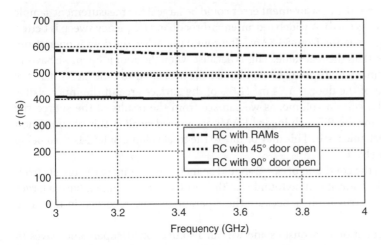

Figure 7.110 Measured τ in three scenarios.

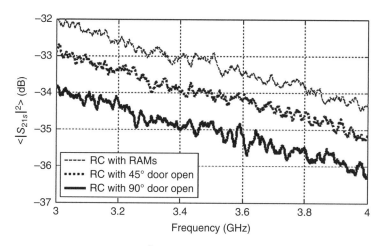

Figure 7.111 Measured $\langle |S_{21,s}|^2 \rangle$ in three scenarios.

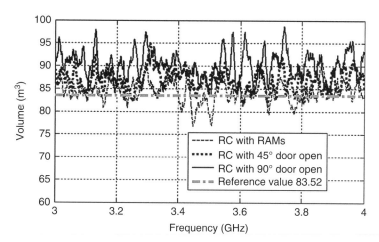

Figure 7.112 Measured volume value in three scenarios using the two-antenna method.

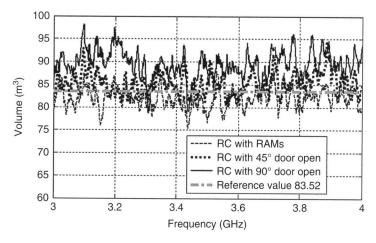

Figure 7.113 Measured volume value in three scenarios using antenna 1 only.

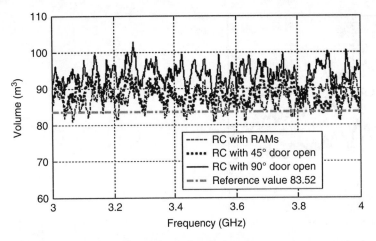

Figure 7.114 Measured volume value in three scenarios using antenna 2 only.

Figure 7.115 Antenna 2 mounted on a rotation platform as a source stirrer.

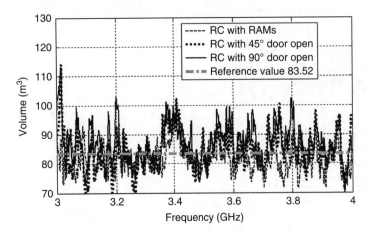

Figure 7.116 Measured volume value in three scenarios using source stir and two-antenna method.

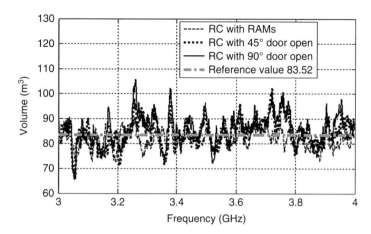

Figure 7.117 Measured volume value in three scenarios using source stir and antenna 2 only.

Table 7.5 Summarised measurements results.

Measurement method		Scenarios	Mean (relative error)	Standard deviation
Mechanical stir	1 and 2[a]	RAMs[a]	84.53 (1.2%)	2.34
		45	86.90 (4.0%)	2.33
		90	90.19 (8.0%)	2.92
	1	RAMs	82.08 (−1.7%)	2.19
		45	85.23 (2.0%)	2.48
		90	89.63 (7.3%)	3.03
	2	RAMs	86.85 (4.0%)	2.71
		45	88.79 (6.3%)	2.57
		90	93.86 (12.4%)	2.84
Source stir	1 and 2[a]	RAMs	82.03 (−1.8%)	4.97
		45	84.34 (1.0%)	6.34
		90	86.87 (4.0%)	6.76
	2	RAMs	82.03 (−1.8%)	3.67
		45	84.42 (1.1%)	5.40
		90	85.66 (2.6%)	6.05

[a] 1 and 2, using the two-antenna method; 1, using antenna 1 only; 2, using antenna 2 only; RAMs, the RC is loaded with RAMs; 45, the door is open at 45°; 90, the door is open at 90°.

$$V = CF \frac{\lambda^2 \langle \tau \rangle}{\langle T_s \rangle} \tag{7.146}$$

where CF is the calibration factor, which can be determined by using a cavity with known volume, λ is the wavelength (electromagnetic or acoustic wave), $\langle T_s \rangle$ is the averaged stirred power transfer function, and $\langle \tau \rangle$ is the averaged decay time. In room acoustics, this can be converted from the 60 dB reverberation time (RT_{60}) with $\tau = RT_{60}/(6 \ln 10)$ [128]. It should be noted that the method has its own limitations. Although the method is independent of Q factor, it is hard to realise a statistical uniform field in low Q cavities. Another issue is if the cavity is partially filled with liquid or dielectric materials that either absorb or reflect waves completely (inhomogeneous cavity), or the wave velocity is changed, which could increase the measurement error. How the inhomogeneous materials affect the results could be very complex. Also if the shape of the cavity is so irregular that the power density is no longer uniform in the cavity, the volume cannot be measured accurately.

7.12   Summary

In this long chapter, a range of applications of the RC (including antenna measurements for radiation efficiency, radiation pattern and DG, measurements of Doppler shift, ACS, material characterisation, shielding effectiveness, and volume measurement) have been introduced and discussed. They are based on the latest developments and advancements using chamber parameters such as Q factor, chamber decay constant, chamber transfer function, etc. The signal in both the FD and the TD has been examined carefully and used to obtain the measurement parameters. With the development of statistical electromagnetics, it is believed that the use of RCs will become increasingly important, and more applications will be developed in the future.

References

1 Hill, D.A. (2009). *Electromagnetic Fields in Cavities: Deterministic and Statistical Theories*. Wiley.
2 IEC 61000-4-21. (2011). Electromagnetic compatibility (EMC) – Part 4-21: Testing and measurement techniques – Reverberation chamber test methods. IEC Standard, Ed 2.0, January.
3 Krauthauser, H.G. (2007). On the measurement of total radiated power in uncalibrated reverberation chambers. *IEEE Transactions on Electromagnetic Compatibility* 49 (2): 270–279.
4 Ladbury, J.M., Johnk, R.T., and Ondrejka, A.R. (1996). *Rapid Evaluation of Mode-Stirred Chambers Using Impulsive Waveforms*. Boulder, CO, NIST Tech. Note 1381: NIST.
5 Hill, D.A., Ma, M.T., Ondrejka, A.R. et al. (1994). Aperture excitation of electrically large, lossy cavities. *IEEE Transactions on Electromagnetic Compatibility* 36 (3): 169–178.
6 Shah, S.M.H.A. (2011). Wireless channel characterization of the reverberation chamber at NIST, MSc thesis, Department of Signals and System, Chalmers University of Technology, Gothenburg.
7 Xu, Q., Huang, Y., Xing, L., and Tian, Z. (2016). Extract the decay constant of a reverberation chamber without satisfying Nyquist criterion. *IEEE Microwave and Wireless Components Letters* 26 (3): 153–155.
8 Chen, X., Kildal, P.-S., Orlenius, C., and Carlsson, J. (2009). Channel sounding of loaded reverberation chamber for over-the-air testing of wireless devices: coherence bandwidth versus average mode bandwidth and delay spread. *IEEE Antennas and Wireless Propagation Letters* 8: 578–681.
9 Arnaut, L.R. (2005). On the maximum rate of fluctuation in mode-stirred reverberation. *IEEE Transactions on Electromagnetic Compatibility* 47 (4): 781–804.
10 Arnaut, L.R. (2007). On the relationship between correlation length and rate of fluctuation of random fields. *IEEE Transactions on Electromagnetic Compatibility* 49 (3): 727–729.
11 Mendes, H.A. (1968). A new approach to electromagnetic field-strength measurements in shielded enclosures, Wescon Technical Papers, Wescon Electronic Show and Convention, Los Angeles.
12 Corona, P., Latmiral, G., Paolini, E., and Piccioli, L. (1976). Use of a reverberating enclosure for measurements of radiated power in the microwave range. *IEEE Transactions on Electromagnetic Compatibility* 18 (2): 54–59.
13 Wilson, P., Koepke, G., Ladbury, J., and Holloway, C.L. (2001). Emission and immunity standards: replacing field-at-a-distance measurements with total-radiated-power measurements, International Symposium on Electromagnetic Compatibility, Montreal, Quebec. vol. 2. pp. 964–969.

14 Xu, Q., Huang, Y., Yuan, S. et al. (2016). Two alternative methods to measure the radiated emission in a reverberation chamber. *International Journal of Antennas and Propagation* 2016, 5291072: 7.

15 Kildal, P.-S., Carlsson, C., and Yang, J. (2001). Measurement of free-space impedances of small antennas in reverberation chambers. *Microwave and Optical Technology Letters* 32 (2): 112–115.

16 Kabacik, P., Byndas, A., Hossa, R., and Bialkowski, M. (2006). A measurement system for determining radiation efficiency of a small antenna, First European Conference on Antennas and Propagation. Nice. pp. 1–6.

17 Pozar, D.M. and Kaufman, B. (1988). Comparison of three methods for the measurement of printed antenna efficiency. *IEEE Transactions on Antennas and Propagation* 36 (1): 136–139.

18 Johnston, R.H. and McRory, J.G. (1998). An improved small antenna radiation-efficiency measurement method. *IEEE Antennas and Propagation Magazine* 40 (5): 40–48.

19 Schantz, H.G. (2002). Radiation efficiency of UWB antennas, IEEE Conference on Ultra Wideband Systems and Technologies. Baltimore, MD. pp. 351–355.

20 Huang, Y. (2015). Radiation efficiency measurements of small antennas. In: *Handbook of Antenna Technologies* (ed. Z. Chen). Springer.

21 Schroeder, W.L. and Gapski, D. (2006). Direct measurement of small antenna radiation efficiency by calorimetric method. *IEEE Transactions on Antennas and Propagation* 54 (9): 2646–2656.

22 Holloway, C.L., Shah, H.A., Pirkl, R.J. et al. (2012). Reverberation chamber techniques for determining the radiation and total efficiency of antennas. *IEEE Transactions on Antennas and Propagation* 60 (4): 1758–1770.

23 Xu, Q., Huang, Y., Zhu, X. et al. (2016). A modified two-antenna method to measure the radiation efficiency of antennas in a reverberation chamber. *IEEE Antennas and Wireless Propagation Letters* 15: 336–339.

24 Boyes, S.J., Zhang, Y., Brown, A.K., and Huang, Y. (2012). A method to de-embed external power dividers in efficiency measurements of all-excited antenna arrays in reverberation chamber. *IEEE Antennas and Wireless Propagation Letters* 11: 1418–1421.

25 Boyes, S. and Huang, Y. (2016). *Reverberation Chambers: Theory and Applications to EMC and Antenna Measurements*. Wiley.

26 Li, C., Loh, T.-H., and Tian, Z. (2017). Evaluation of chamber effects on antenna efficiency measurements using non-reference antenna methods in two reverberation chambers, *IET Microwaves, Antennas & Propagation*, vol.11. pp. 1636–1541.

27 Skrivervik, A. K. (2013)Implantable antennas: The challenge of efficiency, 7th European Conference on Antennas and Propagation (EUCAP). Gothenburg, 8–12 April, pp. 3627–3631.

28 Alrawashdeh, R., Huang, Y., and Cao, P. (2013). A flexible loop antenna for total knee replacement implants in the MedRadio band, Loughborough Antennas and Propagation Conference (LAPC). pp. 225–228.

29 Serafimov, N., Kildal, P.-S., and Bolin, T. (2002). Comparison between radiation efficiencies of phone antennas and radiated power of mobile phones measured in anechoic chambers and reverberation chamber, IEEE Antennas and Propagation Society International Symposium. vol.2. pp. 478–481.

30 Kildal, P.-S., Rosengren, K., Byun, J., and Lee, J. (2002). Definition of effective diversity gain and how to measure it in a reverberation chamber. *Microwave and Optical Technology Letters* 34 (1): 56–59.

31 Rosengren, K. and Kildal, P.S. (2005). Radiation efficiency, correlation, diversity gain and capacity of a six-monopole antenna array for a MIMO system: theory, simulation and measurement in reverberation chamber. *IEE Proceedings - Microwaves, Antennas and Propagation* 152 (1): 7–16.

32 Kildal, P.S. and Rosengren, K. (2004). Correlation and capacity of MIMO systems and mutual coupling, radiation efficiency, and diversity gain of their antennas: simulations and measurements in a reverberation chamber. *IEEE Communications Magazine* 42 (12): 104–112.

33 Mouhamadou, M., Tounou, C.A., Decroze, C. et al. (2010). Active measurements of antenna diversity performances using a specific test-bed, in several environments. *International Journal of RF and Microwave Computer-Aided Engineering* 20: 264–271.

34 Kildal, P.-S. and Rosengren, K. (2003). Electromagnetic analysis of effective and apparent diversity gain of two parallel dipoles. *IEEE Antennas and Wireless Propagation Letters* 2: 9–13.

35 Andersson, M., Orlenius, C., and Franzen, M. (2007). Very fast measurements of effective polarization diversity gain in a reverberation chamber, The Second European Conference on Antennas and Propagation, EuCAP 2007. Edinburgh. pp. 1–4.

36 Holloway, C.L., Hill, D.A., Ladbury, J.M. et al. (2006). On the use of reverberation chambers to simulate a controllable Rician radio environment for the testing of wireless devices. *IEEE Transactions on Antennas and Propagation* 54 (11): 3167–3177.

37 Kurita, D., Okano, Y., Nakamatsu, S., and Okada, T. (2010). Experimental comparison of MIMO OTA testing methodologies, Proceedings of the Fourth European Conference on Antennas and Propagation. Barcelona. pp. 1–5.

38 Hussain, A., Kildal, P.S., and Glazunov, A.A. (2015). Interpreting the total isotropic sensitivity and diversity gain of LTE-enabled wireless devices from over-the-air throughput measurements in reverberation chambers. *IEEE Access* 3: 131–145.

39 Orlenius, C., Kildal, P.S., and G. Poilasne (2005). Measurements of total isotropic sensitivity and average fading sensitivity of CDMA phones in reverberation chamber, IEEE Antennas and Propagation Society International Symposium. vol. 1A, pp. 409–412.

40 Remley, K.A., Dortmans, J., Weldon, C. et al. (2016). Configuring and verifying reverberation chambers for testing cellular wireless devices. *IEEE Transactions on Electromagnetic Compatibility* 58 (3): 661–672.

41 Hussain, A., Por Einarsson, B., and Kildal, P.S. (2015). MIMO OTA testing of communication system using SDRs in reverberation chamber. *IEEE Antennas and Propagation Magazine* 57 (2): 44–53.

42 Hussain, A. and Kildal, P.S. (2014). Estimating TIS of 4G LTE devices from OTA throughput measurements in reverberation chamber, IEEE Conference on Antenna Measurements & Applications (CAMA). Antibes Juan-les-Pins. pp. 1–4.

43 Chen, X., Hussain, A., and Kildal, P.S. (2014). Over-the-air LTE measurements in the reverberation chamber, XXXIth URSI General Assembly and Scientific Symposium (URSI GASS). Beijing. pp. 1–2.

44 Andersson, M., Orlenius, C., and Kildal, P.S. (2008). Three fast ways of measuring receiver sensitivity in a reverberation chamber, International Workshop on Antenna Technology: Small Antennas and Novel Metamaterials. Chiba. pp. 51–54.

45 Alhorr, F., Wiles, M., Fermiñán-Rodríguez, S., and Wehrmann, C. (2013). On the comparison between anechoic and reverberation chambers for wireless OTA testing, 7th European Conference on Antennas and Propagation (EuCAP). Gothenburg. pp. 1852–1856.

46 Genender, E., Holloway, C.L., Remley, K.A. et al. (2010). Simulating the multipath channel with a reverberation chamber: application to bit error rate measurements. *IEEE Transactions on Electromagnetic Compatibility* 52 (4): 766–777.

47 Floris, S.J., Remley, K.A., and Holloway, C.L. (2010). Bit error rate measurements in reverberation chambers using real-time vector receivers. *IEEE Antennas and Wireless Propagation Letters* 9: 619–622.

48 Kildal, P.S. (2007). Overview of 6 years R&D on characterizing wireless devices in Rayleigh fading using reverberation chambers, International Workshop on Antenna Technology: Small and Smart Antennas Metamaterials and Applications. Cambridge. pp. 162–165.

49 Remley, K.A., Fielitz, H., Shah, H.A., and Holloway, C.L. (2011). Simulating MIMO techniques in a reverberation chamber, IEEE International Symposium on Electromagnetic Compatibility. Long Beach, CA. pp. 676–681.

50 Tian, Z., Huang, Y., and Qian, X. (2016). Stirring effectiveness characterization based on Doppler spread in a reverberation chamber, 10th European Conference on Antennas and Propagation (EuCAP). Davos. pp. 1–3.

51 García-Fernández, M.Á., Decroze, C., and Carsenat, D. (2014). Antenna radiation pattern measurements in reverberation chamber using Doppler analysis, IEEE Conference on Antenna Measurements & Applications (CAMA), Antibes Juan-les-Pins. pp. 1–4.

52 García-Fernández, M.Á., Andrieu, G., Decroze, C., and Carsenat, D (2014). Actual antenna radiation pattern measurements in reverberation chamber, XXXIth URSI General Assembly and Scientific Symposium (URSI GASS). Beijing. pp. 1–4.

53 García-Fernández, M.Á., Carsenat, D., and Decroze, C. (2014). Antenna gain and radiation pattern measurements in reverberation chamber using Doppler effect. *IEEE Transactions on Antennas and Propagation* 62 (10): 5389–5394.

54 Jeong, M.H., Park, B.Y., Choi, J.H., and Park, S.O. (2013). Doppler spread spectrum of antenna configurations in a reverberation chamber, Asia-Pacific Microwave Conference Proceedings (APMC). Seoul. pp. 675–677.

55 Karlsson, K., Chen, X., Carlsson, J., and Skårbratt, A. (2013). On OTA test in the presence of Doppler spreads in a reverberation chamber. *IEEE Antennas and Wireless Propagation Letters* 12: 886–889.

56 Chen, X., Kildal, P.S., and Carlsson, J. (2011). Determination of maximum Doppler shift in reverberation chamber using level crossing rate, Proceedings of the 5th European Conference on Antennas and Propagation (EUCAP). Rome. pp. 62–65.

57 Karlsson, K., Chen, X., Kildal, P.S., and Carlsson, J. (2010). Doppler spread in reverberation chamber predicted from measurements during step-wise stationary stirring. *IEEE Antennas and Wireless Propagation Letters* 9: 497–500.

58 Choi, J.H., Lee, J.H., and Park, S.O. (2010). Characterizing the impact of moving mode-stirrers on the Doppler spread spectrum in a reverberation chamber. *IEEE Antennas and Wireless Propagation Letters* 9: 375–378.

59 Hallbjorner, P. and Rydberg, A. (2007). Maximum Doppler frequency in reverberation chamber with continuously moving stirrer, Loughborough Antennas and Propagation Conference. Loughborough. pp. 229–232.

60 Barazzetta, M., Michel, D., and Diamanti, R. (2016). Optimization of 4G wireless access network features by using reverberation chambers: application to high-speed train LTE users, 46th European Microwave Conference (EuMC). London. pp. 719–722.

61 Gradoni, G. and Arnaut, L.R. (2013). Transient evolution of eigenmodes in dynamic cavities and time-varying media, International Symposium on Electromagnetic Theory. Hiroshima. pp. 276–279.

62 Arnaut, L.R. (2016). Pulse jitter, delay spread, and Doppler shift in mode-stirred reverberation. *IEEE Transactions on Electromagnetic Compatibility* 58 (6): 1717–1727.

63 Sorrentino, A., Gifuni, A., Ferrara, G., and Migliaccio, M. (2014). Mode-stirred reverberating chamber Doppler spectra: multi-frequency measurements and empirical model. *IET Microwaves, Antennas & Propagation* 8 (15): 1356–1362.

64 Herbert, S., Wassell, I., Loh, T.H., and Rigelsford, J. (2014). Characterizing the spectral properties and time variation of the in-vehicle wireless communication channel. *IEEE Transactions on Communications* 62 (7): 2390–2399.

65 Wright, C. and Basuki, S. (2010). Utilizing a channel emulator with a reverberation chamber to create the optimal MIMO OTA test methodology, Global Mobile Congress. Shanghai. pp. 1–5.

66 Huitema, L., Reveyrand, T., Mattei, J.-L. et al. (2013). Frequency tunable antenna using a magneto-dielectric material for DVB-H application. *IEEE Transactions on Antennas and Propagation* 61 (9): 4456–4466.

67 Xu, Q., Huang, Y., Zhu, X. et al. (2015). A new antenna diversity gain measurement method using a reverberation chamber. *IEEE Antennas and Wireless Propagation Letters* 14: 935–938.

68 Alja'afreh, S., Huang, Y., and Xing, L. (2014). A compact, wideband and low profile planar inverted-L antenna, The 8th European Conference on Antennas and Propagation (EuCAP 2014). The Hague. pp. 3283–3286.

69 Xu, Q. (2015). Anechoic and Reverberation Chamber Design and Measurements, PhD thesis, Department of Electrical Engineering and Electronics, University of Liverpool, Liverpool.

70 Kildal, P.–.S. (2015). *Foundations of Antenna Engineering: A Unified Approach for Line-of-Sight and Multipath*. Artech House.

71 Chen, X., Kildal, P.S., and Carlsson, J. (2011). Comparisons of different methods to determine correlation applied to multi-port UWB eleven antenna, Proceedings of the 5th European Conference on Antennas and Propagation (EUCAP). Rome. pp. 1776–1780.

72 Pierce, J.N. and Stein, S. (1960). Multiple diversity with nonindependent fading. *Proceedings of the IRE* 48 (1): 89–104.

73 Diallo, A., Thuc, P., Luxey, C. et al. Diversity characterization of optimized two-antenna systems for UMTS handsets. *EURASIP Journal on Wireless Communications and Networking* 2007,: 37574: 1–9.

74 Blanch, S., Romeu, J., and Corbella, I. (2003). Exact representation of antenna system diversity performance from input parameter description. *Electronics Letters* 39 (9): 705–707.

75 Jamaly, N., Kildal, P.S., and Carlsson, J. (2010). Compact formulas for diversity gain of two-port antennas. *IEEE Antennas and Wireless Propagation Letters* 9: 970–973.

76 Dama, Y., Abd-Alhameed, R., Jones, S. et al. An envelope correlation formula for (N, N) MIMO antenna arrays using input scattering parameters, and including power losses. *International Journal of Antennas and Propagation* 2011, 421691: 1–7.

77 Chen, X. Spatial correlation and ergodic capacity of MIMO channel in reverberation chamber. *International Journal of Antennas and Propagation* 2012, 939104: 1–7.

78 Koh, J., De, A., Sarkar, T.K. et al. (2012). Free space radiation pattern reconstruction from non-anechoic measurements using an impulse response of the environment. *IEEE Transactions on Antennas and Propagation* 60 (2): 821–831.

79 Fiumara, V., Fusco, A., Matta, V., and Pinto, I.M. (2011). Free-space antenna field-pattern retrieval in reverberation environments. *IEEE Antennas and Wireless Propagation Letters* 4: 329–332.

80 Besnier, P., Lemonie, C., Sol, J., and Floc'h, J.-M. (2014). Radiation pattern measurements in reverberation chamber based on estimation of coherent and diffuse electromagnetic fields, IEEE Conference on Antenna Measurements & Applications (CAMA), pp.1–4.

81 Lemoine, C., Amador, E., Besnier, P. et al. (2013). Antenna directivity measurement in reverberation chamber from Rician K-factor estimation. *IEEE Transactions on Antennas and Propagation* 61 (10): 5307–5310.

82 García-Fernández, M.Á., Carsenat, D., and Decroze, C. (2013). Antenna radiation pattern measurements in reverberation chamber using plane wave decomposition. *IEEE Transactions on Antennas and Propagation* 61 (10): 5000–5007.

83 Moussa, H., Cozza, A., and Cauterman, M. (2009). A novel way of using reverberation chambers through time reversal, ESA Workshop on Aerospace EMC (ESA'09), pp. 10–12.

84 Cozza, A. and Abou el-Aileh, A. (2010). Accurate radiation-pattern measurements in a time-reversal electromagnetic chamber. *IEEE Antennas and Propagation Magazine* 52 (2): 186–193.

85 Monsef, F., Cozza, A., Meteon, P., and Djedidi, M. (2014). Preliminary results on antenna testing in reverberating environments, IEEE Conference on Antenna Measurements & Applications (CAMA). pp. 1–4.

86 Xu, Q., Huang, Y., Xing, L. et al. (2017). 3-D antenna radiation pattern reconstruction in a reverberation chamber using spherical wave decomposition. *IEEE Transactions on Antennas and Propagation* 65 (4): 1728–1739.

87 Jackson, J. (1983). *Classical Electrodynamics*. New York: Wiley.

88 Chen, Y. and Simpson, T. (1991). Radiation pattern analysis of arbitrary wire antennas using spherical mode expansions with vector coefficients. *IEEE Transactions on Antennas and Propagation* 39 (12): 1716–1721.

89 Galdo, G.D., Lotze, J., Landmann, M., and Haardt, M. (2006). Modelling and manipulation of polarimetric antenna beam patterns via spherical harmonics, 14th European Signal Processing Conference. pp.1–5.

90 Rahola, J., Belloni, F., and Richter, A. (2009). Modelling of radiation patterns using scalar spherical harmonics with vector coefficients, 3rd European Conference on Antennas and Propagation (EuCAP). pp.3361–3365.

91 Mhedhbi, M., Avrillon, S., and Uguen, B. (2013). Comparison of vector and scalar spherical harmonics expansions of UWB antenna patterns, International Conference on Electromagnetics in Advanced Applications (ICEAA). pp.1040–1043.

92 Yang, J., Pivnenko, S., and Laitinen, T. (2010). Measurements of diversity gain and radiation efficiency of the eleven antenna by using different measurement techniques, 4th European Conference on Antennas and Propagation (EuCAP), pp.1–5.

93 Gimbutas, Z. and Greengard, L. (2009). A fast and stable method for rotating spherical harmonic expansions. *Journal of Computational Physics* 228 (16): 5621–5627.

94 Hallbjorner, P. (2007). Accuracy in reverberation chamber antenna correlation measurements, International workshop on Antenna Technology: Small and Smart Antennas Metamaterials and Applications, Cambridge. pp. 170–173.

95 Marquardt, D.W. (1963). An algorithm for least-squares estimation of nonlinear parameters. *Journal of the Society for Industrial and Applied Mathematics* 11 (2): 431–441.

96 Hill, D.A. (1994). Electronic mode stirring for reverberation chambers. *IEEE Transactions on Electromagnetic Compatibility* 36 (4): 294–299.

97 Huang, Y. and Edwards, D.J. (1992). A novel reverberating chamber: the source-stirred chamber, Eighth International Conference on Electromagnetic Compatibility. Edinburgh. pp. 120–124.

98 Rosengren, K., Kildal, P.-S., Carlsson, C., and Carlsson, J. (2001). Characterization of antennas for mobile and wireless terminals in reverberation chambers: improved accuracy by platform stirring. *Microwave and Optical Technology Letters* 39 (6): 391–397.

99 IEEE. (2011). IEEE Standard Test Procedures for Antennas, IEEE Standard, 149-1979, January 2011.

100 Jensen, F. and Frandsen, A. (2005). *On the Number of Modes in Spherical Wave Expansions*. Copenhagen, Denmark: TICRA [Online]. Available: www.ticra.com.

101 Melia, G.C.R., Robinson, M.P., Flintoft, I.D. et al. (2013). Broadband measurement of absorption cross section of the human body in a reverberation chamber. *IEEE Transactions on Electromagnetic Compatibility* 55 (6): 1043–1050.

102 Bamba, A., Gaillot, D.P., Tanghe, E. et al. (2015). Assessing whole-body absorption cross section for diffuse exposure from reverberation chamber measurements. *IEEE Transactions on Electromagnetic Compatibility* 57 (1): 27–34.

103 Tian, Z., Huang, Y., Shen, Y., and Xu, Q. (2016). Efficient and accurate measurement of absorption cross section of a lossy object in reverberation chamber using two one-antenna methods. *IEEE Transactions on Electromagnetic Compatibility* 58 (3): 686–693.

104 Dunlap, C.R. (2013). Reverberation chamber characterization using enhanced backscatter coefficient measurements, PhD dissertation, Department of Electrical, Computer and Energy Engineering, University of Colorado, Boulder.

105 Flintoft, I.D., Melia, G.C.R., Robinson, M.P. et al. (2015). Rapid and accurate broadband absorption cross-section measurement of human bodies in a reverberation chamber. *Measurement Science and Technology* 26 (6): 65701–65709.

106 Burger, W.T.C., Remley, K. A., Holloway, C. L., and Ladbury, J. M. (2013). Proximity and antenna orientation effects for large-form-factor devices in a reverberation chamber, IEEE International Symposium on Electromagnetic Compatibility. Denver, CO. pp. 671–676.

107 Gifuni, A., Sorrentino, A., and Ferrara, G. (2011). Measurements on the reflectivity of materials in reverberating chamber, Loughborough Antennas & Propagation Conference (LAPC), pp. 1–4.

108 Gifuni, A. (2009). On the measurement of the absorption cross section and material reflectivity in a reverberation chamber. *IEEE Transactions on Electromagnetic Compatibility* 51 (4): 1047–1050.

109 Gifuni, A., Khenouchi, H., and Schirinzi, G. (2015). Performance of the reflectivity measurement in a reverberation chamber. *Progress In Electromagnetic Research (PIER)* 154: 87–100.

110 Carlberg, U., Kildal, P.S., Wolfgang, A. et al. (2004). Calculated and measured absorption cross sections of lossy objects in reverberation chamber. *IEEE Transactions on Electromagnetic Compatibility* 46 (2): 146–154.

111 Hallbjorner, P., Carlberg, U., Madsen, K., and Andersson, J. (2005). Extracting electrical material parameters of electrically large dielectric objects from reverberation chamber measurements of absorption cross section. *IEEE Transactions on Electromagnetic Compatibility* 47 (2): 291–303.

112 Xu, Q., Huang, Y., Xing, L. et al. (2016). Average absorption coefficient measurement of arbitrarily shaped electrically large objects in a reverberation chamber. *IEEE Transactions on Electromagnetic Compatibility* 58 (6): 1776–1779.

113 Rius, J.M., Ferrando, M., and Jofre, L. (1993). High-frequency RCS of complex radar targets in real-time. *IEEE Transactions on Antennas and Propagation* 41 (9): 1308–1319.

114 Vouk, V. (1948). Projected area of convex bodies. *Nature* 162: 330–331.

115 Z. Slepian (2012). The average projected area theorem-generalization to higher dimensions, arXiv:1109.0595.

116 Holloway, C.L., Shah, H.A., Pirkl, R.J. et al. (2012). Early time behavior in reverberation chambers and its effect on the relationships between coherence bandwidth, chamber decay time, RMS delay spread, and the chamber buildup time. *IEEE Transactions on Electromagnetic Compatibility* 54 (4): 714–725.

117 Dortmans, J.N.H., Remley, K.A., Senić, D. et al. (2016). Use of absorption cross section to predict coherence bandwidth and other characteristics of a reverberation chamber setup for wireless-system tests. *IEEE Transactions on Electromagnetic Compatibility* 58 (5): 1653–1661.

118 Bohren, C.F. and Huffman, D.R. (1983). *Absorption and Scattering of Light by Small Particles*. New York: Wiley.

119 Holloway, C.L., Hill, D.A., Ladbury, J.M., and Koepke, G. (2006). Requirements for an effective reverberation chamber: unloaded or loaded. *IEEE Transactions on Electromagnetic Compatibility* 48 (1): 187–194.

120 Matzler, C. (2002). Matlab functions for Mie scattering and absorption, research report, Institute of Applied Physics, University of Bern, no. 2002-11.

121 Xu, Q., Huang, Y., and Zhu, X. (2014). Permittivity measurement of spherical objects using a reverberation chamber, Loughborough Antennas and Propagation Conference (LAPC). Loughborough. pp. 44–47.

122 Flintoft, I.D., Bale, S.J., Parker, S.L. et al. (2016). On the measurable range of absorption cross section in a reverberation chamber. *IEEE Transactions on Electromagnetic Compatibility* 58 (1): 22–29.

123 Holloway, C.L., Hill, D.A., Ladbury, J. et al. (2003). Shielding effectiveness measurements of materials using nested reverberation chambers. *IEEE Transactions on Electromagnetic Compatibility* 45 (2): 350–356.

124 Holloway, C.L., Hill, D.A., Sandroni, M. et al. (2008). Use of reverberation chambers to determine the shielding effectiveness of physically small, electrically large enclosures and cavities. *IEEE Transactions on Electromagnetic Compatibility* 50 (4): 770–782.

125 Xu, Q., Huang, Y., Zhu, X. et al. (2015). Shielding effectiveness measurement of an electrically large enclosure using one antenna. *IEEE Transactions on Electromagnetic Compatibility* 57 (6): 1466–1471.

126 Xu, Q., Huang, Y., Xing, L. et al. (2015). A fast method to measure the volume of a large cavity. *IEEE Access* 3: 1555–1561.

127 Flanagin, V.L., Schörnich, S., Schranner, M. et al. (2017). Human exploration of enclosed spaces through echolocation. *Journal of Neuroscience* 37: 1614–1627.

128 Kleiner, M. and Tichy, J. (2014). *Acoustics of Small Rooms*. CRC Press.

8

Measurement Uncertainty in the Reverberation Chamber

Xiaoming Chen[1], Yuxin Ren[2], and Zhihua Zhang[2]

[1] School of Electronic and Information Engineering, Xi'an Jiaotong University, China
[2] Hwa-Tech, Beijing, China

8.1 Introduction

Due to the complicated and time-varying test conditions, reverberation chamber (RC) measurement data are usually analysed from a statistical point of view [1]. It has been used for electromagnetic compatibility (EMC) tests [1–15] as well as over-the-air (OTA) measurements [16–25], and other applications discussed in previous chapters. Therefore, it is of importance to characterise the measurement uncertainty for these applications. In this chapter, we investigate the measurement uncertainty in the RC.

8.2 Procedure for Uncertainty Characterisation

The measurement uncertainty can be assessed by repeating the reference measurement (i.e. measurement of the power transfer function using a reference antenna) several times with different positions and orientations of the reference antenna inside the RC. Due to limitations of the RC space and the measurement time, three heights and three orientations (vertically, horizontally, and at 45° relative to the vertical) of the reference antenna are chosen. Therefore, there are in total nine reference measurements for assessing the measurement uncertainty for a certain loading at each frequency point. The uncertainty, for this loading condition, is characterised by the standard deviation (STD) of the nine average power transfer functions normalised by their mean [25]. To evaluate the uncertainties for different loading conditions, one needs to repeat the uncertainty assessment procedure with the corresponding loading conditions of the RC.

8.3 Uncertainty Model

According to the statistics, the RC measurement uncertainty (in terms of the normalised STD σ) depends on its independent sample numbers N_{ind} [1]

$$\sigma = \frac{1}{\sqrt{N_{\text{ind}}}} \tag{8.1}$$

Different methods have been proposed in the literature for estimating the number of independent samples, N_{ind}. The most commonly used method for estimating N_{ind} is the autocorrelation function (ACF) method

Anechoic and Reverberation Chambers: Theory, Design, and Measurements, First Edition. Qian Xu and Yi Huang.
© 2019 John Wiley & Sons Ltd. Published 2019 by John Wiley & Sons Ltd.

as discussed in [2–6]. It has been recognised that the ACF method gives only coarse estimations [9]. Modified ACF methods were proposed in [7, 8], which suffer from high computational complexity. A simple yet accurate N_{ind} estimator rooted from majorisation theory [26] was adopted in [10, 11] for uncertainty characterisation in RCs. This estimator was first derived in 1969 [27] and termed spatial degrees of freedom (DoF) later in [28]. For this reason, it is referred to as the DoF method in this book. These two methods are presented in the following sections.

8.3.1 ACF Method

The estimated N_{ind} using the ACF method can be written as [2–6]

$$N_{ind} = \frac{N_{meas}}{\Delta} \tag{8.2}$$

where N_{meas} is the number of measured samples and Δ represents the offset at which the autocorrelation coefficient of the measured samples drops to $1/e \approx 0.37$, i.e.

$$ACF(\Delta) = \frac{\sum_{n=0}^{N_{meas}-\Delta-1} x_{n+\Delta}\, x_n^*}{\sum_{n=0}^{N_{meas}-1} x_n\, x_n^*} = 1/e \tag{8.3}$$

where the superscript $*$ is the conjugate operator. Note that the threshold (0.37) actually depends on the number of samples, which is omitted in the standard ACF method. More advanced ACF methods taking the number of samples into account exist in the literature. These methods suffer from high complexity and are out of the scope of this book. Interested readers can refer to [7, 8] for the more advanced ACF methods.

For better measurement accuracy, the RC can employ several stirring mechanisms, e.g. mode stirrers, turn-table platform stirring, and antenna stirring (see Figure 8.1) [25]. Note that platform stirring and antenna stirring are equivalent to so-called source stirring [14]. When several stirring mechanisms are employed in an RC measurement, a straightforward way of estimating the number of independent samples is to apply the ACF

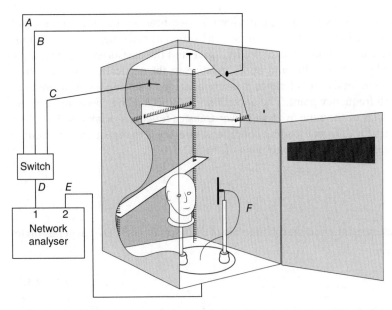

Figure 8.1 Drawing of an RC with two translating plate stirrers, a turntable platform, and three antennas mounted on the walls. Picture from [25].

method to each of the stirring sequences to determine the number of independent antennas $N_{ind,ant}$, the number of independent mode-stirring positions $N_{ind,st}$, and the number of independent platform-stirring positions $N_{ind,pf}$, respectively. In this case, the obtained offset Δ is a function of the stirring sequence of interest. For example, if $N_{ind,st}$ is of interest, then the number of realisations N is the product of the numbers of the other stirring sequences and the corresponding Δ is a function of the mode-stirring position. To improve the estimation accuracy, we take an average of Δ over all the N_{st} mode-stirring positions before substituting it in (8.2). Another way of improving the estimation accuracy is to calculate $N_{ind,st}$ as a function of the mode-stirring sequence and then average $N_{ind,st}$ over all the N_{st} mode-stirring positions. One can easily verify by simulations that the latter way (i.e. averaging $N_{ind,st}$) results in a slower convergence rate than the former way (i.e. averaging Δ). Therefore, the former way for the ACF method is used in this book.

8.3.2 DoF Method

When several stirring mechanisms are used in RC measurement, one can divide the whole stirring sequence into three subsets and estimate the N_{ind} of each subset while treating the other two subsets as observations. Specifically, the number of independent samples in an RC can be decomposed as

$$N_{ind} = N_{ind,ant} N_{ind,st} N_{ind,pf} \tag{8.4}$$

Note that (8.4) holds when the three stirring sequences are independent. When there are correlations between the different stirring sequences, (8.4) serves as the upper bound of the number of independent samples. Nevertheless, due to the different physical mechanisms of the three stirring sequences, (8.4) holds in practice. Each term at the right-hand side of (8.4) $N_{ind,l}$ (l represents ant, st, or pf) can be estimated using the following procedure [10]:

1) Denote the complex field samples at the mth antenna (l = ant), mode-stirring position (l = st), or platform-stirring position (l = pif) as a column vector $\mathbf{x}_m$, $m = 1 \dots N_l$.
2) Concatenate $\mathbf{x}_m$ into a matrix $\mathbf{X} = [\mathbf{x}_1 \ \dots \ \mathbf{x}_{N_l}]$.
3) Estimate the correlation matrix of $\mathbf{X}$ as

$$\mathbf{R} = \mathbf{X}\mathbf{X}^H \tag{8.5}$$

4) The independent antenna (l = ant), mode stirrer (l = st), or platform (l = pf) sample number can be estimated as [26, 27]

$$N_{ind,l} = \frac{tr(\mathbf{R})^2}{tr(\mathbf{R}^2)} = \frac{\left(\sum_i \lambda_i\right)^2}{\sum_i \lambda_i^2} \tag{8.6}$$

The superscript ^{H} denotes conjugate transpose, tr represents the trace operator, and λ_i represents the ith eigenvalue of $\mathbf{R}$. Equation (8.6) determines the dimensionality of the column space of $\mathbf{X}$ and therefore the corresponding independent sample number.

Note that although $\mathbf{R}$ in (8.5) is the maximum likelihood estimation of the true correlation matrix $\mathbf{R}_0$, the eigenvalues of $\mathbf{R}$, i.e. λ_i, are biased estimates of the true eigenvalues [29]. When the number of observations is comparable in magnitude to the number of samples of a certain stirring sequence N_l, estimates of small eigenvalues are biased down, while estimates of large eigenvalues are biased up. Therefore, the estimated eigenvalue vector majorises the true eigenvalue vector [26] and consequently $\mathbf{R}$ is more correlated than $\mathbf{R}_0$. This results in underestimation of $N_{ind,l}$ for data with little correlation. The underestimation can be alleviated using an improved eigenvalue estimator (at the expense of increased complexity) [10]. Interested readers may refer to

[29] for the improved eigenvalue estimator. Nevertheless, the improved eigenvalue estimator is only useful for uncorrelated samples. For oversampled (correlated) data that is required for estimating the maximum available N_{ind} [3], the improved eigenvalue estimator is less effective.

8.3.3 Comparison of ACF and DoF Methods

Comparing the DoF method with the ACF method, it can be seen that the latter requires more computation because of the ACF (8.3) and the inverse mapping needed to determine the offset Δ. The accuracies of both the ACF and DoF methods depend on the number of samples. Because it uses a finite number of samples, the estimate $\mathbf{R} = \mathbf{X}\mathbf{X}^H$ differs from $\mathbf{R}_0$ with high probability, but as N increases the estimated value approaches its true value. For a large number of samples, it approaches asymptotically their corresponding true values.

For the sake of easy exposition and without loss of generality, we focus on estimating the number of independent stirrer positions $N_{\text{ind,st}}$. We first numerically generate an $N_{\text{st}} \times N$ random matrix $\mathbf{X}_w$ consisting of independent and identically distributed (i.i.d.) Gaussian elements. Given the true correlation matrix $\mathbf{R}_0$, the correlations can be introduced to the mode-stirring positions by

$$\mathbf{X} = \mathbf{R}_0^{1/2}\mathbf{X}_w \tag{8.7}$$

where the superscript $^{1/2}$ can be implemented as, for example, the Cholesky decomposition [30].

Simulations were run to compare the estimation performance of the ACF and DoF methods, assuming 10 mode-stirring positions, i.e. $N_{\text{st}} = 10$. We first compared the biasness of the two methods by plotting the estimated $N_{\text{ind,st}}$ as a function of the number of samples with uniform correlations, as shown in Figure 8.2. It can be seen that (i) the DoF method (solid curves) tends to underestimate $N_{\text{ind,st}}$, (ii) the ACF method has larger estimation bias than the DoF method, and (iii) estimates of the two methods tend to converge to the true value as the number of realisations increases and, for correlated samples, the convergence rate of the DoF method is much faster than that of the ACF method.

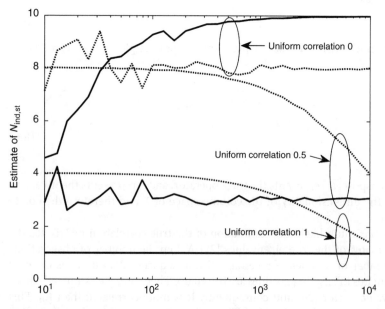

Figure 8.2 Estimated $N_{\text{ind,st}}$ using the DoF and ACF methods as a function of the number of samples with $N_{\text{st}} = 10$. Solid (dotted) curves represent the estimated $N_{\text{ind,st}}$ using the DoF (ACF) method.

To study the variance of the two methods we then repeated the simulation 1000 times to calculate the STD of the estimates of $N_{ind,st}$. For a realistic comparison, an estimated correlation matrix from the measured samples at 50 mode-stirring positions in an RC is used as $\mathbf{R}_0$ for this simulation. The measured correlation matrix is shown in Figure 8.3, from which it can be seen that the measured data are correlated (because of oversampling). The calculated STDs of the estimates of both methods are shown in Figure 8.4. As can been seen, the accuracy of the DoF method depends on the number of realisations N and it has a smaller variance than the ACF method when N is larger than 85. (This sample number is not difficult to obtain for practical RC measurements.)

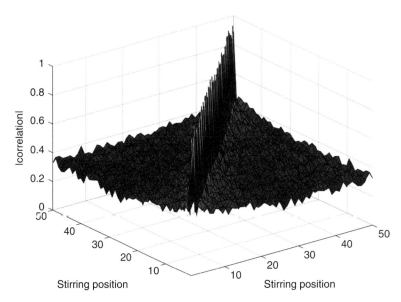

Figure 8.3 Magnitudes of correlation coefficients between measured samples at different mode-stirring positions in the RC.

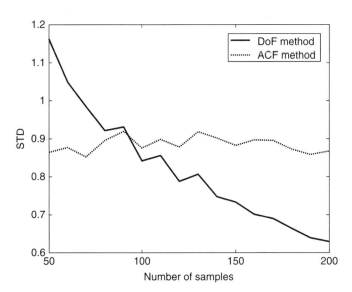

Figure 8.4 STDs of the estimated $N_{ind,st}$ using DoF and ACF methods with the measured correlation matrix.

Moreover, based on the observation of the simulations, the computation time of the ACF method for one realisation is larger than that of the DoF method by a factor of 10 and this factor increases linearly with an increasing number of samples.

To further compare the performances of the two methods, measurements were performed from 500 to 3000 MHz in an RC located at Chalmers University of Technology, Sweden, which has a size of $1.80 \times 1.75 \times 1.25$ m^3 (a drawing of this is shown in Figure 8.1). The RC has two mode-stirring plates, a turntable platform (on which a wideband discone antenna is mounted), and three antennas mounted on three orthogonal walls (referred to as wall antennas hereafter). The wall antennas are actually wideband half-bow-tie antennas. During the measurement, the turntable platform was stepwisely moved to 20 platform stirring positions evenly distributed over one complete platform rotation; at each platform-stirring position the two plates were simultaneously and stepwisely moved to 50 positions (equally spanning the total distances that they can travel along two RC walls). At each stirrer position and for each wall antenna a full frequency sweep was performed by a vector network analyser (VNA) with a frequency step of 1 MHz, during which the scattering parameter (S-parameter) is sampled as a function of frequency and stirring position. Hence, for this measurement setup, we have three wall antennas, 50 plate-stirring positions, and 20 platform-stirring positions, i.e. $N_{\text{ant}} = 3$, $N_{\text{st}} = 50$, and $N_{\text{pf}} = 20$.

To be able to estimate the measurement uncertainty (in terms of STD of the average power level) and to see the effects of $N_{\text{ind,ant}}$, $N_{\text{ind,st}}$, and $N_{\text{ind,pf}}$ on the measurement uncertainty, the same measurement sequence is repeated nine times, each time with a different height/orientation of the antenna on the platform, i.e. the antenna on the platform was placed at three different heights and at each height it was placed with three different orientations (cf. Section 8.2). The nine-measurement procedure were repeated for three loading conditions: *load0* (unloaded RC), *load1* (head phantom equivalent to a human head in terms of microwave absorption), and *load2* (the head phantom plus three polyvinyl chloride (PVC) cylinders filled with electromagnetic (EM) absorbers cut into small pieces). Hereafter measured data from these different loading configurations are simply referred to as *load0*, *load1*, or *load2* data.

We calculated the STD of the average power over the nine measurements for each loading. For clear exhibitions, the calculated STD is plotted using the following dB-transformation [25]

$$\sigma_{\text{dB}} = 5\log_{10}\frac{(1+\sigma)}{(1-\sigma)} \tag{8.8}$$

The dB transformation defined in (8.8) ensures a close-to-linear mapping when σ is small, which helps the visualisation of the STD plots. Note that a similar dB transformation was used in [31] to represent the STD of the fading signal.

After obtaining the estimated total number of independent samples N_{ind} using the ACF and DoF methods, the STD based on the estimated N_{ind} was calculated using (8.1).

Figure 8.5 shows the measured and estimated STDs using the ACF and DoF methods, respectively. Figure 8.5a corresponds to the ACF method, whereas Figure 8.5b corresponds to the DoF method. As can be seen, the DoF method is more accurate than the ACF method. The estimated STDs based on the DoF method agree well with the estimated STDs based on measurements except for the *load2* case. The small discrepancy for the *load2* case is reasonable because the field is more spatially correlated for such heavy loading, rendering the nine measurements (that were described earlier in this section for the uncertainty assessment) less independent. In other words, the measured STD for *load2* is less accurate than the other two loading cases. Since the theoretical STD based on the estimated N_{ind} using the DoF method is not affected by the independence of the nine measurements, it is believed that the estimated STD using the DoF method is more accurate for the *load2* case.

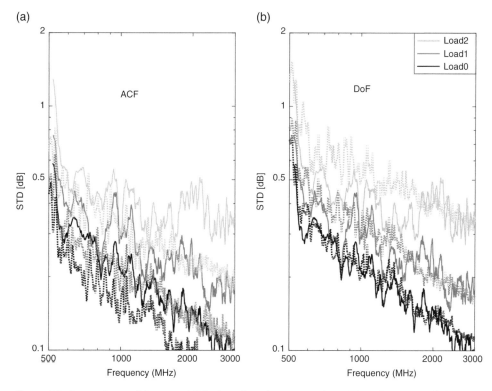

Figure 8.5 Comparisons of theoretical STDs based on the estimated N_{ind} (dotted curves) with the measured STDs (solid curves): (a) ACF method and (b) DoF method.

8.3.4 Semi-empirical Model

The uncertainty models using the ACF method or DoF method relies on accurate estimation of the number of independent samples. The number of independent samples should reflect the effects of various stirring mechanisms as well as the RC loading.

Unlike the ACF and DoF methods, a semi-empirical uncertainty model based on the (Rician) K-factor was proposed in [25]. The K-factor was used as an objective function (which is to be minimised) to optimise their RC mode stirrers [32]. It was pointed out in [25] that the K-factor (which increases with increasing RC loading [33, 34]) represents a residual error in RC measurements, meaning that the RC accuracy cannot be improved by simply improving the stirring mechanisms and that the K-factor must be taken into consideration. It should be noted, however, that the uncertainty model (8.1) together with the DoF or ACF method is not in contradiction with the K-factor based uncertainty model because the estimated N_{ind} using the DoF or ACF method has taken the effect of the RC loading into account.

The K-factor based uncertainty model is given as [25]

$$\sigma = \sqrt{\left(\sigma_{\text{NLOS}}\right)^2 + K_{\text{av}}{}^2\left(\sigma_{\text{LOS}}\right)^2} \Big/ \sqrt{1 + K_{\text{av}}{}^2} \tag{8.9}$$

where σ_{NLOS} is the STD contribution from the stirred electromagnetic fields, σ_{LOS} is the STD from the random line-of-sight (LOS) EM fields, and K_{av} is the average K-factor. The parameters in (8.9) are discussed separately in the following.

Ideally, EM fields in the RC can be expanded into orthogonally perturbed mode functions. These perturbed modes satisfy the boundary condition on mode stirrers as well as on cavity walls. The mode stirrers have the

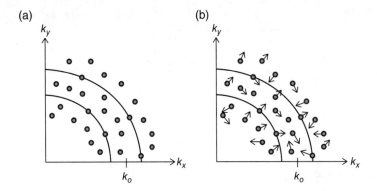

Figure 8.6 Mode distribution near resonating frequency in an RC: (a) stationary mode stirrer and (b) moving mode stirrer.

effect of shifting the eigen-frequencies of the EM modes, as illustrated in Figure 8.6. Figure 8.6a is a two-dimensional (2D) view of the mode distribution (represented by dots) when mode stirrers in the RC are stationary, where $k_0 = 2\pi f_0/c$ is the wave number corresponding to resonant frequency f_0, with speed of light denoted as c. In practice, RCs always have finite Q factors due to different losses [35], corresponding to a non-zero average mode bandwidth Δf [36]. With one excitation at f_0 all the modes falling within the spherical shell Δk will be excited. Δf is related to the thickness of Δk as $\Delta k = 2\pi\Delta f/c$. With moving mode stirrers in the RC, the eigenmodes shift as shown in Figure 8.6b. The amount of shift of the eigenmodes depends on the size, geometry, and moving sequence of the mode stirrers. An effective mode stirrer means the shift of the mode is large enough for it to be shifted in or out of the shell. In this way, the number of independent samples increases. As Wu [37] postulated, the key mechanism behind an effective mode stirrer lies in the shifting of the eigen-frequencies. Therefore, it is necessary to define a mechanical stirring bandwidth, B_{mech} (also in Section 5.5.3), to quantify the effectiveness of the mode stirrers. Due to imperfect EM stirring of the mode stirrers, the EM fields in an RC can be partitioned as stirred and unstirred components. Obviously, the STD from stirred components in the RC, which is Gaussian distributed [1], can be expressed as

$$\sigma_{\text{NLOS}} = 1/\sqrt{N_{\text{th}}} \tag{8.10}$$

The number of theoretically independent samples N_{th} can be determined primarily by the number of independent modes, which is related to the average mode bandwidth (Δf) [36] and the mechanical bandwidth (B_{mech}) in the RC,

$$N_{\text{mode}} = \frac{Vf^2 8\pi}{c^3}(B_{\text{mech}} + \Delta f) \tag{8.11}$$

where the factor in front of the bracket is the well-known Weyl's formula, which will approximate the mode density [38]. However, it is non-trivial to determine B_{mech} in that it is believed to be sensitive to the chamber loading and the frequency, and it depends on mode-stirring methods and sequences, shapes of mode stirrers, and the shape of the chamber. As a result, a detailed study of B_{mech} is not available in the literature yet (except for [7], where the objective function for the stirrer is chosen to be the lowest frequency with certain uncorrelated samples, which does not fit into the present uncertainty model). Instead, a simplified number of theoretically independent sample models is employed

$$N_{\text{th}} = M_{\text{plate}}M_{\text{pf,ind}}M_{\text{ant}} \tag{8.12}$$

where M_{plate} is to the number of plate positions (or the plate-stirring number), M_{ant} is the number of antennas used for the polarisation (or multiprobe) stirring, and $M_{\text{pf,ind}}$ is the number of independent platform positions that was found to be bounded according to

$$M_{\text{pf,ind}} = \max\left\{8, \min\left[M_{\text{pf}}, \frac{2R\sin\left(\pi/M_{\text{pf}}\right)M_{\text{pf}}}{\alpha\lambda_c/2}\right]\right\} \tag{8.13}$$

where M_{pf} is the number of platform positions, R is the radius of the circle along which the reference antenna moves during the measurement, λ_c is the wavelength, and α is a parameter that needs to be determined empirically when the far-field function of the AUT is unknown. The number of theoretically independent samples (8.12) is different from the number of independent samples (8.4). The former does not take the RC loading into account and, therefore, is usually larger than the later. The interpretation of (8.13) is that the independent platform position number is first upper bounded by the maximum independent platform number, and then lower bounded by a factor of 8. The maximum independent platform number is obtained by dividing the arc length that the reference antenna travelled by the coherence distance [39] or correlation length [40]. It is shown that the coherence distance for an isotropic antenna in a well-stirred RC is $\lambda_c/2$ [40], but it differs for practical antennas with non-isotropic far-field functions [41]. Hence α is introduced to take the reference antenna and imperfect stirring into account. Note that since the reference discone antenna is not very directive, the corresponding α was found to be 0.7 by curve fitting against the measured STDs. The factor of 8 in (8.13) comes from the fact that most of the EM wave in the RC can be decomposed into eight plane waves [38].

RCs can be loaded to emulate Rician fading environments, while an unloaded RC with properly designed mode stirrers has a negligible K-factor when the transmit and receive antennas are not pointed towards each other [34] or non-directive [33]. For active OTA tests, it is often necessary to load the RC in order to have a desired coherence bandwidth [36]. Therefore, K-factors (small as they may be) may exist in various OTA tests. The Rician K-factor is defined as the power ratio of the unstirred components to the stirred components [42],

$$K = \frac{|E[H]|^2}{\text{var}[H]} \tag{8.14}$$

where H is the channel transfer function in the RC, and var denotes the variance. Note that the unstirred components consist of both LOS components and unstirred multipath components (UMCs) [34], since the UMC has similar effect as the LOS component in RC measurements; both of them are referred to as LOS components in this chapter. Also note that in the literature various moment-based approaches (that utilise the signal magnitude only), for example [43], were used for the K-factor estimation; the Cramer–Rao lower bound (CRLB) [44] on the estimation error of this kind of estimators diverges as the K-factor approaches zero [45] (meaning that a reliable K-factor estimation using the moment-based approach is infeasible for small K-factors). On the other hand, the CRLB for the estimator (8.14) goes to zero as the K-factor vanishes, and it is strictly smaller than that of the moment-based estimator [45], meaning that the performance of (8.14) surpasses that of its moment-based counterpart. This is quite intuitive since the estimator (8.14) uses both the magnitude and phase information of the signal, which is probably better than the case where only signal magnitude is available. Since the magnitude and phase information can be obtained readily in a VNA measurement, (8.14) is used for the K-factor estimation in this chapter.

Because of the traditional definition of the K-factor, i.e. (8.14), the LOS components are usually regarded as deterministic. However, due to the (stepwise) rotation of the turntable platform in the RC, at each platform position there is an associated K-factor that is different from the other ones with high probability. In other words, the LOS component associated with a moving antenna can be regarded as a random variable. For this reason, an average K-factor is proposed,

$$K_{av} = \frac{\sum\limits_{n=1}^{M_{pf}M_{ant}} |E_n[H]|^2}{\sum\limits_{n=1}^{M_{pf}M_{ant}} var_n[H]} \tag{8.15}$$

where the subscript $_n$ in both E and var denotes an ensemble average over the subset (enumerated by n, i.e. the random channel associated with a fixed platform position and a fixed wall antenna) of the total sample set. The STD contribution from the random LOS components can be expressed as

$$\sigma_{LOS} = 1/\sqrt{M_{LOS,ind}} \tag{8.16}$$

where $M_{LOS,ind}$ is the product of the number of platform-stirring samples and the number of wall antennas, which is given as

$$M_{LOS,ind} = M_{pf}M_{ant} \tag{8.17}$$

The same measurement data as that in Section 8.3.3 are used to verify the K-factor based uncertainty model. Figure 8.7 shows the average K-factors measured in the RC. Note that for better illustrations, a 20 MHz frequency smoothing is performed to the average K-factors before plotting.

Figure 8.8 compares modelled STDs with measured ones for different loading conditions. As can be seen, the K-factor based uncertainty model can predict the measurement well. We next study the effect of different stirring mechanisms and compare them with the uncertainty model in more detail.

Figure 8.9 shows how sensitive the STD is to the number of platform positions, wall antennas, and plate positions for *load0*. Note that any subset of the total measured samples can be chosen for STD calculations at the post-processing stage. Here the subsets of various platform positions, wall antennas, and plate positions are chosen. The curves in the figure were obtained from the measured samples by selecting every second, third or fourth collected sample. (For better illustrations, a 50 MHz frequency smoothing is performed to the STDs

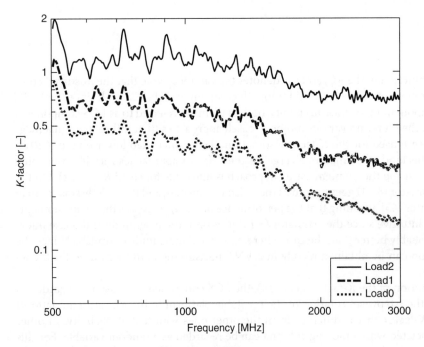

Figure 8.7 Measured average K-factors in the RC.

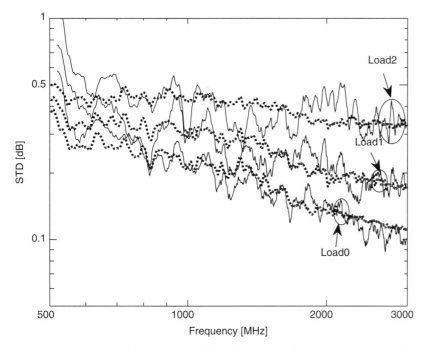

Figure 8.8 Comparison of measured (solid) and modelled (dotted) STDs mode, platform, and antenna stirring.

before plotting.) Reasonable agreements are observed, in particular for the cases with many stirrer positions and for small loads. By comparison, it seems that platform stirring is more effective than the other two stirring mechanisms. Finally, it can be seen that the uncertainty model can predict the measured STD well when $M_{\mathrm{LOS,ind}}$ (8.17) is large, and that it overestimates the STD when $M_{\mathrm{LOS,ind}} = 1$ (i.e. without platform or antenna stirring). This observation makes sense in that when $M_{\mathrm{LOS,ind}}$ is small, (8.16) is not a good approximation of the STD contribution from the random LOS components, which results in mismatches in the prediction. Nevertheless, it is shown that with a sufficient $M_{\mathrm{LOS,ind}}$ number, the uncertainty model approximates the true values with great accuracy. Another interesting observation is that the platform and antenna stirring (i.e. source stirring) play an important role in reducing the measurement uncertainty in the RC.

8.4 Measurement Uncertainty of Antenna Efficiency

In this section, we characterise the measurement uncertainty of antenna efficiency. As discussed in Chapter 7, there are several methods for measuring antenna efficiency in the RC [46–48]. The standard method involving a reference antenna is chosen as an example for uncertainty characterisation. All OTA tests in the RC involve a reference measurement for chamber calibration in similar way to the standard method for measuring antenna efficiency. Hence, the uncertainty analysis in this section also offers insight into OTA tests in the RC.

For the sake of completeness and to facilitate the uncertainty analysis, we briefly repeat the standard method for antenna efficiency measurement in this chapter with slightly different notations. The total efficiency of the antenna under test (AUT) is estimated as

$$\hat{\eta}_{\mathrm{AUT}} = \frac{\hat{P}_{\mathrm{AUT}}}{\hat{P}_{\mathrm{ref}}/\eta_{\mathrm{ref}}} \tag{8.18}$$

(a)

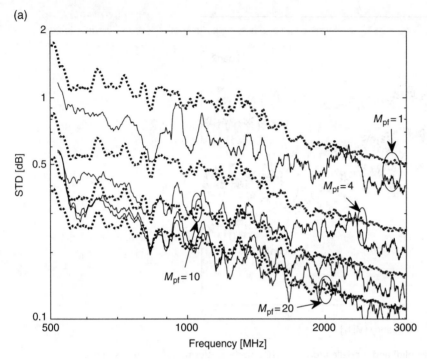

(b)

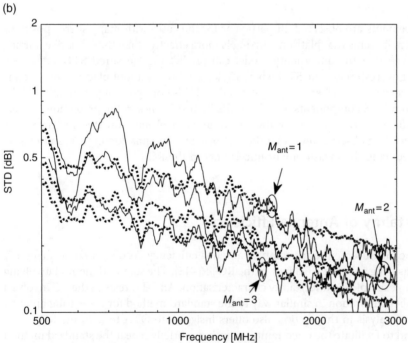

Figure 8.9 Comparison of measured (solid) and modelled (dotted) STDs of the RC with (a) different number of platform positions, (b) different number of wall antennas, and (c) different number of plate positions.

(c)

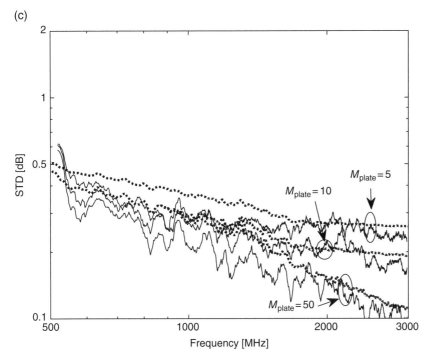

Figure 8.9 (Continued)

where the hat denotes the estimate, e.g. $\hat{\eta}_{\text{AUT}}$ is the estimate of the antenna efficiency η_{AUT}, and $\hat{P}_{\text{AUT}}$ and $\hat{P}_{\text{ref}}$ are the estimates of the average power transfer functions of received by the AUT P_{AUT} and the reference antenna P_{ref}, respectively.

Denoting the net average power transfer functions as

$$\hat{G}_{\text{AUT}} = \hat{P}_{\text{AUT}}/\eta_{\text{AUT}}$$
$$\hat{G}_{\text{ref}} = \hat{P}_{\text{ref}}/\eta_{\text{ref}} \tag{8.19}$$

respectively, (8.18) can be rewritten as

$$\hat{\eta}_{\text{AUT}} = \frac{\eta_{\text{AUT}}\hat{G}_{\text{AUT}}}{\hat{G}_{\text{ref}}} \tag{8.20}$$

Since η_{AUT} is a constant, we are interested in the distribution of the random variable $G_{\text{AUT}}/G_{\text{ref}}$ only. Assuming that there are N independent random variables of the net power transfer functions,

$$\hat{G}_{\text{AUT}} = \frac{1}{N}\sum_{i=1}^{N} G_{\text{AUT},i},$$
$$\hat{G}_{\text{ref}} = \frac{1}{N}\sum_{i=1}^{N} G_{\text{ref},i}, \tag{8.21}$$

$$\frac{\hat{G}_{\text{AUT}}}{\hat{G}_{\text{ref}}} = \frac{\sum_{i=1}^{N} G_{\text{AUT},i}}{\sum_{i=1}^{N} G_{\text{ref},i}}. \tag{8.22}$$

Note that $E[G_{\mathrm{AUT}}] = E[G_{\mathrm{AUT},i}] = E[G_{\mathrm{ref}}] = E[G_{\mathrm{ref},i}] = G_0$, where E denotes the expectation. Since $G_{\mathrm{AUT},i}$ and $G_{\mathrm{ref},i}$ are i.i.d. random variables that are exponentially distributed [1], both the numerator and the denominator on the right-hand side of (8.22) follow Gamma distribution, Gamma(N, G_0) [49]. The probability density function (PDF) of Gamma(N, G_0) is

$$f(x) = \frac{x^{N-1}\exp(-x/G_0)}{G_0^N \Gamma(N)} \tag{8.23}$$

where Γ is the Gamma function. Since N is an integer, $\Gamma(N) = (N-1)!$, where ! represents the factorial operator. Note that this distribution approaches Gaussian as N grows large due to the central limit theorem.

For notational convenience, we denote $X = \sum_{i=1}^N G_{\mathrm{AUT},i}$, $Y = \sum_{i=1}^N G_{\mathrm{ref},i}$, and $Z = X/Y$. Hence, we are interested in the distribution of the random variable Z. For distinction, we denote the PDF of Z as f_Z. Thus, (8.23) is denoted as f_X hereafter. In order to determine f_Z we need an auxiliary equation $U = Y$. Hence the group of equations for the multivariate transformation is

$$\begin{cases} Z = g_1(X,Y) = X/Y \\ U = g_2(X,Y) = Y \end{cases} \tag{8.24}$$

whose inverse map is

$$\begin{cases} X = g_1^{-1}(Z,U) = ZU \\ Y = g_2^{-1}(Z,U) = U \end{cases} \tag{8.25}$$

The Jacobian transform is

$$J = \begin{vmatrix} \dfrac{\partial g_1^{-1}}{\partial z} & \dfrac{\partial g_2^{-1}}{\partial z} \\[2mm] \dfrac{\partial g_1^{-1}}{\partial u} & \dfrac{\partial g_2^{-1}}{\partial u} \end{vmatrix} = u \tag{8.26}$$

The joint PDF of Z and U is [49]

$$f_{Z,U}(z,u) = f_{X,Y}\left(g_1^{-1},g_2^{-1}\right)|J| = f_{X,Y}(zu,u)u \tag{8.27}$$

Since X and Y are i.i.d.,

$$f_{X,Y}(zu,u) = f_X(zu)f_X(u) = \frac{z^{N-1}u^{2N-2}\exp\left(-\dfrac{zu+u}{G_0}\right)}{\left(G_0^N\Gamma(N)\right)^2} \tag{8.28}$$

Substitute (8.28) into (8.27),

$$f_{Z,U}(z,u) = \frac{z^{N-1}u^{2N-1}\exp(-(zu+u)/G_0)}{\left(G_0^N\Gamma(N)\right)^2} \tag{8.29}$$

In order to obtain f_Z, we integrate both sides of (8.29) over u,

$$f_Z(z) = \int_0^\infty f_{Z,U}(z,u)du = \frac{z^{N-1}\displaystyle\int_0^\infty u^{2N-1}\exp\left(-\dfrac{(z+1)u}{G_0}\right)du}{\left(G_0^N\Gamma(N)\right)^2}$$

$$= \frac{\Gamma(2N)}{(\Gamma(N))^2} \frac{z^{N-1}}{(1+z)^{2N}} \tag{8.30}$$

Once the distribution is known, one can readily calculate the mean and the variance (and other moments) of the random variable Z,

$$E[Z] = \int_0^\infty z f_Z(z,u) du = \frac{N}{N-1}$$

$$Var[Z] = \int_0^\infty (z-E[Z])^2 f_Z(z,u) du = \frac{N(2N-1)}{(N-2)(N-1)^2} \tag{8.31}$$

where Var denotes the variance operator. Note that Z is related to the measured antenna efficiency by $\hat{\eta}_{AUT} = \eta_{AUT} Z$. Therefore, the mean and the variance of $\hat{\eta}_{AUT}$ are [50]

$$E[\hat{\eta}_{AUT}] = \frac{N}{N-1} \eta_{AUT}$$

$$Var[\hat{\eta}_{AUT}] = \frac{N(2N-1)}{(N-2)(N-1)} \eta_{AUT}^2 \tag{8.32}$$

respectively. It can be seen from (8.32) that the efficiency estimator (8.18), i.e. the standard method for measuring antenna efficiency in an RC, is asymptotically unbiased (as N goes to infinity), and that its variance goes to zero as N goes to infinity. Thus, the antenna efficiency estimator (8.18) is consistent [44]. Note that, although the estimator (8.18) is asymptotically unbiased, it is biased for finite N. Nevertheless, it is straightforward to derive an unbiased estimator of the antenna efficiency based on (8.32)

$$\hat{\eta}_{AUT}^{unbiased} = \frac{N-1}{N} \frac{\hat{P}_{AUT}}{\hat{P}_{ref}/\eta_{ref}} \tag{8.33}$$

The mean and the variance of the unbiased estimator (8.33) are

$$E\left[\hat{\eta}_{AUT}^{unbiased}\right] = \eta_{AUT}$$

$$Var\left[\hat{\eta}_{AUT}^{unbiased}\right] = \frac{2N-1}{N(N-2)} \eta_{AUT}^2 \tag{8.34}$$

respectively.

The root mean squares (RMSs) of $\hat{\eta}_{AUT}$ and $\hat{\eta}_{AUT}^{unbiased}$ can be easily derived from (8.32) and (8.34), respectively,

$$RMS[\hat{\eta}_{AUT}] = \eta_{AUT} \sqrt{\frac{N(N+1)}{(N-2)(N-1)}} \tag{8.35}$$

$$RMS\left[\hat{\eta}_{AUT}^{unbiased}\right] = \eta_{AUT} \sqrt{\frac{N^2-1}{N(N-2)}} \tag{8.36}$$

An approximation of $Var[\hat{\eta}_{AUT}]$ was given as [50]

$$Var[\hat{\eta}_{AUT}] \approx \frac{2}{N} \eta_{AUT}^2 \tag{8.37}$$

Comparing (8.32) and (8.37), it can be seen that the approximation (8.37) is accurate for large N.

As expected, the statistics of $\hat{\eta}_{AUT}$ and $\hat{\eta}_{AUT}^{unbiased}$ are functions of the number of independent samples N, from which it can be seen that estimators (8.18) and (8.33) converge to η_{AUT} with a convergence rate of about $\sqrt{1/N}$. Hence, it is important to have many independent samples for an accurate measurement of the antenna efficiency. Since (8.18) is a biased estimator, a suitable performance metric of it is the mean square error (MSE), i.e. $E\left[(\hat{\eta}_{AUT} - \eta_{AUT})^2\right]$, which can be easily derived from (8.32) as

$$MSE[\hat{\eta}_{AUT}] = \frac{2(N+1)}{(N-2)(N-1)}\eta_{AUT}^2 \tag{8.38}$$

Since (8.33) is unbiased, its MSE equals its variance.

In practice, the measured samples in an RC may be correlated. One can estimate the number of independent samples N from the measurements using the DoF method (cf. Section 8.3.2). For simplicity, we assume that the number of independent samples is the same for the measurements of the reference antenna and the AUT. The statistics of the antenna efficiency measurement with different numbers of independent samples of the measurements of the reference antenna and the AUT have been derived in [51], which is out of the scope of this book. The interested reader is encouraged to refer to [51] for this matter. Note that the absolute measurement error is assumed in the above derivation, therefore the STD, variance, and MSE of the antenna efficiency estimator contain the true antenna efficiency η_{AUT}. For relative measurement error, the corresonponding statistics will not contain η_{AUT}.

We resort to simulations for verifying the derived statistics of the efficiency estimator. For simplicity and without loss of generality, we assume the antenna efficiency is unity (i.e. $\eta_{AUT} = 1$). For each independent sample number N, we numerically generate 1000 samples (that follow i.i.d. exponential distribution) for both $\hat{G}_{AUT}$ and $\hat{G}_{ref}$. As a result, we have 1000 realisations of $\hat{\eta}_{AUT}$ for each N, based on which we can obtain the empirical PDF, mean, and variance, as well as other moment-based statistics.

Figure 8.10 shows the comparison between the empirical PDF and the analytical PDF (8.30) for $N = 10$, 30, and 50, respectively. The good agreement between the analytical PDF and the empirical PDF verifies the derived PDF.

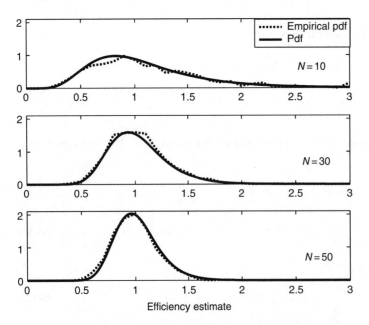

Figure 8.10 Comparisons of the analytical PDF (8.30) and the corresponding empirical PDF for $N = 10$, 30, and 50, respectively.

Figure 8.11 shows the analytical and empirical RMSs of estimators (8.18) and (8.33) as a function of N, respectively. It can also be seen that the unbiased estimator (8.33) offers slight improvement over the standard one (8.18) not only in estimation bias but also in estimation variance for small N. Nevertheless, the performances of the two estimators are indistinguishable for large N.

In order to further verify the derived statistics of the antenna efficiency estimators, extensive measurements were performed from 700 to 3000 MHz in the RC (see Figure 8.1). During the measurement, the turntable platform was moved stepwise to 20 platform-stirring positions; at each platform-stirring position the two plates were moved simultaneously and stepwise to 50 positions. At each stirrer position and for each wall antenna, a full frequency sweep was performed by a VNA with a frequency step of 1 MHz, during which the S-parameters were sampled. Hence, for each measurement, we have three wall antennas, 50 plate-stirring positions, and 20 platform-stirring positions. To facilitate the characterisation of the antenna efficiency estimators, the same measurement sequence is repeated 12 times, each time with a different height/orientation of the reference antenna on the platform, i.e. the reference antenna was placed at four different heights and at each height it is placed with one vertical and two horizontal orientations (in radial and tangential directions of the platform), respectively. The heights and orientations are chosen to ensure independent measurements. In postprocessing, arbitrary pairs of antenna heights/orientations are chosen as the AUTs and the reference antennas, respectively, for estimating η_{AUT}, and we introduce 0- and 3-dB attenuators to the AUT (whose negative value in dB is η_{AUT}). Note that it is nontrivial to find the maximum number of independent samples of $\hat{\eta}_{\text{AUT}}$ and $\hat{\eta}_{\text{AUT}}^{\text{unbiased}}$ from the 12 measurements. To be safe, we chose six pairs of distinct measurements to obtain six independent antenna efficiency estimates, which are used for evaluating the performance metric of the antenna efficiency estimator. In order to see the RC loading effect on the estimation performance of the antenna efficiency, the measurement procedure was repeated for two loading configurations: load 0 and load 1. Since the standard method for measuring antenna efficiency results in a biased estimate, the MSE is used as the performance metric for characterising estimation bias and variance together.

Figure 8.12 shows the empirical MSE of $\hat{\eta}_{\text{AUT}}$ (for $\eta_{\text{AUT}} = 0$ and -3 dB) estimated based on independent measurements (solid) and the analytical MSE model with N estimated from a single measurement (dotted)

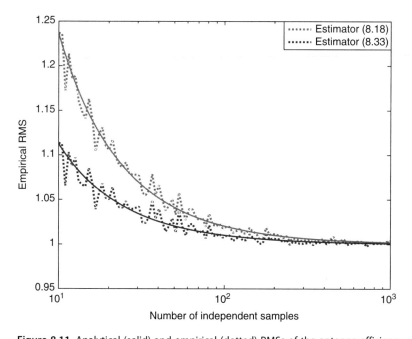

Figure 8.11 Analytical (solid) and empirical (dotted) RMSs of the antenna efficiency estimators (8.18) and (8.33).

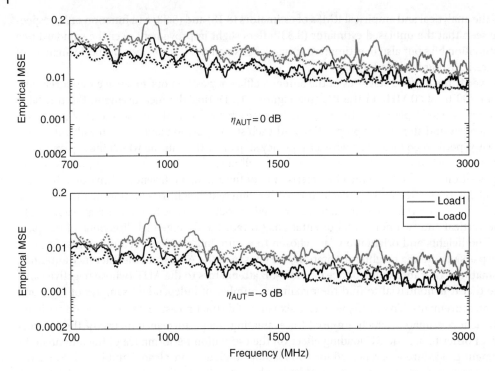

Figure 8.12 Empirical MSE of $\hat{\eta}_{AUT}$ based on independent measurements (solid) and the analytical MSE model (dotted).

using the DoF method, respectively. Note that 40 MHz frequency smoothing is applied to the empirical MSE curves before plotting. As can be seen, there are good agreements between the MSE expressions (8.38) (with N estimated using the DoF method) and the direct MSE estimate (based on the independent measurements). It can also be seen that the MSE of $\hat{\eta}_{AUT}$ decreases with decreasing η_{AUT}. This observation can be readily explained from the analytical MSE expression (8.38). Note that the estimation performances of $\hat{\eta}_{AUT}$ and $\hat{\eta}_{AUT}^{unbiased}$ are similar for large N. Therefore, for the sake of conciseness, the MSE of $\hat{\eta}_{AUT}^{unbiased}$ (estimated using all stirring samples) is not shown here.

8.5 Summary

In this chapter, various measurement uncertainty models of the RC together with an uncertainty characterisation procedure were introduced. The measurement uncertainty models can be divided into two categories. The first category relies on modelling the effective number of independent samples (which is inversely proportional to the measurement STD) [8, 10], whereas the second category is a K-factor based semi-empirical model [25]. There are different methods for estimating the effective number of independent samples, such as the ACF method [8] and the DoF method [10]. The DoF method outperforms the ACF method when the number of samples exceeds 100 (which is a typical value for most RC measurements). One can accurately predict measurement uncertainty based on one measurement using the DoF-based uncertainty model. The K-factor based semi-empirical model allows more insight into the RC uncertainty. These models were verified by extensive measurements. The measurement uncertainty of the standard antenna efficiency measurement was also analysed. The distribution of the measured antenna efficiency using the standard method has been derived, based on which statistics of the measured antenna efficiency were derived. It was shown that the standard method of antenna efficiency measurement results in biased estimation. Nevertheless, the standard method becomes asymptotically unbiased when the number of samples becomes large.

References

1 Kostas, J. G. and Boverie, B. (1991). Statistical model for a mode-stirred chamber. *IEEE Transactions on Electromagnetic Compatibility* 33 (4): 366–370.

2 Clegg, J., Marvin, A. C., Dawson, J. F., and Porter, S. J. (2005). Optimization of stirrer designs in a reverberation chamber. *IEEE Transactions on Electromagnetic Compatibility* 47 (4): 824–832.

3 Hallbjörner, P. (2006). Estimating the number of independent samples in reverberation chamber measurements from sample differences. *IEEE Transactions on Electromagnetic Compatibility* 48 (2): 354–358.

4 Arnaut, L. R. (2001). Effect of local stir and spatial averaging on measurement and testing in mode-stuned and mode-stirred reverberation chambers. *IEEE Transactions on Electromagnetic Compatibility* 43 (3): 305–325.

5 Delangre, O., Doncker, P. D., Lienard, M., and Degauque, P. (2009). Analytical angular correlation function in mode-stirred reverberation chamber. *Electronics Letters* 45 (2): 90–91.

6 Moglie, F. and Primiani, V. M. (2011). Analysis of the independent positions of reverberation chamber stirrers as a function of their operating conditions. *IEEE Transactions on Electromagnetic Compatibility* 53 (2): 288–295.

7 Wellander, N., Lunden, O., and Bäckström, M. (2007). Experimental investigation and methematical modeling of design parameters for efficient stirrers in mode-stirred reverberation chambers. *IEEE Transactions on Electromagnetic Compatibility* 49 (1): 94–103.

8 Krauthäuser, H.G., Winzerling, T., and Nitsch, J. (2005). Statistical interpretation of autocorrelation coefficients for fields in mode-stirred chambers. In Proceedings of the IEEE International Symposium on EMC. Chicago, FL. Volume 2, pp. 550–555.

9 Lemoine, C., Besnier, P., and Drissi, M. (2008). Estimating the effective sample size to select independent measurements in a reverberation chamber. *IEEE Transactions on Electromagnetic Compatibility* 50 (2): 227–236.

10 Chen, X. (2013). Experimental investigation of the number of independent samples and the measurement uncertainty in a reverberation chamber. *IEEE Transactions on Electromagnetic Compatibility* 55 (6): 816–824.

11 Pirkl, R. J., Remley, K. A., and Patané, C. S. L. (2012). Reverberation chamber measurement correlation. *IEEE Transactions on Electromagnetic Compatibility* 54 (3): 533–544.

12 Gradoni, G., Mariani Primiani, V., and Moglie, F. (2013). Reverberation chamber as a multivariate process: FDTD evaluation of correlation matrix and independent positions. *Progress in Electromagnetics Research* 133: 217–234.

13 Gifuni, A., Flintoft, I. D., Bale, S. J. et al. (2016). A theory of alternative methods for measurements of absorption cross section and antenna radiation efficiency using nested and contiguous reverberation chambers. *Transactions on Electromagnetic Compatibility* 58 (3): 678–685.

14 Huang, Y. and Edwards, D.J. (1992). A novel reverberating chamber: source-stirred chamber. In Proceedings of the International Conference on EMC. Edinburgh. pp.120–124.

15 Chen, X. (2013). Using Akaike information criterion for selecting the field distribution in a reverberation chamber. *IEEE Transactions on Electromagnetic Compatibility* 55 (4): 664–670.

16 Ferrara, G., Migliaccio, M., and Sorrentino, A. (2007). Characterization of GSM non-line-of-sight propagation channels generated in a reverberating chamber by using bit error rates. *IEEE Transactions on Electromagnetic Compatibility* 49 (3): 467–473.

17 Sorrentino, A., Ferrara, G., and Migliaccio, M. (2015). The reverberating chamber as a line-of-sight wireless channel emulator. *IEEE Transactions on Antennas and Propagation* 14: 935–938.

18 Xu, Q., Huang, Y., Zhu, X. et al. (2015). A new antenna diversity gain measurement method using a reverberation chamber. *IEEE Antennas and Wireless Propagation Letters* 14: 935–938.

19 Remley, K. A., Wang, C. -M. J., Williams, D. F. et al. (2016). A significance test for reverberation-chamber measurement uncertainty in total radiated power of wireless devices. *IEEE Transactions on Electromagnetic Compatibility* 58 (1): 207–219.

20 Genender, E., Holloway, C. L., Remley, K. A. et al. (2010). Simulating the multipath channel with a reverberation chamber: application to bit error rate measurements. *IEEE Transactions on Electromagnetic Compatibility* 52: 766–777.

21 Zhou, X., Zhong, Z., Bian, X. et al. (2017). Impacts of absorber loadings on simulating the multipath channel in a reverberation chamber. *International Journal of Antennas and Propagation* 2017: 1–7.

22 Micheli, D., Barazzetta, M., Moglie, F., and Primiani, V. M. (2015). Power boosting and compensation during OTA testing of a real 4G LTE base station in reverberation chamber. *IEEE Transactions on Electromagnetic Compatibility* 57 (4): 623–634.

23 Chen, X. (2014). Throughput modeling and measurement in an isotropic-scattering reverberation chamber. *IEEE Transactions on Antennas and Propagation* 62 (4): 2130–2139.

24 Sanchez-Heredia, J. D., Valenzuela-Valdes, J. F., Martinez-Gonzalez, A. M., and Sanchez-Hernandez, D. A. (2011). Emulation of MIMO Rician fading environments with mode-stirred reverberation chambers. *IEEE Transactions on Antennas and Propagation* 59 (2): 654–660.

25 Kildal, P. -S., Chen, X., Orlenius, C. et al. (2012). Characterization of reverberation chambers for OTA measurements of wireless devices: physical formulations of channel matrix and new uncertainty formula. *IEEE Transactions on Antennas and Propagation* 60 (8): 3875–3891.

26 Jorswieck, E. and Boche, H. (2006). Majorization and matrix-monotone functions in wireless communications. *Foundations and Trends in Communications and Information Theory* 3 (6): 553–701.

27 Bagrov, N. A. (1969). On the equivalent number of independent data. *Trudy Gidrometeotsentra* 44: 3–11.

28 Bretherton, C. S., Widmann, M., Dymnikov, V. P. et al. (1999). The effective number of spatial degrees of freedom of a time-varying field. *Journal of Climate* 12 (7): 1990–2009.

29 Mestre, X. (2008). Improved estimation of eigenvalues and eigenvectors of covariance matrices using their sample estimates. *IEEE Transactions on Information Theory* 54 (11): 5113–5129.

30 Laub, A. J. (2005). *Matrix Analysis for Scientists and Engineers*. Philadelphia, PA: SIAM.

31 Vaughan, R. G. (1990). Polarization diversity in mobile communications. *IEEE Transactions on Vehicle Technology* 39 (3): 177–186.

32 Lemoine, C., Amador, E., and Besnier, P. (2011). On the K-factor estimation for Rician channel simulated in reverberation chamber. *IEEE Transactions on Antennas and Propagation* 59 (3): 1003–1012.

33 Chen, X., Kildal, P.-S., and Lai, S.-H. (2011). Estimation of average Rician K-factor and average mode bandwidth in loaded reverberation chamber. *IEEE Antennas and Wireless Propagation Letters* 10: 1437–1440.

34 Holloway, C. L., Hill, D. A., Ladbury, J. M. et al. (Nov. 2006). On the use of reverberation chamber to simulate a Rician radio environment for the testing of wireless devices. *IEEE Transactions on Antennas and Propagation* 54 (11): 3167–3177.

35 Hill, D. A., Ma, M. T., Ondrejka, A. R. et al. (1994). Aperture excitation of electrically large, lossy cavities. *IEEE Transactions on Electromagnetic Compatibility* 36 (3): 169–178.

36 Chen, X., Kildal, P.-S., Orlenius, C., and Carlsson, J. (2009). Channel sounding of loaded reverberation chamber for over-the-air testing of wireless devices - coherence bandwidth and delay spread versus average mode bandwidth. *IEEE Antennas and Wireless Propagation Letters* 8: 678–681.

37 Wu, D. I. and Chang, D. C. (1989). The effect of an electrically large stirrer in a mode-stirred chamber. *IEEE Transactions on Electromagnetic Compatibility* 31 (2): 164–169.

38 Rosengren, K. (2005). Characterization of Terminal Antennas for Diversity and MIMO Systems by Theory, Simulations and Measurements in Reverberation Chamber. PhD thesis, Chalmers University of Technology.

39 Paulraj, A., Nabar, R., and Gore, D. (2003). *Introduction to Space-Time Wireless Communication*. Cambridge University Press.

40 Hill, D. A. (1998). Plane wave integral representation for fields in reverberation chambers. *IEEE Transactions on Electromagnetic Compatibility* 40 (3): 209–217.

41 Hill, D. A. (1999). Linear dipole response in a reverberation chamber. *IEEE Transactions on Electromagnetic Compatibility* 41 (4): 365–368.

42 Rappaport, T. S. (2002). *Wireless communications—principles and practice*, 2e, 196–202. Prentice Hall PTR.

43 Greenstein, L. J., Michelson, D. G., and Erceg, V. (1999). Moment-method estimation of the Ricean K-factor. *IEEE Communications Letters* 3: 175–176.

44 Kay, S. M. (1993). *Fundamentals of Statistical Signal Processing: Estimation Theory, Volumn I: Estimation Theory*. Prentice Hall.

45 Tepedelenlioglu, C., Abdi, A., and Giannakis, G. B. (2003). The Ricean K-factor: estimation and performance analysis. *IEEE Transactions on Wireless Communication* 2 (4): 799–810.

46 IEC 61000-4-21. (2011). Electromagnetic compatibility (EMC) – Part 4–21: Testing and measurement techniques – Reverberation chamber test methods. 2nd edition. IEC Standard. Ed 2.0. IEC 61000-4-21. January 2011.

47 Holloway, C. L., Shah, H., Pirkl, R. J. et al. (2012). Reverberation chamber techniques for determining the radiation and total efficiency of antennas. *IEEE Transactions on Antennas and Propagation* 60 (4): 1758–1770.

48 Xu, Q., Huang, Y., Zhu, X. et al. (2016). A modified two-antenna method to measure the radiation efficiency of antennas in a reverberation chamber. *IEEE Antennas and Wireless Propagation Letters* 15: 336–339.

49 Grimmett, G. and Stirzaker, D. (2001). *Probability and Random Processes*, 3e. Oxford University Press.

50 Chen, X. (2013). On statistics of the measured antenna efficiency in a reverberation chamber. *IEEE Transactions on Antennas and Propagation* 61 (11): 5417–5424.

51 Chen, X. (2014). Generalized statistics of antenna efficiency measurement in a reverberation chamber. *IEEE Transactions on Antennas and Propagation* 62 (3): 1504–1507.

9

Inter-Comparison Between Antenna Radiation Efficiency Measurements Performed in an Anechoic Chamber and in a Reverberation Chamber

Tian-Hong Loh[1] and Wanquan Qi[2]

[1] *National Physical Laboratory, Engineering, Materials & Electrical Science Department, 5G & Future Communications Technology Group, Teddington, United Kingdom*
[2] *Beijing Institute of Radio Metrology and Measurements, Division of EMC, Beijing, China*

9.1 Introduction

Antenna radiation efficiency is an important attribute of antennas [1, 2] as it has a significant effect on the performance, reliability, and efficiency of wireless communications systems [3–6]. It has been defined in the Institute of Electrical and Electronics Engineers (IEEE) Standard 145 [2] as the ratio of the total power radiated by an antenna to the net power accepted by the antenna from the connected transmitter. It takes into account losses at the input terminals and within the structure of the antenna. However, accurately and effectively measuring it has been a challenge for many decades [6–11].

The conventional way of obtaining the efficiency of an antenna under test (AUT) is to measure the maximum gain and maximum directivity of the antenna in an anechoic chamber (AC) and to take the ratio of the former over the latter [1, 12]. There are several techniques to obtain antenna gain and directivity [13–15] and they require precise alignment between transmitting and receiving antennas, which is usually time-consuming and expensive. Alternatively, other methods such as the Wheeler cap method [16–18] and the reverberation chamber (RC) method [19–23], which do not require alignment, have been proposed. The Wheeler cap method is intrinsically band limited and usually used for measuring electrically small antennas; on the other hand, the RC method has much broader applications.

In this chapter, we have chosen antenna radiation efficiency to inter-compare the measured results obtained in an AC and an RC. The fundamental operating principles of the RC and the AC are different. The main differences are:

- The alignment of the test configuration in an AC needs to be precisely controlled in order to get repeatable results, but the data processing is straightforward since there are only line of sight (LoS) signals to consider. Also, a reference antenna with a known gain is required if the antenna substitution method is employed for the antenna gain measurement.
- The position of the AUT in an RC does not need to be precise, but the data processing is considerably more involved because complex field statistics need to be interpreted correctly in order to perform limit tests on the AUT. Also, a reference antenna with a known efficiency is required if the reference antenna method is employed for the efficiency measurement of the AUT.

Anechoic and Reverberation Chambers: Theory, Design, and Measurements, First Edition. Qian Xu and Yi Huang.
© 2019 John Wiley & Sons Ltd. Published 2019 by John Wiley & Sons Ltd.

In an electromagnetic compatibility (EMC) context, to inter-compare the measured results obtained in an AC and an RC, the RC data will represent an average total emitted power (emission testing), or some defined average and maximum field (susceptibility testing). During emission testing the measured total power from the AUT or device under test (DUT) in the RC is converted to an equivalent received power obtained from the receiving antenna, P_{AC}, which would be measured in an AC. This is done by equating field components from one facility to the other [24]. The RC effectively averages the DUT directivity, D, over all directions, thus it is assumed to be unity for the RC calculations. In order to estimate the equivalent field which would be measured in an AC an estimated value of DUT directivity is required. Since the DUT will radiate in specific directions when placed in an AC, the choice of directivity value is often taken to be the maximum directivity of a radiator of the same electrical size as the DUT. When the DUT dimensions are of the order of a wavelength a dipole directivity value is often used, but for DUTs with slots and other features the real value can be much higher. A common approach is to equate the squared magnitude of the rectangular field component generated in the RC with the equivalent in the AC, which generates the following relationship:

$$\frac{P_{AC}}{\langle P_{RC} \rangle} = \frac{2}{3}D \tag{9.1}$$

where $\langle P_{RC} \rangle$ is the average power received by an ideal antenna in a RC. In [24] it is recommended to equate the rectangular components of E-field as the basis for radiated susceptibility (RS) testing, thus:

$$\frac{P_{AC}}{\langle P_{RC} \rangle} = \frac{\pi D}{6} \cdot N \cdot \left(\frac{\Gamma(N)}{\Gamma(N + 1/2)} \right)^2 \tag{9.2}$$

where N is the finite number of samples and $\Gamma(N)$ is the gamma or factorial function evaluated at N. For large N the received power radio in (9.2) tends towards a value of $0.52D$ [24]. P_{AC} may be converted to E-field for comparison with other facilities such as semi-ACs in which the receive antenna is height scanned over a conducting ground plane [25]. Here the E-field recorded would be a maximum value measured over the height scan, which may be estimated from knowledge of the test configuration. When operating in the RCs the DUT is more likely to receive the maximum power during the RS stirring process, whereas in an AC the maximum directivity may not be aligned with the source so the DUT may not receive such a harsh test. This is one major benefit of the RC facilities. The standards on testing in transverse electromagnetic (TEM) waveguides (International Electrotechnical Commission (IEC) 61000-4-20 [26]) contain expressions to approximate radiated far-field from total radiated power, and they make assumptions about the directivity of the DUT. This approximation is repeated in Annex E of IEC 61000-4-21 [27].

9.2 Measurement Facilities and Setups

In this chapter, all the measurements were performed in the UK National Physical Laboratory (NPL) AC and RC facilities with controlled temperature typically at 23 ± 1 °C. Both facilities have control rooms with a network analyser, computer, and software to enable automated measurements.

9.2.1 Anechoic Chamber

The NPL AC is a fully anechoic screened room and has dimensions of 7 m × 6.2 m × 6.2 m with two low permittivity mounts for the transmitting (Tx) and receiving (Rx) antennas (see Figure 9.1) [28]. The anechoic absorber has a −40 dB reflectivity from 400 MHz to 110 GHz [29]. The absorbers at the walls and ceiling of the chamber are covered with white radio frequency (RF) transparent polystyrene tiles for improved room illumination by the four-corner ceiling light. The parts of the absorbers on the floor are covered with polystyrene walkover to allow easy access to the measurement setup.

Figure 9.1 NPL anechoic chamber. The mount on the left is the roll-over-azimuth positioner system for the Rx AUT whereas the mount on the right is for the Tx antenna.

As depicted in Figure 9.1, the chamber contains a roll-over-azimuth positioner system at the receiving end with dielectric belt driven rotator model tower [28, 30]. This provides a lower-scattering alternative to conventional metal towers and is generally used for omni-directional antenna measurements. The positioner system enables the three-dimensional (3D) radiation pattern of the Rx AUT located at the centre of rotation to be acquired over a spherical surface with a Tx probe antenna located at a fixed distance oriented transversely to the spherical surface, with a particular polarisation angle where vertical polarisation (VP) and horizontal polarisation (HP) are often chosen. A source tower can be positioned to give a range length from 0.1 to 3.5 m between the Tx and Rx antennas. The antenna supports have very low reflectivity to achieve the low uncertainty measurements [31–33]. For antenna calibration, one could refer to useful standard document such as CISPR 16–1–6 [34], which provides complete information on procedures, configurations, and types of test instrumentation used.

9.2.2 Reverberation Chamber

The NPL RC operates from 170 MHz to 18 GHz and has dimensions 6.5 m × 5.85 m × 3.5 m with a full-height floor to ceiling stirrer which can be controlled to move either stepped (for mode-tuned operation) or continuously (for mode-stir operation). The volume of the chamber is approximately 134 m^3 and it has the option of an additional wall-mounted stirrer. Figure 9.2 shows a photo of the RC with an antenna setup.

The RC has spangle-galvanized mild stainless-steel walls and aluminium inner door skins. Three circular metallic plates, each with diameter of 1.2 m, are horizontally fixed to the vertical rotation paddle axis and serve as support decks for various rectangular aluminium tuner blades. For the RC operation, one could refer to standard documents such as IEC 61000–4–21 [27], which provides complete information on construction of the RC and stirrer, and guidance on setting up the source antenna and evaluating the relevant test volume. This standard also allows both mode-tuned and mode-stirred methods. Note that, for the relevant RC measured results presented throughout this chapter, only mode-tuned operation is considered where the relevant

Figure 9.2 NPL reverberation chamber with paddle stirrer and an antenna setup.

stationary S-parameters are acquired for each of N paddle steps (i.e. the paddle rotates one-Nth of a complete revolution) using a vector network analyser (VNA). Also, for antenna efficiency measurement, the reference antenna method is employed where the AUT and reference (REF) antenna are positioned to minimise proximity effects from other conductors and the chamber walls, and to minimise LoS correlation with the transmit antenna.

9.3 Antenna Efficiency Measurements

Until now antenna radiation efficiency measurement has not been a standard service provided by most accredited laboratories. Due to the cost-effective nature of establishing an RC, methods for antenna efficiency measurement using an RC have been developed rapidly over recent years [7, 8, 19]. The RC is perhaps one of the best ways to experimentally characterise the antenna radiation efficiency [35], but these methods usually require a reference antenna with known efficiency or require measurement of the quality factor of the RC [22, 36].

9.3.1 Theory

The following presents the formula used for the radiation efficiency evaluation using the AC and the RC, respectively.

9.3.1.1 Radiation Efficiency Using the Anechoic Chamber

For the AC, the antenna radiation efficiency of AUT, $\eta_{\text{AUT}_{\text{AC}}}$, can be calculated using the antenna's directivity, D_{AUT}, and gain, G_{AUT}. The directivity of AUT is a measure of the concentration of the radiation in a desired direction $(\theta_0, \emptyset_0)$. In the following, the IEEE definition of gain is used: the gain of an antenna in a given direction is defined as 'the ratio of the radiation intensity, in a given direction, to the radiation intensity that would be obtained if the power accepted by the antenna were isotropically radiated' [2]. These are defined as [1, 14]:

$$\eta_{\text{AUT}_{\text{AC}}} = G_{\text{AUT}}(\theta_0, \emptyset_0) / D_{\text{AUT}}(\theta_0, \emptyset_0) \tag{9.3}$$

$$D_{\text{AUT}}(\theta_0, \emptyset_0) = F_{\text{Nor}}(\theta_0, \emptyset_0) / F_{\text{Nor}_{\text{Av}}} \tag{9.4}$$

$$F_{\text{Nor}_{\text{Av}}} = \frac{1}{4\pi} \iint\limits_{4\pi} F_{\text{Nor}}(\theta, \emptyset) d\Omega \tag{9.5}$$

where $F_{\text{Nor}}(\theta_0, \emptyset_0)$ is the normalised radiation intensity for direction $(\theta_0, \emptyset_0)$, and $F_{\text{Nor}_{\text{Av}}}$ is the average value of normalised radiation intensity over 4π space. Based on the Friis transmission formula [1], the realised gain of AUT can be obtained by using the two-antenna method (i.e. with a reference antenna of known gain where the method is also known as the gain transfer method) or the three-antenna method [1, 13, 14]. The gain of AUT can then be obtained by correcting the realised gain of AUT so as to take into account the mismatch that occurred during the through measurement. The directivity of AUT can be obtained by measuring the far-field radiation pattern of the AUT over a full sphere in the AC. As shown in (9.4) and (9.5), the ratio of the radiation intensity in a given direction from the AUT to the radiation intensity average over all directions where the average radiation intensity is equal to the total power radiated by the antenna divided by 4π.

9.3.1.2 Radiation Efficiency Using the Reverberation Chamber

The radiation efficiency measurement method in the RC follows the guidance given in IEC 61000–4–21 [27] where a reference antenna with known efficiency, η_{REF}, was employed. It is noted that, in this chapter, the efficiency of all the reference antennas are obtained using the NPL AC. The antenna radiation efficiency of AUT, $\eta_{\text{AUT}_{\text{RC}}}$, can be calculated using the following formula:

$$\eta_{\text{AUT}_{\text{RC}}} = F_{\text{AUT}} . F_{\text{REF}} . \eta_{\text{REF}} \tag{9.6}$$

$$F_{\text{AUT}} = \frac{\left\langle \left| S_{21_{\text{AUT}}} \right|^2 \right\rangle}{\left(1 - \left| \left\langle S_{11_{\text{AUT}}} \right\rangle \right|^2\right) \left(1 - \left| \left\langle S_{22_{\text{AUT}}} \right\rangle \right|^2\right)} \tag{9.7}$$

$$F_{\text{REF}} = \frac{\left(1 - \left| \left\langle S_{22_{\text{REF}}} \right\rangle \right|^2\right) \left(1 - \left| \left\langle S_{11_{\text{REF}}} \right\rangle \right|^2\right)}{\left\langle \left| S_{21_{\text{REF}}} \right|^2 \right\rangle} \tag{9.8}$$

where S_{11}, S_{21}, and S_{22}, are the stationary S-parameters that are acquired for each paddle step under mode-tuned operation using a VNA. The averaged power transmission was measured using a fixed transmitting antenna and receiving with both the AUT and a reference antenna in turn. The triangular brackets, $\langle \blacksquare \rangle$, denote the average over all paddle steps. The average power is obtained from the average of S_{21} squared, which is adjusted by the receive antenna match, S_{22}. Note that the average of the complex S_{11} and S_{22} data is calculated before taking the square of the magnitude. In order to present the same load to the chamber both the AUT and reference antenna were present in the chamber at the same time. This regularises the Q-factor of the chamber. The antenna not connected to the receive cable was terminated with a 50 Ω load.

9.3.2 Comparison Between the AC and the RC

In this section, we present a comparison of antenna efficiency measurements performed in the AC and the RC at NPL where several omni- and directional antennas are considered.

9.3.2.1 Biconical Antenna

Biconical antennas are dipole antennas that are formed by placing two roughly conical conductive elements together, which offers an omnidirectional or dipole-like pattern and operates over a broad range of frequencies [1, 37]. The biconical elements allow a continuous sweep over the complete frequency range hence there is no need for a time-consuming change of the antenna elements as required when operating with tuned half-wave dipoles. Typical applications include wide bandwidth evaluation of test sites, measurement of field strength, generation of a defined field strength, and electromagnetic interference and immunity testing.

   A commercially available broadband biconical AUT was measured, namely, the Schwarzbeck Model SBA 9112 [38], in both the AC and the RC between 3 and 8 GHz. The AUT is designed with balun, which is envisaged to attenuate the common mode current that flows on the outside of the coaxial feed line. An ETS-Lindgren's Model 3117 double-ridged waveguide horn antenna was used as the Tx antenna in the AC whereas another Schwarzbeck Model SBA 9112 biconical antenna was used as the Tx antenna in the RC.

In the AC, the distance between the Tx antenna and the Rx AUT was 3.25 m and the height of the antennas above the chamber ground was 2.53 m. Both the VP and HP radiation pattern measurements were performed with an angular resolution of 5° in the roll and azimuth spherical axes. The two-antenna method was considered where an EMCO Model 3115 dual-ridge horn antenna was used as the reference gain antenna. For the RC measurement, an ETS-Lindgren's Model 3117 double-ridged waveguide horn antenna was used a reference efficiency antenna where the reference efficiency value was employed in (9.6) for obtaining the AUT radiation efficiency.

Figures 9.3 and 9.4 show the measurement setup in the AC and the RC, respectively. The measured 3D $E_\emptyset$ and E_θ radiation patterns of the AUT operating at 4 and 8 GHz, obtained from the AC, are shown in Figure 9.5.

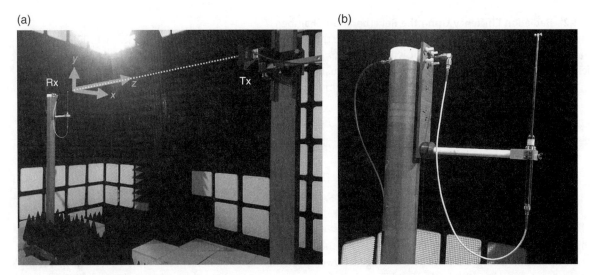

Figure 9.3 Schwarzbeck SBA 9112 Biconical AUT radiation efficiency measurement: (a) the AC measurement setup and (b) AUT.

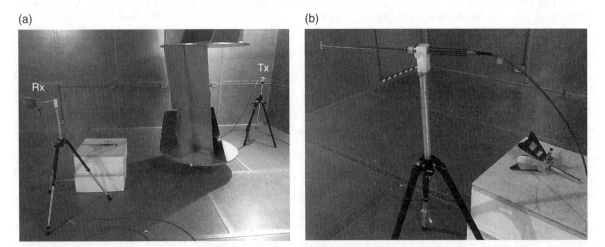

Figure 9.4 Schwarzbeck SBA 9112 Biconical AUT radiation efficiency measurement: (a) the RC measurement setup and (b) AUT. Note that REF antenna is terminated with a 50 Ω load.

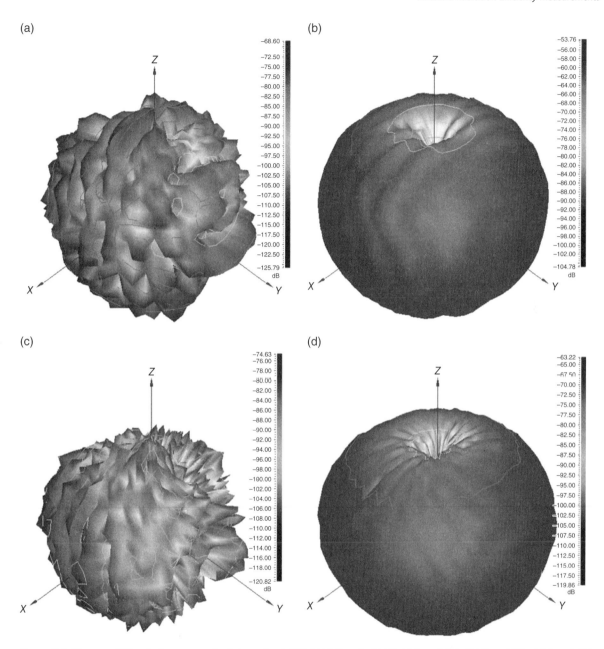

Figure 9.5 Measured 3D radiation pattern for Schwarzbeck SBA 9112 Biconical AUT: (a) E_ϕ at 4 GHz, (b) E_θ at 4 GHz, (c) E_ϕ at 8 GHz, and (d) E_θ at 8 GHz.

Note that the correspondent rectangular coordinations of the measurement are shown in Figure 9.3a where the endfire direction of the biconical antenna is aligned to the *z*-axis (i.e. the roll axis).

Figure 9.6 shows the comparison between the measured radiation efficiency of the Schwarzbeck biconical AUT in both the AC and the RC. The comparison shows reasonable cross-correlation between the AC and the RC.

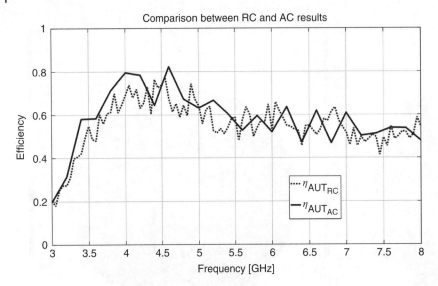

Figure 9.6 Antenna radiation efficiency comparison between the AC and the RC for Schwarzbeck SBA 9112 Biconical AUT.

9.3.2.2 Horn Antenna

Horn antennas consist of a flared metal waveguide shaped like a horn to direct radio waves into a beam and are used to transmit radio waves from a waveguide out into space or collect radio waves propagating in space into a waveguide for reception [1, 39]. They are widely used as feed antennas for larger antenna structures such as parabolic antennas, as standard antennas of known gain to measure the gain of other antennas, and as directive antennas. Horns can have different flare angles as well as different profiles in the E-field and H-field directions, making possible a wide variety of different beam shapes.

A commercially available double-ridged waveguide horn AUT was measured, namely, ETS-Lindgren's Model 3117, in both the AC and the RC between 1 and 8 GHz with 50 MHz frequency resolution. An EMCO Model 3115 and another ETS-Lindgren's Model 3117 double-ridged waveguide horn antenna were used, respectively, as the reference and Tx antennas in both the AC and the RC.

In the AC, the distance between the Tx antenna and the Rx AUT was 3.19 m and the height of the antennas above the chamber ground was 3.06 m. Both the VP and HP radiation pattern measurements were measured with an angular resolution of 5° in the roll and azimuth spherical axes. Note that in this subsection, the three-antenna method was considered. A detailed discussion of the effects of influence factors on two- and three-antenna methods in an AC for antenna gain and directivity measurement, and hence the antenna radiation efficiency are given in [40].

In the RC, the stationary S-parameters were acquired for each of 1000 paddle steps (i.e. the paddle rotates one-thousandth of a complete revolution between S-parameter measurements). Figures 9.7 and 9.8 show the measurement setup in the AC and the RC, respectively. The measured 3D $E_\emptyset$ and E_θ radiation patterns of the AUT operating at 1, 4, and 8 GHz, obtained in the AC, are shown in Figure 9.9.

As depicted in Figure 9.9, more side lobes are observed as the frequency increases. Note that the rectangular coordinates used in the measurements are shown in Figure 9.7a. In Figure 9.10, the measured radiation efficiency results obtained in the AC and the RC are compared for the ETS-Lindgren's Model 3117 double-ridged waveguide horn AUT. The comparison shows reasonable cross-correlation between the AC and the RC.

9.3.2.3 MIMO Antenna

The use of multiple-input-multiple-output (MIMO) antennas, coupled with modulation formats such as orthogonal frequency division multiple access (OFDMA), can provide both increased channel capacity and

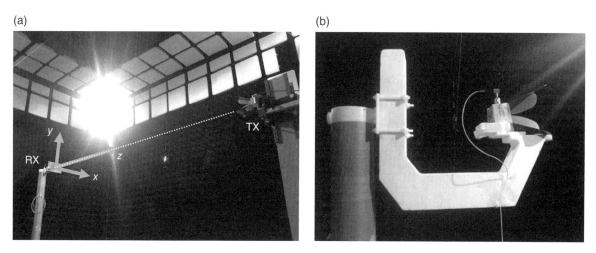

Figure 9.7 Double-ridged waveguide horn AUT radiation efficiency measurement: (a) the AC measurement setup and (b) AUT.

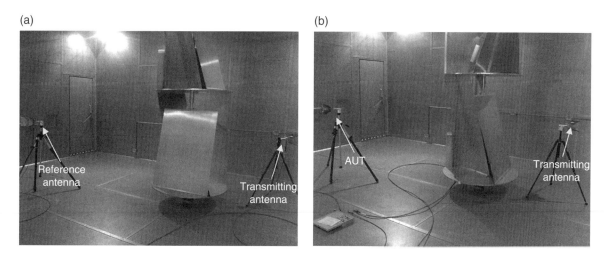

Figure 9.8 The RC measurement setup: (a) Tx-Reference antenna pair and (b) Tx-AUT antenna pair.

protection against multipath fading [41]. To date, these technologies have been embedded in fourth-generation (4G) wireless communication systems such as 3GPP Long-Term-Evolution (LTE), WiMAX, and Wi-Fi [42]. This technology will play a key role in the development of fifth-generation (5G) wireless mobile communication systems [43–45]. Furthermore, massive MIMO communication is an exciting area of 5G wireless research and it is envisaged that it will become one of the major drivers for new radio access technologies in 5G [46–48].

Antenna radiation efficiency is an important parameter for a MIMO antenna system [9, 49]. It is decreased by small array spacing and the reduction of efficiency reduces the channel capacity [50]. There are particular challenges in assessing the radiation efficiency of MIMO antennas as it has been demonstrated that both diversity gain and MIMO capacity depend upon the number of antennas, signal-to-noise ratio (SNR), and radiation efficiency [51].

To date, there are only a limited number of publications [49, 52] presenting work relating to characterising MIMO antennas and comparing them between the AC and the RC. The main focus in [49, 52] is on the characterisation of the diversity gain and the capacity of MIMO antennas. The relevant AC diversity evaluations

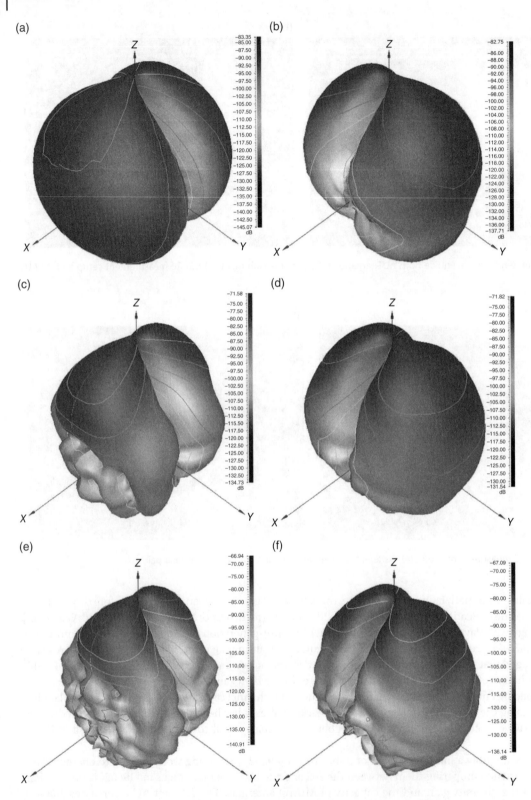

Figure 9.9 Measured 3D radiation pattern for ETS-Lindgren's Model 3117 double-ridged waveguide horn AUT: (a) E_ϕ at 1 GHz, (b) E_θ at 1 GHz, (c) E_ϕ at 4 GHz, (d) E_θ at 4 GHz, (e) E_ϕ at 8 GHz, and (f) E_θ at 8 GHz.

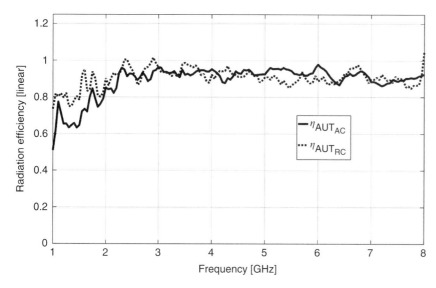

Figure 9.10 Antenna radiation efficiency comparison between the AC and the RC for ETS-Lindgren's Model 3117 double-ridged waveguide horn AUT.

require measurement of far-field radiation patterns and radiation efficiencies at every antenna port. However, considering the time consuming radiation pattern measurements in the AC, the authors in [49, 52] assume the MIMO antenna is symmetric and only measure radiation efficiencies at one port.

In this section, we present a study into comparing the measured results obtained in the AC and the RC for two different directional dual polarised full LTE band MIMO antennas [43]. The work aims to evaluate the antenna radiation efficiency at every antenna port as well as the antenna radiation efficiency when the ports are combined using a combiner.

Two commercially available directional dual polarised full LTE band two-port MIMO AUTs were measured, namely, Laird PAS69278 and Poynting XPOL-A0002, in both the AC and the RC between 1 and 3 GHz. An ETS-Lindgren 3117 double-ridged waveguide horn antenna was used as the Tx antenna in both the AC and the RC. To assess the MIMO AUT as a single-port AUT a power combiner Mini-Circuits ZN2PD2-50-S+ was used. A four-port Rohde & Schwarz ZVB8 VNA was used with output power of 0 dBm.

In the AC, the distance between the Tx antenna and the Rx AUT was 2.8 m and the height of the antennas above the chamber ground was 3.06 m. Both the VP and HP radiation pattern measurements were performed with an angular resolution of 5° in the roll and azimuth spherical axes. An ETS-Lindgren 3117 double-ridged waveguide horn antenna was used as the reference gain antenna.

In the RC, an ETS-Lindgren's model 3117 double-ridged waveguide horn antenna was used as the reference antenna of known efficiency. The efficiency of the reference antenna was measured in the AC using (9.3)–(9.5) (see Section 9.3.1.1). The stationary S-parameters were acquired for each of 1000 paddle steps (i.e. the paddle was rotated one-thousandth of a complete revolution between S-parameter measurements). Figures 9.11 and 9.12 show the measurement setup in the AC and the RC, respectively.

The measured 3D E_ϕ and E_θ radiation patterns obtained in the AC for both AUTs operating at 2.6 GHz are shown in Figures 9.13 and 9.14. Note that the rectangular coordinates used in the measurement are shown in Figure 9.11a. In Figures 9.15 and 9.16, the measured radiation efficiency results obtained in the AC and the RC are compared for the Laird PAS69278 and Poynting XPOL-A0002, respectively.

Note that in Figures 9.15 and 9.16 'Total' means that the two ports of the MIMO AUT are combined into one port to connect to the VNA (i.e. Port 1 of the VNA was connected to the Tx antenna and Port 2 of the VNA was connected to the combined port of the Rx antenna using the combiner). On the other hand, 'Channel One' and

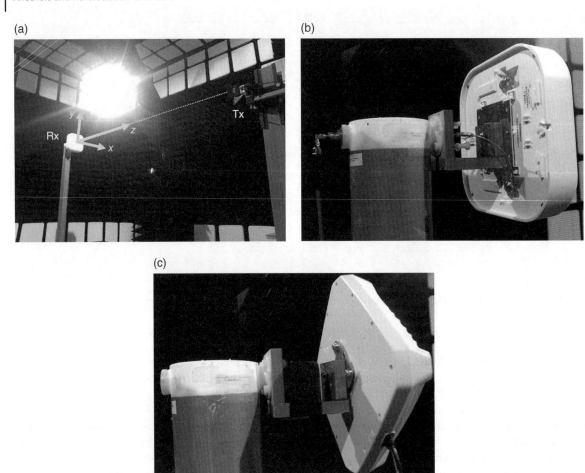

Figure 9.11 MIMO AUT radiation efficiency measurement: (a) the AC measurement setup, (b) the Laird PAS69278 MIMO AUT, and (c) the Laird PAS69278 MIMO AUT.

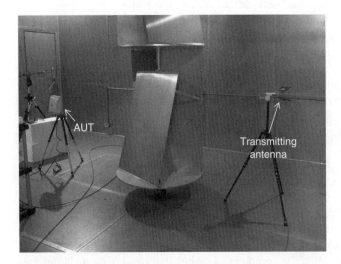

Figure 9.12 The NPL RC setup for MIMO AUT radiation efficiency measurement.

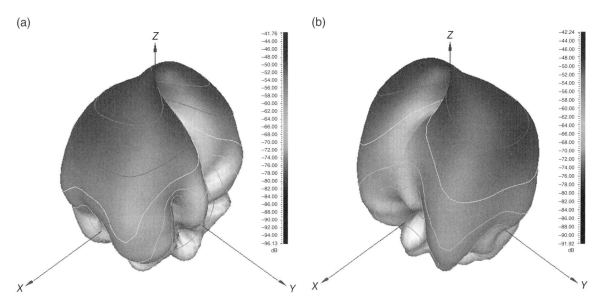

Figure 9.13 The measured 3D radiation pattern for Laird PAS69278 operating at 2.6 GHz: (a) E_ϕ and (b) E_θ field components.

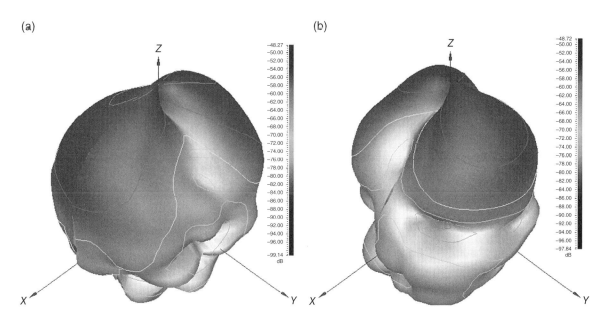

Figure 9.14 The measured 3D radiation pattern for Poynting XPOL-A0002 operating at 2.6 GHz: (a) E_ϕ and (b) E_θ field components.

'Channel Two' mean that the two ports of the MIMO AUT are each connected to a separate port of the VNA (i.e. Port 1 at the VNA was connected to the Tx antenna and Ports 2 and 3 of the VNA were connected to the two ports of the Rx antenna). The measured radiation efficiency results show reasonable agreement between the AC and the RC. Nevertheless, the 'Total', 'Channel One', and 'Channel Two' radiation efficiencies are all different.

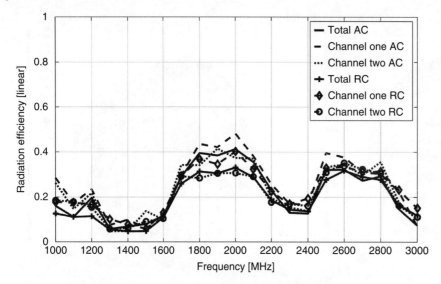

Figure 9.15 Antenna radiation efficiency comparison between the AC and the RC for Poynting XPOL-A0002 (operating frequency: 650–960 MHz, 1710–2170 MHz, and 2500–2700 MHz) AUT.

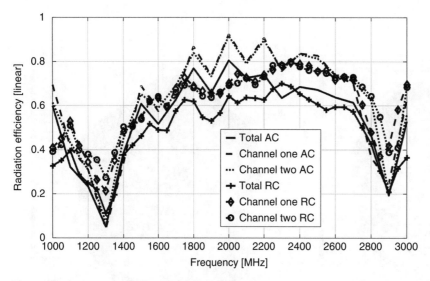

Figure 9.16 Antenna radiation efficiency comparison between the AC and the RC for Laird PAS69278 (operating frequency: 698–960 MHz and 1710–2700 MHz) AUT.

9.4 Summary

In this chapter, an inter-comparison of antenna radiation efficiency measurements between the AC and the RC was presented. Radiation efficiencies for both an omni-directional antenna (biconical antenna) and a directional antenna (horn antenna) measured in an AC were compared with that measured in an RC. The radiation efficiency of a MIMO antenna was evaluated at each antenna port as well as when the ports were combined and different efficiencies were obtained in each case. All the comparisons show reasonable cross-correlation between the AC and the RC. The RC measurement uncertainties for antenna radiation efficiency are covered in Chapter 8, and a review of different antenna efficiency measurement methods has also been given in [5] with typical uncertainty values. Note that the measurement uncertainty for antenna radiation efficiency has not been assessed in this chapter but it agrees well with the typical uncertainties reported in [5].

Acknowledgement

The work of T. H. Loh was supported by the 2014–2016 and the 2017–2020 National Measurement System Programme of the UK government's Department for Business, Energy, and Industrial Strategy (BEIS), under Science Theme Reference EMT of that Programme, and W. Qi was supported by the China State Foundation under a Studying Abroad Programme.

References

1 Balanis, C. A. (2016). *Antenna Theory: Analysis and Design*, 4e. Wiley-Blackwell.

2 IEEE Standards (2014). IEEE Standard for Definitions of Terms for Antennas. IEEE Std 145-2013 (Revision of IEEE Std 145-1993). pp. 1–50.

3 Liu, D., Hong, W., Rappaport, T. S. et al. (2017). What will 5G antennas and propagation be? *IEEE Transactions on Antennas and Propagation* 65 (12): 6205–6212.

4 Loh, T.H., Matthews, J., and Alexander, M. et al. (2011). Measurements of body wearable antennas. IET Seminar on Antenna and Propagation for Body-Centric Wireless Communications.

5 Huang, Y. (2014). Radiation efficiency measurements of small antennas. In: *Handbook of Antenna Technologies*, vol. 3 (ed. Z.N. Chen), 2165–2189. Springer.

6 Del Barrio, S.C. (2013). Tunable Antennas to Address the LTE Bandwidth Challenge on Small Mobile Terminals: One World, One Radio. PhD Thesis, Aalborg University.

7 Piette, M. (2004). Antenna radiation efficiency measurements in a reverberation chamber, Asia-Pacific Radio Science Conference. pp. 19–22.

8 Senic, D., Williams, D. F., Remley, C. A. et al. (2017). Improved Antenna Efficiency Measurement Uncertainty in a Reverberation Chamber at Millimeter-Wave Frequencies. *IEEE Transactions on Antennas and Propagation* 65 (8): 4209–4219.

9 Moongilan, D. (2017). Wireless MIMO base station near and far field measurement challenges. the 2017 IEEE International Symposium on Electromagnetic Compatibility & Signal/Power Integrity (EMCSI). Washington, DC.

10 Huang, Y., Lu, Y., Boyes, S., and Loh, T. H. (2011). Source-stirred method for antenna efficiency measurements. 5th European Conference on Antennas and Propagation. (EuCAP 2011), Rome.

11 Newman, E., Bohley, P., and Walter, C. (1975). Two methods for the measurement of antenna efficiency. *IEEE Transactions on Antennas and Propagation* 23 (4): 457–461.

12 Ulaby, F. T., Ravaioli, U., and Michielssen, E. (2014). *Fundamentals of Applied Electromagnetics*, 7e. Pearson Inc.

13 Holllis, J. S., Lyong, T. J., and Clayton, L. (1985). *Microwave Antenna Measurements*. Scientific-Atlanta Inc.

14 IEEE Standard Test Procedures for Antennas, IEEE Std 149-1979. IEEE Inc., distributed by Wiley-Interscience. (1979).

15 Foegelle, M. D. (2002). Antenna Pattern Measurement: Concepts and Techniques. *Compliance Engineering* 19 (3): 22–33.

16 Pozar, D. M. and Kaufman, B. (1988). Comparison of three methods for the measurement of printed antenna efficiency. *IEEE Transactions on Antennas and Propagation* 36: 136–139.

17 McKinzie, W.E. (1997). A modified Wheeler cap method for measuring antenna efficiency. IEEE Antennas and Propagation Society International Symposium 1997. Montreal, Canada. Volume 1. July 1997. pp. 542–545.

18 Huang, Y., Loh, T. H., and Foged, L.J. et al. (2010). Broadband antenna measurement comparisons. 4th European Conference on Antennas and Propagation (EuCAP 2010). Barcelona. 11–16.

19 Kildal, P. S. and Rosengren, K. (2004). Correlation and capacity of MIMO systems and mutual coupling, radiation efficiency, and diversity gain of their antennas: simulations and measurements in a reverberation chamber. *IEEE Communications Magazine* 42: 104–112.

20 Conway, G. A., Scanlon, W. G., and Orlenius, C. (2008). In situ measurement of UHF wearable antenna radiation efficiency using a reverberation chamber. *IEEE Antennas and Wireless Propagation Letters* 7: 271–274.

21 Boyes, S. J., Soh, P. J., Huang, Y. et al. (2013). Measurement and performance of textile antenna efficiency on a human body in a reverberation chamber. *IEEE Transactions on Antennas and Propagation* 61 (2): 871–881.

22 Li, C., Loh, T. H., Tian, Z. et al. (2017). Evaluation of chamber effects on antenna efficiency measurements using non-reference antenna methods in two reverberation chambers. *IET Microwaves, Antennas and Propagation* 11 (11): 1536–1541.

23 Li, C., Loh, T. H., Tian, Z. H., and Xu, Q. et al. (2015). A Comparison of Antenna Efficiency Measurements Performed in Two Reverberation Chambers Using Non-reference Antenna Methods. 11th Loughborough Antennas and Propagation Conference (LAPC 2015). Loughborough, 2–3 November 2015. Best Paper Award.

24 Ladbury, J.M. and Koepke, G. H. (1999). Reverberation chamber relationships: corrections and improvements or three wrongs can (almost) make a right. IEEE International Symposium on Electromagnetic Compatability. Symposium Record (Cat. No. 99CH36261), Seattle, WA. Volume 3. pp. 1–6.

25 Wilson, P., Holloway, C. L., and Koepke, G. (2004). A review of dipole models for correlating emission measurements made at various EMC test facilities. International Symposium on Electromagnetic Compatibility (IEEE Cat. No. 04CH37559), Volume 3. pp. 898–901.

26 IEC Standards (2010). *IEC 61000-4-20: 2010, Electromagnetic Compatibility, Part 4–20: Testing and Measurement Techniques – Emission and Immunity Testing in Transverse Electromagnetic (TEM) Waveguides.* BSI Standards Publication.

27 IEC Standards (2011). *IEC 61000-4-21: 2011, Electromagnetic Compatibility, Part 4–21: Testing and Measurement Techniques – Reverberation Chamber Test Methods.* BSI Standards Publication.

28 NPL SMART Anechoic Chamber, Available at: www.npl.co.uk/electromagnetics/rf-microwave/products-and-services/smart-antenna-testing-range.

29 Loh, T.H., Alexander, M., and Widmer, F. (2009). Validation of a new small-antenna radiated testing range. 3rd European Conference on Antennas and Propagation (EuCAP 2009). Berlin, Germany. pp. 699–703.

30 Dielectric Belt Driven Rotator model tower. Available at: http://www.orbitfr.com/sites/http://www.orbitfr.com/files/DBDR-ModelTowers_DataSheet.pdf.

31 Loh, T.H. (2013). Non-invasive measurement of electrically small ultra-wideband and smart antennas, Invited Paper, Loughborough Antennas and Propagation Conference (LAPC 2013). Loughborough, 11–12. pp. 456–460.

32 Loh, T.H., Alexander, M., Miller, P., and Betancort, A.L. (2010). Interference minimisation of antenna-to-range interface for pattern testing of electrically small antennas. 4th European Conference on Antennas and Propagation (EuCAP 2010). Barcelona. 11–16.

33 Loh, T. H., Meng, D., and Huang, Y. (2011). Effects of antenna mount on a spherical scanning measurement of the radiation efficiency of a wire antenna. 13th International Conference on Electromagnetics in Advanced Applications. (ICEAA 2011). Torino, Italy, 12-16. pp. 1176–1179.

34 CISPR Standards (2014). *CISPR 16–1-6, Specification for Radio Disturbance and Immunity Measuring Apparatus and Methods - Part 1–6: Radio Disturbance and Immunity Measuring Apparatus - EMC Antenna Calibration.* CISPR.

35 Rosengren, K., Kildal, P.S., Carlsson, C., and Carlssonm, J. (2001). Characterization of antennas for mobile and wireless terminals by using reverberation chambers: improved accuracy by platform stirring. IEEE International Symposium of Antennas and Propagation Society. Boston, MA. Volume 3. pp. 350–353.

36 Krauthäuser, H.G. and Herbrig, M. (2010). Yet another antenna efficiency measurement method in reverberation chambers. 2010 IEEE International Symposium on Electromagnetic Compatibility. Fort Lauderdale, FL.

37 Biconical antenna. Available at: https://en.wikipedia.org/wiki/Biconical_antenna.

38 Schwarzbeck Microwave Biconical Broadband Antenna. Available at: http://schwarzbeck.de/Datenblatt/k9112.pdf.

39 Horn Antenna. Available at: https://en.wikipedia.org/wiki/Horn_antenna.

40 Qi, W. and Loh, T. H. (2015). A study on the effects of influence factors for antenna radiation efficiency measurements in anechoic chamber. Antenna Measurement Techniques Association 37th Annual Meeting Symposium. (AMTA 2015). Long Beach, California, 11–16.

41 Brown, T., Kyritsi, P., and Carvalho, E. D. (2012). *Practical Guide to MIMO Radio Channel: With MATLAB Examples*. Wiley.

42 Rumney, M. (2013). *LTE and the Evolution to 4G Wireles: Design and Measurement Challenges*, 2e. Wiley-Blackwell.

43 Loh, T.H. and Qi, W. (2015). A comparison of MIMO antenna efficiency measurements performed in anechoic chamber and reverberation chamber. 85th ARFTG Microwave Measurement Conference. (ARFTG 2015). Phoenix, Arizona, 22.

44 Loh, T.H., Cheadle, D., and Miller, P. (2017). A millimeter wave MIMO testbed for 5G communications. 89th ARFTG Microwave Measurement Conference. (ARFTG 2017). Honolulu, Hawaii, 9th June, 2017. pp. 1–4.

45 Loh, T.H., Hudlicka, M., Brown, T. et al. (2016). Literature review of wireless link quality metrics. EURAMET EMPIR JRP 14IND10 Project entitled – Metrology for 5G Communications (MET5G), Activity A1.1.1 under project EURAMET EMPIR, 2 November, 2016.

46 Osseiran, A., Monserrat, J. F., and Marsch, P. (2016). *5G Mobile and Wireless Communications Technology*. Cambridge University Press.

47 Hu, F. (2016). *Opportunities in 5G Networks: A Research and Development Perspective*, 1e. CRC Press.

48 NetWorld2020. (2014). NetWorld2020 ETP (European Technology Platform for Communications Networks and Services). Expert Working Group on 5G. Challenges, Research Priorities, and Recommendations, Joint White Paper (Draft version published for public consultation). Available: http://networld2020.eu/wp-content/uploads/2014/02/NetWorld2020_Joint-Whitepaper-V8_public-consultation.pdf.

49 Chen, X. (2012). Characterization of MIMO Antennas and Terminals: Measurements in Reverberation Chambers. PhD thesis, Chalmers University of Technology.

50 Arai, H. (2013). An optimum design of handset antenna for MIMO systems. 7th European Conference on Antennas and Propagation. (EuCAP 2013). pp. 224–227.

51 Valenzuela-Valdes, J. F., Garcia-Fernandez, M. A., Martinez-Gonzalez, A. M., and Sanchez-Hernandez, D. A. (2008). The influence of efficiency on receive diversity and MIMO capacity for Rayleigh-fading channels. *IEEE Transactions on Antennas and Propagation* 56 (5): 1444–1450.

52 Chen, X., Kildal, S., Carlsson, J., and Yang, J. (2011). Comparison of ergodic capacities from wideband MIMO antenna measurements in reverberation chamber and anechoic chamber. *IEEE Antennas and Wireless Propagation Letters* 10: 466–469.

10

Discussion on Future Applications

10.1 Introduction

In this chapter, we discuss the possible future development and applications of the anechoic chamber (AC) and the reverberation chamber (RC). Although many applications have already been explored, we still face a lot of unsolved problems not only from the AC and RC themselves but also from possible applications.

10.2 Anechoic Chambers

It has been shown that the computer-aided design (CAD) tool is efficient and accurate for the AC simulation and suitable for real-world chamber design. It is important to note that geometric optics (GO) is a high frequency approximation method. In the measurement part we have compared the simulation and measurement results which were in good agreement, and the error was smaller than ±2 dB.

A potential problem is that at higher frequencies (the trend is moving to higher frequencies), the tip scattering of radio absorbing material (RAM) becomes significant and it is not considered in the model. When the tip scattering becomes the major contribution for the unexpected field (in millimetre wave), it will limit the boundary of the high frequency of this method. This could be in the region of statistical electromagnetics but not deterministic electromagnetics. Is there a way to investigate it in statistical electromagnetics and characterise the scattering field statistically?

Another issue is the diffraction or scattering of the rays, if the tip scattering effect is considered, the ray number will be huge and outside the ability of standard computers. For example, if the number of scattered rays is 1000, every time the ray is reflected 1000 extra rays need to be generated, which will increase the simulation time drastically (1000 times) and make the ray tracing algorithm useless. There has been some research [1] using an inverse algorithm to consider part (but not all) of the diffracted waves. How to deal with the scattered field is still a challenging problem not just in the electromagnetic (EM) community but also in the computer graphics community.

In terms of the applications, as massive multiple-input multiple-output (MIMO) antenna systems are becoming a key technology for future radio communication systems, how to evaluation such an antenna system has become a new topic. At the moment, the multiprobe method and the two-stage method [2, 3] have been proposed to be conducted in an AC. How these methods will evolve remains to be seen.

10.3 Reverberation Chambers

In this book, the RC is in a regular shape (an example is shown in Figure 10.1a). We have not discussed the general coupled cavities shown in Figure 10.1b. Existing work has given the mathematical model to describe the statistical property inside each small cavity [4–10]. It would be interesting to decouple the system and

Anechoic and Reverberation Chambers: Theory, Design, and Measurements, First Edition. Qian Xu and Yi Huang.
© 2019 John Wiley & Sons Ltd. Published 2019 by John Wiley & Sons Ltd.

(a)

(b)

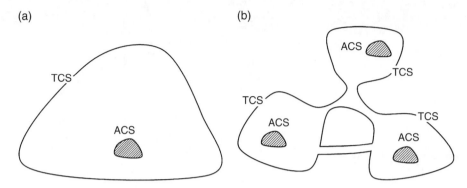

Figure 10.1 Generally coupled electrically large cavity: (a) normal shape and (b) general topology.

investigate the frequency domain and the time domain response in a generally coupled electrically large cavity (the time domain response no longer decays exponentially). This could be very useful for some practical electromagnetic compatibility (EMC) engineering problems.

In the permittivity measurement when the conductivity is very large, the absorption cross section (ACS) value will become less dependent on ε_r, which will increase the uncertainty for the measured ε_r. How to measure the object with a high conductivity could be a problem. Also, the proposed method is limited to the spherical object; is it possible to extend it to an arbitrarily shaped object? Currently, we have characterised the material property in the RC mainly by using the ACS. It would be possible to explore new information from the measured data. How to characterise a general material system in the RC has not been fully studied [11].

For the shielding effectiveness (SE) measurement using one antenna, when the equipment under test (EUT) is well shielded the uncertainty of the measured SE increases quickly. How to measure the decay time constant τ with a very small uncertainty/high resolution could be challenging, or finding the measurable range of the τ. This could be useful for many applications, as with ACS measurement [12]. For the volume measurement, if the cavity is partially filled with liquid or dielectric materials that either absorb or reflect waves completely (inhomogeneous cavity), the wave velocity is changed, which could increase the measurement error. How the inhomogeneous materials affect the Hill's equation, and how to generalise this equation into an inhomogeneous environment could be an interesting problem.

We have demonstrated the use of the RC to characterise the global absorption coefficient of an arbitrarily shaped object by using the ACS in Section 7.9.2. The total/average scattering cross section has also been investigated in Section 6.5. They are all global properties. The RC has its own advantages to measure global properties like radiation efficiency, total isotropic sensitivity, SE, total radiated power, average ACS, and average scattering cross section. An interesting problem is: is it possible to measure the angle dependency variables in the RC? We have shown the radiation pattern extraction in the RC, how about the scattering cross section in a specific direction or in a range of angles? It could be possible to realise the 'passive sensing' in an RC, which extracts the impulse response (Green's function) between two points by using the cross-correlation of signals captured from them [13–16]. This technique has been used in the seismology, and it would be useful to apply it in the RC.

On the statistical distribution of variables, there is an alternative way to obtain the probability distribution function (PDF) – the maximum (differential) entropy principle. The differential entropy can be interpreted as a functional of the PDF: by applying certain constraints and maximise the entropy, the PDF can be obtained [17]. The Gaussian distribution can be obtained from the maximum entropy method by applying the mean and standard deviation of the random variable in the unbounded interval of $(-\infty, +\infty)$ [18]. The Gaussian distribution has also been obtained by using the minimum energy and maximum entropy principles in [19]. The concept of Rényi entropy has been used to describe the stirring performance of the RC in the frequency domain

[20] and in the time domain [21]. It is interesting to use the techniques in information theory to understand the physical phenomenon in the RC, and the relationships among many parameters are still unknown. There could be more understanding from this distinctive perspective.

With the development of 5G communication, an important application is the RC in the millimetre wave range [22]. It has been found that in a hyper-sized RC (electrical volume > $(100\lambda)^3$) [19], the enhanced backscatter constant e_b deviates from 2 [23]. This phenomenon could lead to the concept of the 'highest usable frequency' (HUF), although the 'lowest usable frequency' (LUF) is normally of interest. We have to admit the HUF must exist, as a laser beam cannot be well diffused in the RC due to absorption or other reasons. Although there is an unexpected phenomenon at high frequencies, it does not affect the use of the RC at some applications when the cavity is not that large. Experimental verifications have been performed to demonstrate the use of a small volume RC at millimetre wave [24]; the RC works well as expected. Recently, the millimetre RC has been used in the over-the-air (OTA) test [25–30], and it has been found that when the cavity is not that large, the condition of the non-reference two-antenna method can be satisfied ($e_{b1} = e_{b2}$) and can be used as well [25]. Practical engineering has demonstrated the validity of the small volume RC (still electrically large but not hyper-size), but what factors affect the HUF and how to predict the HUF have not been fully studied. We have discussed some typical MIMO OTA measurements in this book and more applications could be possible; since an RC is already a multipath environment and very different from an AC, it may offer unique advantages over an AC in this important area.

The RC technique has been developed for around 50 years, and has become more and more popular. There is still undiscovered knowledge that we are not aware of, and it is believed that, with the effort of scientists and engineers, the RC will be employed for more applications in both academia and industry.

References

1 Choudhury, B., Singh, H., Bommer, J. P., and Jha, R. M. (2013). RF field mapping inside a large passenger-aircraft cabin using a refined ray-tracing algorithm. *IEEE Antennas and Propagation Magazine* 55 (1): 276–288.

2 Rumney, M., Kong, H., and Jing, Y. (2014). Practical active antenna evaluation using the two-stage MIMO OTA measurement method. 8th European Conference on Antennas and Propagation (EuCAP). The Hague. pp. 3500–3503.

3 Glazunov, A. A., Kolmonen, V. -M., and Laitinen, T. (2012). MIMO over-the-air testing. In: *LTE-Advanced and Next Generation Wireless Networks Channel Modelling and Propagation* (ed. G. de la Roche, A.A. Glazunov and B. Allen). Wiley.

4 Tian, Z., Huang, Y., and Xu, Q. Efficient methods of measuring shielding effectiveness of electrically large enclosures using nested reverberation chambers with only two antennas. *IEEE Transactions on Electromagnetic Compatibility* doi: 10.1109/TEMC.2017.2696743.

5 Romero, S. F., Gutieerrez, G., and Gonzalez, I. (2015). A shielding effectiveness prediction method for coupled reverberant cavities validated on a real object. *Journal of Electromagnetic Waves and Applications* 29 (14): 1829–1840.

6 Tait, G. B., Richardson, R. E., Slocum, M. B. et al. (2011). Reverberant microwave propagation in coupled complex cavities. *IEEE Transactions on Electromagnetic Compatibility* 53 (1): 229–232.

7 Giuseppe, J. S., Hager, C., and Tait, G. B. (2011). Wireless RF energy propagation in multiply-connected reverberant spaces. *IEEE Antennas and Wireless Propagation Letters* 10: 1251–1254.

8 Richardson, R. (2008). Reverberant microwave propagation. Naval Surface Warfare Center, Dahlgren Division, Dahlgren, VA, USA, Technical Report NSWCDD/TR-08/127.

9 Davis, A. H. (1925). Reverberation equations for two adjacent rooms connected by an incompletely soundproof partition. *Philosophical Magazine Series 6* 50 (295): 75–80.

10 Pierce, A. D. (1981). *Acoustics: An Introduction to its Physical Principles and Applications*. New York, NY, USA: McGraw-Hill, ch. 6.

11 Arnaut, L. R. (2000). *Complex Media in Complex Fields: A Statistical Approach*. Defense Technical Information Center.

12 Flintoft, I. D., Bale, S. J., Parker, S. L. et al. (2016). On the measurable range of absorption cross section in a reverberation chamber. *IEEE Transactions on Electromagnetic Compatibility* 58 (1): 22–29.

13 Weaver, R. L. and Lobkis, O. I. (2001). Ultrasonics without a source: thermal fluctuation correlations at MHz frequencies. *Physical Review Letters* 87 (13): 134301.

14 Davy, M., Fink, M., and Rosny, J. (2013). Green's function retrieval and passive imaging from correlations of wideband thermal radiations. *Physical Review Letters* 110: 203901.

15 Snieder, R. (2004). Extracting the Green's function from the correlation of coda waves: a derivation based on stationary phase. *Physical Review E* 69: 046610.

16 Weaver, R. and Lobkis, O. (2002). On the emergence of the Green's function in the correlations of a diffuse field: pulse-echo using thermal phonons. *Ultrasonics* 40: 435–439.

17 Naus, H. W. L. (2008). Statistical electromagnetics: complex cavities. *IEEE Transactions on Electromagnetic Compatibility* 50 (2): 316–324.

18 Hill, D. A. (2009). *Electromagnetic Fields in Cavities: Deterministic and Statistical Theories*. Wiley.

19 Demoulin, B. and Besnier, P. (2011). *Electromagnetic Reverberation Chambers*. Wiley.

20 Pirkl, R. J., Remley, K. A., and Patane, C. S. L. (2012). Reverberation chamber measurement correlation. *IEEE Transactions on Electromagnetic Compatibility* 54 (3): 533–545.

21 Gradoni, G., Primiani, V. M., and Moglie, F. (2012). Reverberation chamber as a statistical relaxation process: entropy analysis and fast time domain simulations. International Symposium on Electromagnetic Compatibility – EMC EUROPE. Rome. pp. 1–6.

22 Fall, A. K., Besnier, P., Lemoine, C. et al. (2015). Design and experimental validation of a mode-stirred reverberation chamber at millimeter waves. *IEEE Transactions on Electromagnetic Compatibility* 57 (1): 12–21.

23 Dunlap, C.R. (2013). Reverberation chamber characterization using enhanced backscatter coefficient measurements. PhD Dissertation, Department of Electrical, Computer and Engineering, University of Colorado, Boulder.

24 Fall, A. K., Besnier, P., Lemoine, C. et al. (2015). Design and experimental validation of a mode-stirred reverberation chamber at millimeter waves. *IEEE Transactions on Electromagnetic Compatibility* 57 (1): 12–21.

25 Senic, D., Williams, D. F., Remley, K. A. et al. Improved antenna efficiency measurement uncertainty in a reverberation chamber at millimeter-wave frequencies. *IEEE Transactions on Antennas and Propagation* doi: 10.1109/TAP.2017.2708084.

26 Senic, D., Remley, K. A., Wang, C. M. et al. (2016). Estimating and reducing uncertainty in reverberation-chamber characterization at millimeter-wave frequencies. *IEEE Transactions on Antennas and Propagation* 64 (7): 3130–3140.

27 Lötbäck, C.S.P. (2017). Extending the frequency range of reverberation chamber to millimeter waves for 5G over-the-air testing. 11th European Conference on Antennas and Propagation (EUCAP). Paris. pp. 3012–3016.

28 Fall, A. K., Besnier, P., Lemoine, C. et al. (2016). Experimental dosimetry in a mode-stirred reverberation chamber in the 60-GHz band. *IEEE Transactions on Electromagnetic Compatibility* 58 (4): 981–992.

29 Senic, D., Remley, K. A., and Williams, D. F. et al. (2016). Radiated power based on wave parameters at millimeter-wave frequencies for integrated wireless devices. 88th ARFTG Microwave Measurement Conference (ARFTG). Austin, TX. pp. 1–4.

30 Santoni, F., Pastore, R., and G. Gradoni et al. (2016). Experimental characterization of building material absorption at mmWave frequencies: by using reverberation chamber in the frequency range 50–68 GHz. IEEE Metrology for Aerospace (MetroAeroSpace). Florence. pp. 166–171.

Appendix A

Code Snippets

A.1 Import STL File into Matlab

STL (STereoLithography) is a file format native to the STL computer-aided design (CAD) software created by 3D Systems. STL files describe only the surface geometry of a three-dimensional object without any representation of colour, texture, or other common CAD model attributes. This code snippet imports STL file with name STLFile.stl into Matlab®. The program checks the STL file line by line and records the coordinates of each triangle. A progress bar appears during the reading process. After importing the STL file, it is plotted using the 'patch' command.

```
STL_FILE_PATH='D:\STLFile.stl';    % The location of STLFile.stl
t.Start = tic;                     % Start timer
fidin=fopen(STL_FILE_PATH);        % Open STL file
Ind_Facet=1;                       % The index of the face
Ind_Vert=1;                        % The index of the point of the face 1,2,3.
fprintf('> STL file path: \n');
fprintf('> %s \n', STL_FILE_PATH);
fprintf('> Reading STL file.....\n');
h = waitbar(0,'Reading STL file...'); % Initialise the waitbar
LINE_NUM=0;
while ~feof(fidin)
  fgetl(fidin);
  LINE_NUM=LINE_NUM+1;                 % Count the line number
end
fclose('all');
fidin=fopen(STL_FILE_PATH);
Lin_Ind=0;Perct1=0;

while ~feof(fidin)              % Check if it is the end of file
  tline=fgetl(fidin);          % Read line by line
  tline=strtrim(tline);        % Trim the space of this line
  Lin_Ind=Lin_Ind+1;Perct2=Lin_Ind/LINE_NUM;
  if Perct2-Perct1>0.01
    waitbar(Lin_Ind/LINE_NUM,h,sprintf('Reading STL
file...%6.2f%%',Lin_Ind/LINE_NUM*100));     % Update wait bar
    Perct1=Perct2;
  end
```

Anechoic and Reverberation Chambers: Theory, Design, and Measurements, First Edition. Qian Xu and Yi Huang.
© 2019 John Wiley & Sons Ltd. Published 2019 by John Wiley & Sons Ltd.

```matlab
    if strcmp(tline(1:5),'solid')        % Check if this line start with 'solid'
      fprintf('> Solid name: %s.\n',tline(7:end));    % Print the name of the solid
      SOLID_NAME=tline(7:end);
    elseif strcmp(tline(1:5),'facet')    % Check if this line start with 'facet'
      temp=str2num(tline(13:end));
      Fac_NormX(Ind_Facet)=temp(1);      % Read the normal vector of the face
      Fac_NormY(Ind_Facet)=temp(2);
      Fac_NormZ(Ind_Facet)=temp(3);
    elseif min(tline(1:6)=='vertex') && Ind_Vert==1   % Read the 1st point of the face
      temp=str2num(tline(7:end));
      VertX1(Ind_Facet)=temp(1);
      VertY1(Ind_Facet)=temp(2);
      VertZ1(Ind_Facet)=temp(3);
      Ind_Vert=2;
    elseif min(tline(1:6)=='vertex')&&Ind_Vert==2    % Read the 2nd point of the face
      temp=str2num(tline(7:end));
      VertX2(Ind_Facet)=temp(1);
      VertY2(Ind_Facet)=temp(2);
      VertZ2(Ind_Facet)=temp(3);
      Ind_Vert=3;
    elseif min(tline(1:6)=='vertex')&&Ind_Vert==3    % Read the 3rd point of the face
      temp=str2num(tline(7:end));
      VertX3(Ind_Facet)=temp(1);
      VertY3(Ind_Facet)=temp(2);
      VertZ3(Ind_Facet)=temp(3);
      Ind_Vert=1;
      Ind_Facet=Ind_Facet+1;
    elseif min(tline(1:6)=='endsol')
      tElapsed = toc(tStart);
      NUM_FACES=Ind_Facet-1;
      fprintf('> Reading STL file finished with %g triangle faces. \n', NUM_FACES);
      fprintf('> Reading .stl time consuming %g secs. \n', tElapsed);
    else
    end
end
waitbar(1,h,sprintf('Reading STL file finished.'));
fclose('all'); delete(h);       %Finished reading, close open file, delete waitbar

WholeXmin=min([min(VertX1),min(VertX2),min(VertX3)]);
WholeYmin=min([min(VertY1),min(VertY2),min(VertY3)]);
WholeZmin=min([min(VertZ1),min(VertZ2),min(VertZ3)]);
WholeXmax=max([max(VertX1),max(VertX2),max(VertX3)]);
WholeYmax=max([max(VertY1),max(VertY2),max(VertY3)]);
WholeZmax=max([max(VertZ1),max(VertZ2),max(VertZ3)]);
%=====================2.Display .stl file=====================
STL_disp_x=[VertX1(1:NUM_FACES);
  VertX2(1:NUM_FACES);
  VertX3(1:NUM_FACES)];
STL_disp_y=[VertY1(1:NUM_FACES);
```

```
  VertY2(1:NUM_FACES);
  VertY3(1:NUM_FACES)];
STL_disp_z=[VertZ1(1:NUM_FACES);
  VertZ2(1:NUM_FACES);
  VertZ3(1:NUM_FACES)];
figure('Name','Imported STL Model', 'NumberTitle','off')
  p=patch(STL_disp_x,STL_disp_y,STL_disp_z,[173,235,255]./255);
  view(3);
  axis([WholeXmin WholeXmax WholeYmin WholeYmax WholeZmin WholeZmax]);
  xlabel('X axis');ylabel('Y axis');zlabel('Z axis');
  grid on;box on; set(gcf,'Color','w');
 figure(gcf);
  fprintf('> Display the .stl model..... \n');
```

A.2 Possible Intersection Check Between a Triangle and a Hexahedron

This Matlab function checks if there is any potential intersection between a triangle and a hexahedron. Because it is not necessary to apply a strict check in the octree allocation for the triangles, this intersection check function provides a necessary, but not sufficient, condition.

```
function ReturnValue=TriCubeInterTest(TriVert1,TriVert2,TriVert3,XYZList)
% TriVert1=[VertX1 VertY1 VertZ1]
% TriVert2=[VertX2 VertY2 VertZ2]
% TriVert3=[VertX3 VertY3 VertZ3]
% Tolerance; 1e-3 m
CubeXmin=XYZList(1);CubeXmax=XYZList(2);
CubeYmin=XYZList(3);CubeYmax=XYZList(4);
CubeZmin=XYZList(5);CubeZmax=XYZList(6);

TriBdXmin=min([TriVert1(1),TriVert2(1),TriVert3(1)])-1e-3;
TriBdXmax=max([TriVert1(1),TriVert2(1),TriVert3(1)])+1e-3;
TriBdYmin=min([TriVert1(2),TriVert2(2),TriVert3(2)])-1e-3;
TriBdYmax=max([TriVert1(2),TriVert2(2),TriVert3(2)])+1e-3;
TriBdZmin=min([TriVert1(3),TriVert2(3),TriVert3(3)])-1e-3;
TriBdZmax=max([TriVert1(3),TriVert2(3),TriVert3(3)])+1e-3;

OverlapX=(TriBdXmin>CubeXmin && TriBdXmin<CubeXmax) ||...
    (TriBdXmax>CubeXmin && TriBdXmax<CubeXmax) ||...
    (CubeXmin>TriBdXmin && CubeXmin<TriBdXmax) ||...
    (CubeXmax>TriBdXmin && CubeXmax<TriBdXmax);
OverlapY=(TriBdYmin>CubeYmin && TriBdYmin<CubeYmax) ||...
    (TriBdYmax>CubeYmin && TriBdYmax<CubeYmax) ||...
    (CubeYmin>TriBdYmin || CubeYmin<TriBdYmax) ||...
    (CubeYmax>TriBdYmin && CubeYmax<TriBdYmax);
OverlapZ=(TriBdZmin>CubeZmin && TriBdZmin<CubeZmax) ||...
    (TriBdZmax>CubeZmin && TriBdZmax<CubeZmax) ||...
    (CubeZmin>TriBdZmin && CubeZmin<TriBdZmax) ||...
    (CubeZmax>TriBdZmin && CubeZmax<TriBdZmax);
```

```
if OverlapX && OverlapY && OverlapZ
  ReturnValue=true;return;
else
  ReturnValue=false;return;
end
```

A.3 Intersection Check Between a Ray and a Hexahedron in the Octree

This Matlab function checks if a ray intersects a hexahedron, an error tolerance of 10^{-4} is used.

```
function IntFlag=RayCubIntTest(StartPoint,DirLstXYZ,CubeXYZ)
% StartPoint=[StartPointX StartPointY StartPointZ]
% DirLstXYZ=[DirLstX DirLstY DirLstZ]
% CubeXYZ=[CubeXmin CubeXmax CubeYmin CubeYmax CubeZmin CubeZmax] ;
% x=StartPointX+t*DirLstX
% y=StartPointY+t*DirLstY
% z=StartPointZ+t*DirLstZ
CubeXYZ(1)=CubeXYZ(1)-1e-4;CubeXYZ(2)=CubeXYZ(2)+1e-4;
CubeXYZ(3)=CubeXYZ(3)-1e-4;CubeXYZ(4)=CubeXYZ(4)+1e-4;
CubeXYZ(5)=CubeXYZ(5)-1e-4;CubeXYZ(6)=CubeXYZ(6)+1e-4;

% Check if the startpoint is inside the cube
if StartPoint(1)>=CubeXYZ(1) && StartPoint(1)<=CubeXYZ(2) && ...
    StartPoint(2)>=CubeXYZ(3) && StartPoint(2)<=CubeXYZ(4) && ...
    StartPoint(3)>=CubeXYZ(5) && StartPoint(3)<=CubeXYZ(6)
  IntFlag=true;return;
else  % The StartPoint is outside the cube, solve for t
  % Check Xmin
  tXmin=(CubeXYZ(1)-StartPoint(1))/DirLstXYZ(1);
  % Check Xmax
  tXmax=(CubeXYZ(2)-StartPoint(1))/DirLstXYZ(1);
  % Check Ymin
  tYmin=(CubeXYZ(3)-StartPoint(2))/DirLstXYZ(2);
  % Check Ymax
  tYmax=(CubeXYZ(4)-StartPoint(2))/DirLstXYZ(2);
  % Check Zmin
  tZmin=(CubeXYZ(5)-StartPoint(3))/DirLstXYZ(3);
  % Check Zmax
  tZmax=(CubeXYZ(6)-StartPoint(3))/DirLstXYZ(3);

  IntXmin=StartPoint+tXmin*DirLstXYZ;  % Solve for the intersection coordinate
  IntXmax=StartPoint+tXmax*DirLstXYZ;

  IntYmin=StartPoint+tYmin*DirLstXYZ;
  IntYmax=StartPoint+tYmax*DirLstXYZ;

  IntZmin=StartPoint+tZmin*DirLstXYZ;
  IntZmax=StartPoint+tZmax*DirLstXYZ;
```

```
    if tXmin>0 && IntXmin(2)>CubeXYZ(3) && IntXmin(2)<CubeXYZ(4) &&
IntXmin(3)>CubeXYZ(5) && IntXmin(3)<CubeXYZ(6)
        % y,z in the cube area
        IntFlag=true;return;    % if y,z in the range
      elseif tXmax>0 && IntXmax(2)>CubeXYZ(3) && IntXmax(2)<CubeXYZ(4) &&
IntXmax(3)>CubeXYZ(5) && IntXmax(3)<CubeXYZ(6)
        % y,z in the cube area
        IntFlag=true;return;    % if y,z in the range
      elseif tYmin>0 && IntYmin(1)>CubeXYZ(1) && IntYmin(1)<CubeXYZ(2) &&
IntYmin(3)>CubeXYZ(5) && IntYmin(3)<CubeXYZ(6)
        % x,z in the range
        IntFlag=true;return;
      elseif tYmax>0 && IntYmax(1)>CubeXYZ(1) && IntYmax(1)<CubeXYZ(2) &&
IntYmax(3)>CubeXYZ(5) && IntYmax(3)<CubeXYZ(6)
        % x,z in the range
        IntFlag=true;return;
      elseif tZmin>0 && IntZmin(1)>CubeXYZ(1) && IntZmin(1)<CubeXYZ(2) &&
IntZmin(2)>CubeXYZ(3) && IntZmin(2)<CubeXYZ(4)
        % x,y in the range
        IntFlag=true;return;
      elseif tZmax>0 && IntZmax(1)>CubeXYZ(1) && IntZmax(1)<CubeXYZ(2) &&
IntZmax(2)>CubeXYZ(3) && IntZmax(2)<CubeXYZ(4)
        % x,y in the range
        IntFlag=true;return;
      else
        IntFlag=false;return;
      end
end
```

A.4 Intersection Check Between a Ray and a Triangle

This Matlab function checks if a ray intersects a triangle. Note, the code is not easy to read as it has been generated by computer using the CodeGeneration package in Maple®.

```
function [IntFlag,IntPtCoord,Distance]=RayTriIntTest(StartPoint,DirLstXYZ,
  TriXYZ)
% StartPoint=[StartPointX StartPointY StartPointZ]
% DirLstXYZ=[DirLstX DirLstY DirLstZ]
% TriXYZ=[X1 Y1 Z1 X2 Y2 Z2 X3 Y3 Z3]
% Solve Linear Equation Ax=b
% t_A=[-DirLstXYZ(1) TriXYZ(4)-TriXYZ(1) TriXYZ(7)-TriXYZ(1);...
%    -DirLstXYZ(2) TriXYZ(5)-TriXYZ(2) TriXYZ(8)-TriXYZ(2);...
%    -DirLstXYZ(3) TriXYZ(6)-TriXYZ(3) TriXYZ(9)-TriXYZ(3)];
% t_b=[StartPoint(1)-TriXYZ(1);StartPoint(2)-TriXYZ(2);StartPoint(3)-
  TriXYZ(3)];
% [t_x,~]=linsolve(t_A,t_b);
%t_x=t_A\t_b;
```

```
t3=DirLstXYZ(3)*TriXYZ(1);
t4=t3*TriXYZ(8);
t6=DirLstXYZ(3)*TriXYZ(2)*StartPoint(1);
t9=DirLstXYZ(2)*TriXYZ(1);
t10=t9*TriXYZ(9);
t12=DirLstXYZ(2)*TriXYZ(3)*StartPoint(1);
t15=DirLstXYZ(1)*TriXYZ(2);
t16=t15*TriXYZ(9);
t18=DirLstXYZ(1)*TriXYZ(3)*StartPoint(2);
t19=DirLstXYZ(3)*TriXYZ(7);
t21=t19*TriXYZ(2);
t22=t3*StartPoint(2);
t23=DirLstXYZ(2)*TriXYZ(7);
t25=t23*TriXYZ(3);
t26=t9*StartPoint(3);
t27=DirLstXYZ(1)*TriXYZ(8);
t29=t27*TriXYZ(3);
t30=t15*StartPoint(3);
t31=-DirLstXYZ(3)*TriXYZ(8)*StartPoint(1)+t4+t6+DirLstXYZ(2)*TriXYZ(9)
*StartPoint(1)-t10-t12-DirLstXYZ(1)*TriXYZ(9)*StartPoint(2)+t16+t18
+t19*StartPoint(2)-t21-t22-t23*StartPoint(3)+t25+t26+t27*StartPoint(3)-
t29-t30;
t32=DirLstXYZ(3)*TriXYZ(4);
t34=t32*TriXYZ(2);
t36=t3*TriXYZ(5);
t37=DirLstXYZ(2)*TriXYZ(4);
t39=t37*TriXYZ(3);
t41=t9*TriXYZ(6);
t42=DirLstXYZ(1)*TriXYZ(5);
t44=t42*TriXYZ(3);
t46=t15*TriXYZ(6);
t47=t32*TriXYZ(8)-t34-t4-t19*TriXYZ(5)+t21+t36-
t37*TriXYZ(9)+t39+t10+t23*TriXYZ(6)-t25-t41+t42*TriXYZ(9)-t44-t16-
t27*TriXYZ(6)+t29+t46;
t48=0.1e1/t47;
u=-t31*t48;
if u<0 || u>1
  IntFlag=false;
  IntPtCoord=NaN;
  Distance=NaN;
  return;
end
t59=DirLstXYZ(2)*TriXYZ(6)*StartPoint(1)-t41-t12-
DirLstXYZ(3)*TriXYZ(5)*StartPoint(1)+t36+t6+t32*StartPoint(2)-t34-t22-
DirLstXYZ(1)*TriXYZ(6)*StartPoint(2)+t46+t18+t42*StartPoint(3)-t44-t30-
t37*StartPoint(3)+t39+t26;
v=t59*t48;
if v<0 || v>1
```

```
  IntFlag=false;
  IntPtCoord=NaN;
  Distance=NaN;
  return;
end
t60=TriXYZ(8)*TriXYZ(6);
t66=TriXYZ(4)*TriXYZ(9);
t73=TriXYZ(7)*TriXYZ(6);
t80=TriXYZ(4)*TriXYZ(8);
t82=-t60*TriXYZ(1)-TriXYZ(8)*TriXYZ(3)*StartPoint(1)-
TriXYZ(2)*TriXYZ(6)*StartPoint(1)+t66*StartPoint(2)-t66*TriXYZ(2)-
TriXYZ(4)*TriXYZ(3)*StartPoint(2)-TriXYZ(1)*TriXYZ(9)*StartPoint(2)-
t73*StartPoint(2)+t73*TriXYZ(2)+TriXYZ(7)*TriXYZ(3)*StartPoint(2)+TriXYZ(1)
*TriXYZ(6)*StartPoint(2)-t80*StartPoint(3);
t88=TriXYZ(7)*TriXYZ(5);
t95=TriXYZ(5)*TriXYZ(9);
t103=t80*TriXYZ(3)+TriXYZ(4)*TriXYZ(2)*StartPoint(3)+TriXYZ(1)*TriXYZ(8)
*StartPoint(3)+t88*StartPoint(3)-t88*TriXYZ(3)-TriXYZ(7)*TriXYZ(2)
*StartPoint(3)-TriXYZ(1)*TriXYZ(5)*StartPoint(3)-t95*StartPoint(1)
+t95*TriXYZ(1)+TriXYZ(5)*TriXYZ(3)*StartPoint(1)+TriXYZ(2)*TriXYZ(9)
*StartPoint(1)+t60*StartPoint(1);
t=(t82+t103)*t48;
if u+v<=1 && t>1e-4
  IntFlag=true;
  IntPtCoord=StartPoint+t*DirLstXYZ;
  Distance=t;
  return;
else
  IntFlag=false;
  IntPtCoord=NaN;
  Distance=NaN;
  return;
end
```

The Maple code used to generate the above Matlab code is also listed below. As can be seen, after optimisation the multiplication times have been reduced from 228 to 63. Computer code generation is very useful, as it can generate optimised code for expressions and minimise the errors when typing the code manually.

```
restart;with(CodeGeneration):with(LinearAlgebra):with(codegen, cost,
optimize):
t_A:=Matrix([[-rx, v1x-v0x, v2x-v0x], [-ry, v1y-v0y,v2y-v0y],
  [-rz,v1z-v0z,v2z-v0z]]):
b:=Matrix([[p0x-v0x], [p0y-v0y], [p0z-v0z]]):
Sol:=LinearSolve(A,b):
OriginalCode:=[t=Sol[1][1],u=Sol[2][1],v=Sol[3][1]]:
cost(OriginalCode);
          110*additions+228*multiplications+3*divisions+3*assignments
cost(optimize(OriginalCode,'tryhard'));
          63*multiplications+21*assignments+60*additions+divisions
Matlab(OriginalCode,optimize);
```

A.5 Field Uniformity Extraction

This Matlab code extracts the field uniformity with 75% of the measured 16 points. The code can be generalised to calculate the field uniformity with an arbitrary percentage from an arbitrary number of samples. For example, if the field uniformity is required for all the sample points (100%), we just need to modify 0.75 to 1. The *EfieldList* saves all the unsorted measured readings from all sample points.

```matlab
FUPtsNo=numel(EfieldList); % EfieldList saves the measured readings (V/m) from
  all points.
SortE=sort(EfieldList);          % Sort E-field in ascending order.
MinPassNo=fix(0.75*FUPtsNo);     % 75% points are used.
TryNo=FUPtsNo-MinPassNo+1;
FUList=zeros(1,TryNo);

for ti=1:TryNo
  EndInd=ti+MinPassNo-1;
  FUList(ti)=20*log10(SortE(EndInd)/SortE(ti));   % Calculate FU with different
reference
end

FieldUniformity=min(FUList);
```

A.6 Mean and Standard Deviation of the Maximum Received Power with N Independent Samples

```matlab
clc;clear all;close all;
sigma=1;
NList=1:10000;
MaxMeanList=zeros(1,numel(NList));StdList=zeros(1,numel(NList));
for ti=1:numel(NList)
  N=NList(ti);
  F=@(x)N./(2*sigma.^2).*(1-exp(-x./(2*sigma.^2))).^(N-1).
  *exp(-x./(2*sigma.^2));
  MaxMeanList(ti)=quadgk(@(x)F(x).*x,0,inf );
  StdList(ti)=sqrt(quadgk(@(x)(x-MaxMeanList(ti)).^2.*F(x),0,inf ));
end

figure;
StdIndB=10*log10((StdList+MaxMeanList)./MaxMeanList);
plot(NList,StdIndB);
grid on;xlabel('N');ylabel('std (dB)');
set(gcf,'Color','w');set(gca, 'Xscale', 'log');
set(findall(gcf,'-property','FontSize'),'FontSize',14);
figure(gcf );

figure;
MaxMeanIndB=10*log10(MaxMeanList./MaxMeanList(1));
plot(NList,MaxMeanIndB);
```

```
grid on;xlabel('N');ylabel('mean (dB)');
set(gcf,'Color','w');set(gca, 'Xscale', 'log');
set(findall(gcf,'-property','FontSize'),'FontSize',14);
figure(gcf);
```

A.7 Mean and Standard Deviation of the Maximum Rectangular E-Field with N Independent Samples

```
clc;clear all;close all;
sigma=1;
NList=1:10000;
MaxMeanList=zeros(1,numel(NList));StdList=zeros(1,numel(NList));
for ti=1:numel(NList)
  N=NList(ti);
  F=@(x)N*x./sigma.^2.*(1-exp(-x.^2./(2*sigma.^2))).^(N-1).
  *exp(-x.^2./(2*sigma.^2));
  MaxMeanList(ti)=quadgk(@(x)F(x).*x,0,inf);
  StdList(ti)=sqrt(quadgk(@(x)(x-MaxMeanList(ti)).^2.*F(x),0,inf));
end

figure;
StdIndB=20*log10((StdList+MaxMeanList)./MaxMeanList);
plot(NList,StdIndB);
grid on;xlabel('N');ylabel('std (dB)');
set(gcf,'Color','w');set(gca, 'Xscale', 'log');
set(findall(gcf,'-property','FontSize'),'FontSize',14);
figure(gcf);

figure;
MaxMeanIndB=20*log10(MaxMeanList./MaxMeanList(1));
plot(NList,MaxMeanIndB);
grid on;xlabel('N');ylabel('mean (dB)');
set(gcf,'Color','w');set(gca, 'Xscale', 'log');
set(findall(gcf,'-property','FontSize'),'FontSize',14);
figure(gcf);
```

A.8 Monte Carlo Simulation in the RC

These are typical Matlab codes to plot the empirical cumulative distribution function (CDF) of the magnitude of the rectangular E-field in the reverberation chamber (RC) and compare it with the theoretical distribution. The code can be modified to plot other CDFs and used for other purposes such as correlation coefficient, boundary field distributions, etc. The parameters (number of plane waves, simulation times) can also be changed depending on applications.

```
% Monte Carlo Simulation
clc;close all;clear all;
RunTime=1000;  % Run MC simulation for 1000 times
VectorETotX=zeros(1,RunTime);
```

```
VectorETotY=zeros(1,RunTime);
VectorETotZ=zeros(1,RunTime);
for RT=1:RunTime

  N=1000;  %Number of random incident plane waves (incident angles)
  u=rand(1,N);
  v=rand(1,N);
  alpha=pi*rand(1,N);
  theta=acos(2*u-1);
  phi=2*pi*v;
  kx=sin(theta).*cos(phi);  % k is not necessary when r=0 is used,
  but we leave it here.
  ky=sin(theta).*sin(phi);
  kz=cos(theta);

  ex=cos(alpha).*cos(theta).*cos(phi)-sin(alpha).*sin(phi);
  ey=cos(alpha).*cos(theta).*sin(phi)+sin(alpha).*cos(phi);
  ez=-cos(alpha).*sin(theta);

  Phi=2*pi*rand(1,N);
  E0=1;
  VectorETot=0;
  for ti=1:N
    VectorE=E0*[ex(ti) ey(ti) ez(ti)].*exp(1j*Phi(ti))/sqrt(N); %exp(jk dot r)=1
    VectorETot=VectorETot+VectorE; % Superimpose N plane waves
  end

  VectorETotX(RT)=VectorETot(1);  % Record Ex field
  VectorETotY(RT)=VectorETot(2);
  VectorETotZ(RT)=VectorETot(3);

end

[AbsExCDF,Ex]=ecdf(abs(VectorETotX));  % Empirical CDF from MC simulation
x=linspace(0,2,100);sigma=E0/sqrt(6);
TheoCDF=1-exp(-x.^2./(2*sigma^2));   % Theoretical CDF from analytical
expression

figure;
stairs(Ex,AbsExCDF,'LineWidth',2);hold on;
plot(x,TheoCDF,'color','r','LineStyle','none','Marker','o');hold on;grid on;
grid on;xlabel('|Ex|');ylabel('CDF(|Ex|)');
set(gcf,'Color','w');legend('Empirical','Theoretical');
set(findall(gcf,'-property','FontSize'),'FontSize',14);
figure(gcf);

figure;
quiver3(kx,ky,kz,-kx,-ky,-kz,'AutoScaleFactor',0.6); hold on;
quiver3(kx,ky,kz,ex,ey,ez,'AutoScaleFactor',0.4); hold on;
pbaspect([1 1 1]);xlim([-1 1]);ylim([-1 1]);zlim([-1 1]);
figure(gcf);
```

A.9 The Plot of Theoretical Rayleigh CDF

This Matlab code plots the theoretical Rayleigh CDF in diversity gain measurement.

```
% Plot theoretical Rayleigh distribution.
RayleighMagdB = -30:0.1:10;
RayleighMag = 10.^( RayleighMagdB /20);
RayleighCDF= 1 - exp(-RayleighMag.^2);
figure;
semilogy(RayleighMagdB, RayleighCDF,'color','r','LineWidth', 2);
xlabel('dB');ylabel('CDF');
figure(gcf);
```

Appendix B

Reference NSA Values

The NSA (normalised site attenuation) values of an anechoic chamber are plotted from the data in standard [1]. The NSA values when using tunable half-wave dipoles are shown in Figures B.1–B.3 for distances of 3, 10, and 30 m, respectively. The ±4 dB limit is also given. The height of the transmitting (Tx) antenna is 2 m for horizontal polarisation and 2.75 m for vertical polarisation, but the scanning range of the height of the receiving (Rx) antenna is not always from 1 to 4 m. At low frequency, the dipole is very long and the scanning height cannot start from 1 m. Also, for the 30 m distance, the height range of 2–6 m is normally scanned. Note that in the measurement the dipole length needs to be tuned for each measurement frequency and the correction factors need to be used to correct the measured NSA values. More details can be found in [1].

The NSA values when using broadband antennas (such as biconical or log-periodic dipole arrays) are shown in Figures B.4–B.9 for distances of 3, 10, and 30 m.

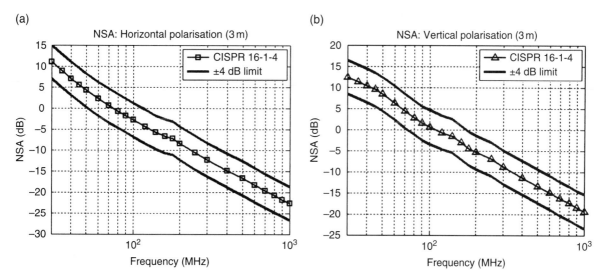

Figure B.1 NSA values for tunable dipoles when the separation distance is 3 m: (a) horizontal polarisation, the height of the Tx antenna is fixed as 2 m, and (b) vertical polarisation, the height of the Tx antenna is fixed as 2.75 m.

Anechoic and Reverberation Chambers: Theory, Design, and Measurements, First Edition. Qian Xu and Yi Huang.
© 2019 John Wiley & Sons Ltd. Published 2019 by John Wiley & Sons Ltd.

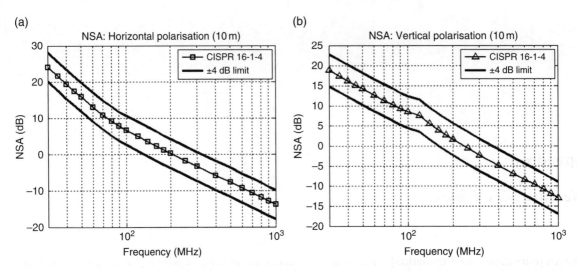

Figure B.2 NSA values for tunable dipoles when the separation distance is 10 m: (a) horizontal polarisation, the height of the Tx antenna is fixed as 2 m, and (b) vertical polarisation, the height of the Tx antenna is fixed as 2.75 m.

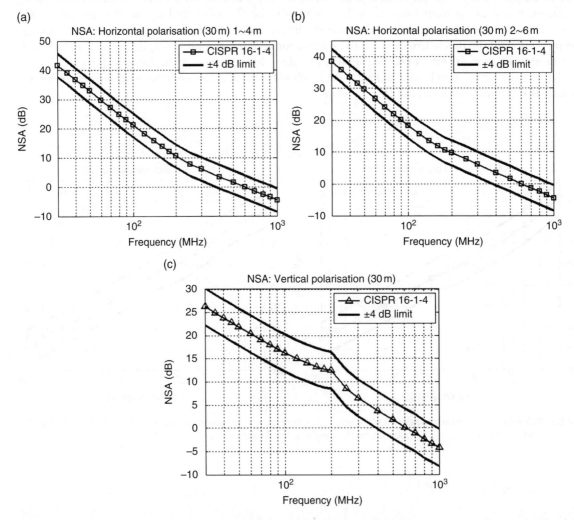

Figure B.3 NSA values for tunable dipoles when the separation distance is 30 m: (a) horizontal polarisation, the height of the Tx antenna is fixed as 2 m and the height of the Rx antenna is scanned from 1 to 4 m, (b) horizontal polarisation, the height of the Rx antenna is scanned from 2 to 6 m, and (c) vertical polarisation, the height of the transmitting antenna is fixed as 2.75 m.

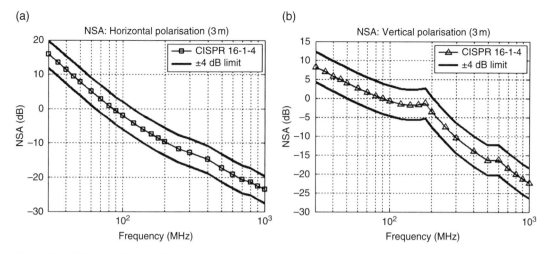

Figure B.4 NSA values for broadband antennas when the separation distance is 3 m, the height of the Tx antenna is fixed as 1 m, and the height of the Rx antenna is scanned from 1 to 4 m: (a) horizontal polarisation and (b) vertical polarisation.

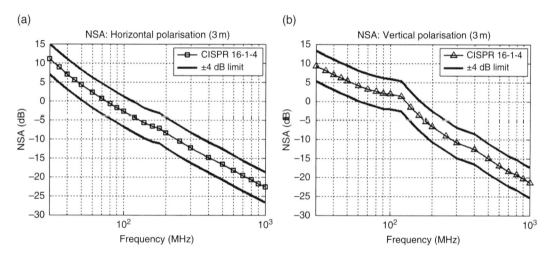

Figure B.5 NSA values for broadband antennas when the separation distance is 3 m and the height of the Rx antenna is scanned from 1 to 4 m. (a) horizontal polarisation, the height of the Tx antenna is fixed as 2 m, (b) vertical polarisation, the height of the Tx antenna is fixed as 1.5 m.

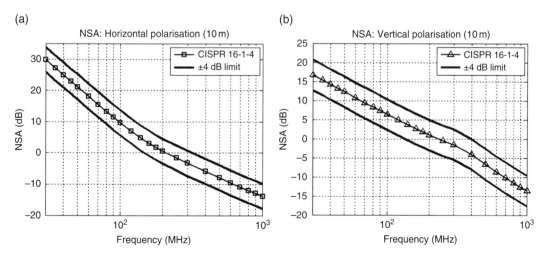

Figure B.6 NSA values for broadband antennas when the separation distance is 10 m, the height of the Rx antenna is scanned from 1 to 4 m, and the height of the Tx antenna is fixed as 1 m: (a) horizontal polarisation and (b) vertical polarisation.

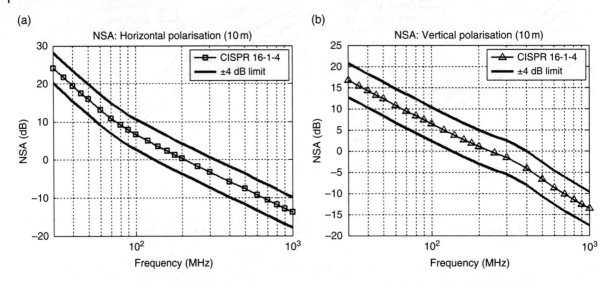

Figure B.7 NSA values for broadband antennas when the separation distance is 10 m and the height of the Rx antenna is scanned from 1 to 4 m: (a) horizontal polarisation, the height of the Tx antenna is fixed as 2 m, and (b) vertical polarisation, the height of the Tx antenna is fixed as 1.5 m.

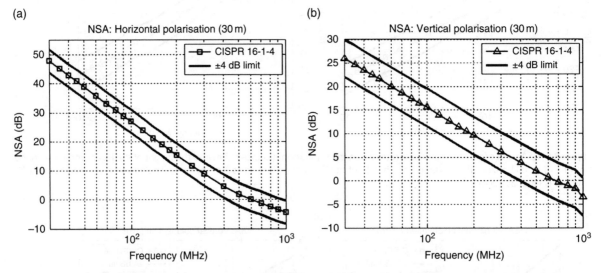

Figure B.8 NSA values for broadband antennas when the separation distance is 30 m, the height of the Rx antenna is scanned from 1 to 4 m, and the height of the Tx antenna is fixed as 1 m: (a) horizontal polarisation and (b) vertical polarisation.

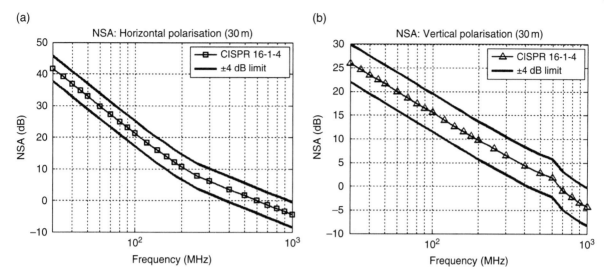

Figure B.9 NSA values for broadband antennas when the separation distance is 30 m and the height of the Rx antenna is scanned from 1 to 4 m: (a) horizontal polarisation, the height of the Tx antenna is fixed as 2 m, and (b) vertical polarisation, the height of the Tx antenna is fixed as 1.5 m.

Reference

1 CISPR 16-1-4. (2012). *Specification for radio disturbance and immunity measuring apparatus and methods – Part 1–4: Radio disturbance and immunity measuring apparatus – Antennas and test sites for radiated disturbance measurements*, 3.1e. IEC Standard. International Electrotechnical Commission, July 2012.

Appendix C

Test Report Template

This appendix gives a template of typical test report of an anechoic chamber including normalised site attenuation (NSA), site voltage-standing-wave ratio (SVSWR), and field uniformity (FU) measurements. The figures and results in the report are omitted. More details can be found in [1]. Obviously, the report can have different forms, depending on the requirement in practice.

Test report no. ∗∗∗

On: site validation

Subject: semi-anechoic chamber at ∗∗∗

Ordered by: ∗∗∗

Internal order no.: ∗∗∗

Technical responsibility: ∗∗∗   Test performed by: ∗∗∗

Data: ∗∗∗

Number of pages: ∗∗∗

C.1 Summary

Tests performed	Setup	Acceptability criterion	Conclusion
NSA	$d = 10$ m, $v = 4$ m	±4 dB	∗∗∗
	$d = 3$ m, $v = 2$ m	±4 dB	∗∗∗
Site VSWR	$d = 3$ m, $v = 4$ m	≤6 dB	∗∗∗
Field uniformity	$d = 3$ m	−0 dB, +6 dB for ≥75%	∗∗∗

C.2 Date and Place of Measurements

Test engineer: ∗∗∗

Date of measurements: from ∗∗∗ to ∗∗∗

Address of the site: ∗∗∗

Anechoic and Reverberation Chambers: Theory, Design, and Measurements, First Edition. Qian Xu and Yi Huang.
© 2019 John Wiley & Sons Ltd. Published 2019 by John Wiley & Sons Ltd.

C.3 Description of the Test Site

Type: Semi-Anechoic Chamber
Dimensions: ✳✳✳
Date of manufacture: ✳✳✳
Shielding provided by: ✳✳✳
Absorbers: Walls and ceiling: ✳✳✳
       Floor for SVSWR: ✳✳✳
       Floor for FU: ✳✳✳
Turntable: ✳✳✳, 4 m diameter
Environmental conditions: Temperature: ✳✳✳, Humidity: ✳✳✳

C.4 Measurements and Results

C.4.1 Normalised Site Attenuation (NSA)

C.4.1.1 Test Description
Standard: CISPR 16-1-4 Amd. 1 Ed. 3.0 2012-07 (clause 5.4.4)
Frequency range:

Frequency range (MHz)	Frequency resolution (MHz)
30–100	1
100–500	5
500–1000	10

Test configuration and acceptability criterion:

Distance	Volume diameter	Axis	Limit
$d = 10\,$m	$v = 4\,$m	$-15°$ off axis	±4 dB
$d = 3\,$m	$v = 2\,$m	In axis	±4 dB

Antenna heights:

Position	Tx Hor (m)	Ver (m)	Rx
Lower	1.0	1.0	Scan 1–4 m
Upper	2.0	1.5	

C.4.1.2 Measurement Equipment

Equipment	Type and identification
Network analyser	∗∗∗, calibration due ∗∗∗
Pair of biconical antennas	∗∗∗, certificate: ∗∗∗, antenna calibration according to CISPR 16-1-4 APR procedure (section 5.4.4.4)
Pair of log. per. antennas	∗∗∗, certificate: ∗∗∗, antenna calibration according to CISPR 16-1-4 APR procedure (section 5.4.4.4)
Preamplifier	∗∗∗
Attenuators	∗∗∗
Masts	∗∗∗

C.4.1.3 Results

Results are given in the figures and also in files in MS-EXCEL format:

Distance (m)	Frequency range	Figure	File	Conclusion
10	30 MHz–200 MHz	Figure ∗∗∗	∗∗.xls	∗∗∗
	200 MHz–1 GHz	Figure ∗∗∗	∗∗.xls	∗∗∗
3	30 MHz–200 MHz	Figure ∗∗∗	∗∗.xls	∗∗∗
	200 MHz–1 GHz	Figure ∗∗∗	∗∗.xls	∗∗∗

C.4.2 Site VSWR

C.4.2.1 Test Description

Standard: CISPR 16-1-4 Ed. 3.0 2010-04 (clause 8)
Frequency range: 1–18 GHz
Acceptability criterion:

Distance	Limit
$d = 3$ m	≤6 dB

Test volume and distance:

Distance	Volume diameter	Volume height	Axis
$d = 3$ m	$v = 4$ m	$h = 2$ m	In axis

Antenna heights:

Position	Tx (m)	Rx
Front, right, left, centre	1.0	Same as Tx
Top	2.0	

C.4.2.2 Measurement Equipment

Equipment	Type and identification
Network analyser	***, calibration due ***
Transmit antenna	***
Receive antenna	***
Low noise amplifier	***
Masts	***

C.4.2.3 Results
Results for the frequency range 1–18 GHz in horizontal and vertical polarisation are given in the following figures and also in files in MS-EXCEL format:

Distance (m)	Frequency range (GHz)	Figure	File	Conclusion
3	1–6	Figure ***	**.xls	***
	6–18	Figure ***	**.xls	***

C.4.3 Field Uniformity (FU) Measurements

C.4.3.1 Test Description
Standard: IEC 61000-4-3 Ed. 3.1 b 2008
Frequency range: 80 MHz to 6 GHz at 1% frequency steps
Acceptability criterion:

Distance	Limit
$d = 3\,m$	−0 dB, +6 dB for ≥75%

Uniform field area: The purpose of this measurement is to characterise the properties of the uniform field area (UFA), which is defined as follows: 16 points in a vertical plane of 1.50 m × 1.50 m, by 0.50 m steps, starting at a height of 0.80 m. The uniform plane is located through the centre of the turntable.

Antenna height: $h = 1.55\,m$ above floor.

Test distance: $d = 3\,m$, between the centre of the uniform plane and the tip of the antenna.

Floor absorbers: 3.0 m × 2.4 m.

Measurement axis: in axis.

C.4.3.2   Measurement Equipment

Equipment	Type and identification
Network analyser	***, calibration due ***
Transmit antenna	***
Field probe	***
Amplifier	***
Directional coupler	***
Antenna mast	***
Probe stand	***

C.4.3.3   Results

Results are given in the following figures and also in files in MS-EXCEL format:

Frequency range	Figure	File	Conclusion
80 MHz–200 MHz	Figure ***	**.xls	***
200 MHz–1 GHz	Figure ***	**.xls	***
1 GHz–2 GHz	Figure ***	**.xls	***
2 GHz–4 GHz	Figure ***	**.xls	***
4 GHz–6 GHz	Figure ***	**.xls	***

Reference

1 https://www.seibersdorf-laboratories.at/en/home.

Appendix D

Typical Bandpass Filters

In order to measure the time domain (TD) response of a reverberation chamber (RC), a frequency domain (FD) measurement is normally conducted because the FD measurement can offer a much larger dynamic range than the TD measurement and the TD response can be obtained from the inverse Fourier transform (IFT) of the measured FD response. In the RC measurement, a bandpass filter (BPF) is used to filter the measured S-parameters, and then the IFT is applied to obtain the TD response of the filtered S-parameters. From the TD response, the decay constant (or Q factor) can be obtained by using the least-squares fit method [1], which is frequency dependent, and the resolution depends on the passband of the BPF used. Elliptic filters were used in [2–5] while rectangular filters were used in [6] without detailed discussion. In this appendix, different types of BPFs are investigated, and results from measurement data are given and compared [7]. It is found that the extraction of chamber decay constant τ_{RC} (or Q factor) is not sensitive to the type of employed filters, but for some applications the BPF selection is important [7].

A schematic plot of a typical RC measurement is given in Figure D.1a and a typical measurement scenario is shown in Figure D.1b. S-parameters can be measured by using a vector network analyser (VNA), and then a BPF is used to select the frequency band of interest to process the collected S-parameter data. If one defines Port 1 as the input port and Port 2 as the output port, as depicted in Figure D.1c, the BPF can be placed at the input port or the output port. Although in measurement the S-parameters are normally measured first and BPF is applied later, they are actually equivalent in theory:

$$F(j\omega) \times S_{21} = S_{21} \times F(j\omega) = \widetilde{S}_{21} \tag{D.1}$$

The BPF is placed at the input port to provide a better understanding. It is shown in Figure D.1c that, when the impulse signal $\delta(t)$ is filtered by a BPF, a modulated impulse signal is obtained. The TD response from the IFT of $\widetilde{S}_{21}$ is actually the response of a modulated impulse excitation.

If we check Figure D.1c carefully, it can be found that the input signal of the RC depends on the impulse response of the BPF. The extracted τ_{RC} is actually the average value in the passband of the BPF. It is expected to have a filter with a narrow passband in the FD and a short impulse response in the TD, so that the TD response of the RC will not be dominated by the input modulated impulse (impulse response of the BPF) and a good resolution of τ_{RC} in the FD can be obtained. It is understandable that these two requirements cannot be ideally satisfied simultaneously as a short time signal in the TD means a wideband spectrum in the FD. An extreme case is that when the passband of the BPF is infinitely small (single frequency), the impulse response of the BPF is a continuous sine wave (infinitely long). The response of the RC will also be a sine wave, which makes it impossible to observe the decay of the signal.

The performance of typical BPFs are compared as it is impossible to enumerate all types of BPFs. S-parameters were measured in the RC at the University of Liverpool (Figure D.1b). 10 001 points were recorded for each stirrer position in the frequency range of 2–3.5 GHz, and 100 stirrer positions were used

Anechoic and Reverberation Chambers: Theory, Design, and Measurements, First Edition. Qian Xu and Yi Huang.
© 2019 John Wiley & Sons Ltd. Published 2019 by John Wiley & Sons Ltd.

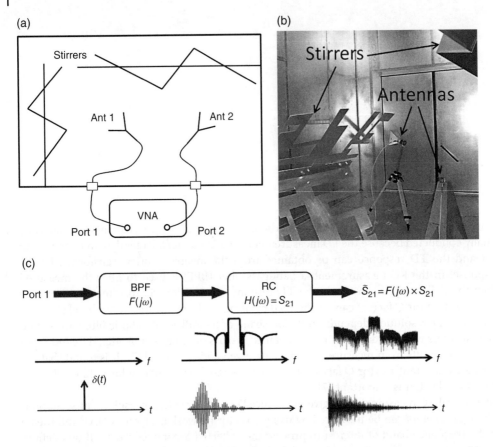

Figure D.1 TD response measurement in an RC: (a) schematic plot of a typical measurement setup, (b) measurement setup at the University of Liverpool, (c) the use of a BPF to obtain the TD response of a modulated impulse input. $F(j\omega)$ is the transfer function of the BPF, S_{21} is the transfer function of the RC. and $\tilde{S}_{21}$ is the total transfer function. FD spectrums are shown with frequency axis f, and the corresponding TD signals are illustrated with time axis t.

with 3.6°/step. Typical measured S-parameters at one stirrer position are shown in Figure D.2. The detailed steps to extract the TD response and the decay constant are given below with different BPFs.

Rectangular filter: The rectangular BPF is a well-known filter which removes all frequency components outside the passband in the FD. Although the impulse response of the rectangular filter is non-causal, it does not affect the use of it, as we use it in the post processing of the measurement data rather than realise such a filter in practice. A rectangular filter with 100 MHz and 1.3 GHz passband is illustrated in Figure D.3a. The corresponding normalised impulse responses can be obtained from the IFT of the FD response and are shown in Figure D.3b. As expected, a wider passband BPF has a faster TD decay but the FD resolution is coarser.

The power delay profile (PDP) can be obtained from $\left\langle \text{IFT}\left(\tilde{S}_{21}\right)^2 \right\rangle$, where $\langle \cdot \rangle$ means averaging for all stirrer positions, and the normalised PDPs are shown in Figure D.3c. The TD aliasing effect [1] can be well observed as the sampling interval in the FD is 150 kHz, which corresponds to $1/(150\,\text{kHz}) = 6666.7$ ns in the figures.

It can be observed in Figure D.3b that, when a 100 MHz BPF is used (the obtained PDP and τ_{RC} have a frequency resolution of 100 MHz), the normalised input modulated impulse signal drops to −56 dB at 4000 ns (normalised to the peak value of 0 dB). Meanwhile, the PDP at 4000 ns is around −15 dB lower than the peak value. Ideally, the response should be purely from the RC and the input signal should decay to 0 (−∞ dB) quickly.

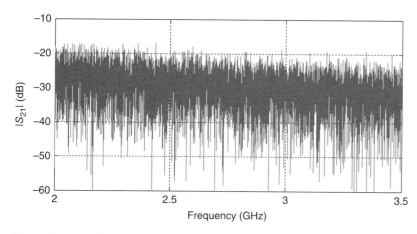

Figure D.2 Typical measured S_{21} at one stirrer position.

However, when the IFT is used to inverse the FD response, the decay of the input signal depends on the used BPF and there is always a finite tail for the input modulated impulse. This interference from the finite tail could lead to errors when we analyse the TD response from the RC. An extreme case is that, when the input signal is a continuous sine wave (no decay or decay extremely slow), the output will also be a sine wave and no RC characteristics can be extracted.

By comparing Figure D.3b and Figure D.3c, if we assume that the peak response (0 dB) of the PDP is from the peak magnitude of the input signal (TD impulse response of the BPF), it can be seen that the interference from the input signal is around 41 dB lower than the RC response at 4000 ns, and the response from the RC dominates the TD response. The interference level from the input signal can be expressed as:

$$\text{Interference Level} = \text{Norm}[\text{IFT}(F(j\omega)]\,(\text{dB}) - \text{Norm}(\text{PDP})\,(\text{dB}) \qquad (D.2)$$

which can be obtained from the difference between Figure D.3b and Figure D.3c and is shown in Figure D.4. As expected, the interference from the input signal is much smaller than the PDP in the whole time range, and the TD response is dominated by the RC but not the input signal.

This rectangular filter has been used in the antenna radiation efficiency measurement in [6] and good results were obtained. Figure D.4 is important, as it quantifies the interference from the input signal against the TD response of the RC. It should be noted that, although the interference shown in Figure D.4 is already very small in the τ_{RC} extraction, it could be significant in other applications, e.g. the extraction of the scattering damping time τ_s, because in the τ_s extraction, the decay speed of $\left|\left\langle \text{IFT}\left(\tilde{S}_{21}\right)\right\rangle\right|^2$ is much faster than $\left\langle \left|\text{IFT}\left(\tilde{S}_{21}\right)\right|^2\right\rangle$, as discussed in Chapter 6 [4, 5]. As long as the interference level is quantified, a trustable region of the TD response of the RC can be identified.

Elliptic filter: The elliptic filter is famous for the sharpest transition between the passband and the stopband when compared with the same order Butterworth or Chebyshev filters. The same procedure is repeated for different orders of elliptic BPFs. As a high resolution in the FD is expected, we fix the passband as 100 MHz (the same as we used in the rectangular BPF). The transfer functions of different order elliptic BPFs are shown in Figure D.5a, and the corresponding interference level from the input signal is given in Figure D.5b. As can be seen, the interference level for low order elliptic filters decays faster than high order ones, and a large *Rs* means a low level tail (>100 ns). Compared to the rectangular filter (Figure D.4), for the same passband bandwidth, the elliptic filter can perform better than the rectangular filter in some applications.

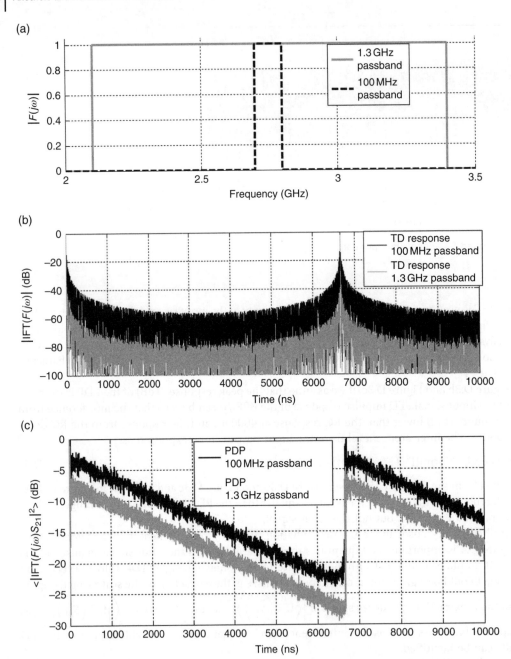

Figure D.3 (a) Rectangular filter transfer functions $|F(j\omega)|$ in the FD, (b) the impulse response of the filter in the TD (IFT$|F(j\omega)|$), only responses in positive time are plotted and normalised to peak value, and (c) PDPs at the centre frequency of 2.75 GHz, normalised to the peak value.

Butterworth filter: The magnitude response of Butterworth filters is maximally flat in the passband and monotonic in the pass- and stopbands, while the sacrifice is the roll-off steepness between the pass- and stopbands. Typical transfer functions of Butterworth filters are shown in Figure D.6a and the interference levels are given in Figure D.6b. As can be seen, Butterworth filters can also provide sharp early time decay and low level tail.

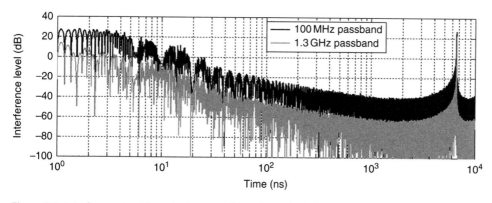

Figure D.4 Interference level from the input modulated signal. The horizontal axis is plotted in log scale to have a good resolution of the early time.

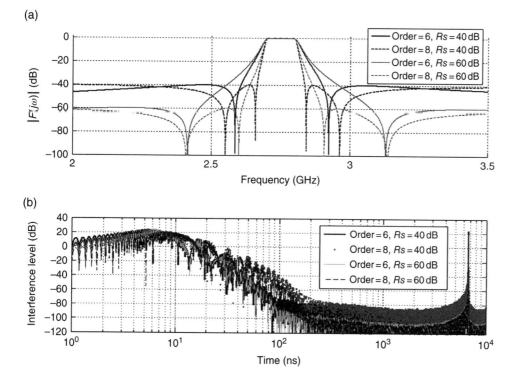

Figure D.5 (a) Transfer functions of elliptic filters with 100 MHz passband, 0.3 dB ripples in the passband, *Rs* means the stopband is *Rs* dB down from the peak value in the passband. (b) Interference level from the input signal with different elliptic filters.

Chebyshev filter: Chebyshev BPFs (100 MHz passband, 0.3 dB ripples) have also been studied and the interference level is comparable with the Butterworth filter in Figure D.6b. Typical transfer functions of Chebyshev BPFs are shown in Figure D.7a with 100 MHz passband. As can be seen in Figure D.7b, when the order is 6, the interference level is comparable with the Butterworth filter in Figure D.6b. When the order becomes 8, the transfer functions are still similar, but the interference level rolls off slower than the Butterworth filter.

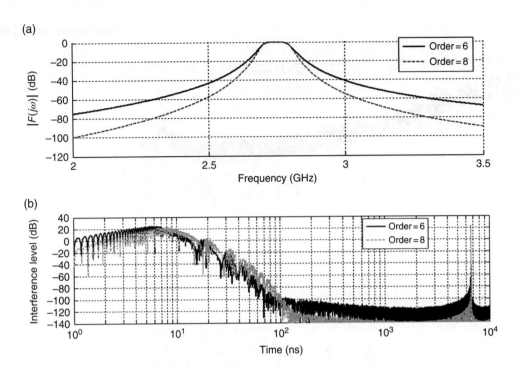

Figure D.6 (a) Transfer functions of Butterworth filters with 100 MHz passband. (b) Interference level from the input signal with different Butterworth filters.

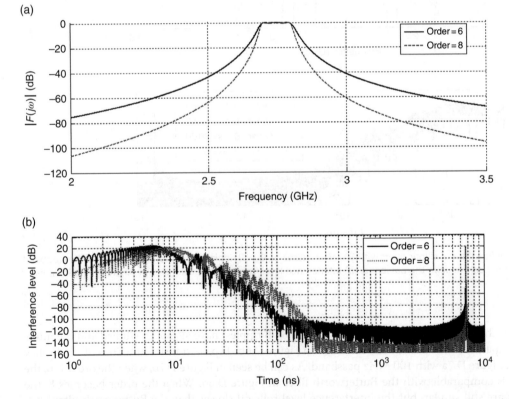

Figure D.7 (a) Transfer functions of Chebyshev filters with 100 MHz passband, 0.3 dB ripples in the passband. (b) Interference level from the input signal with different Chebyshev filters.

Gaussian filter: Gaussian pulse has been widely used in the TD simulation in computational electromagnetics [8]; it can also be used in the inversion of the TD signal in the RC (the RC is excited by a Gaussian pulse). A typical Gaussian BPF is shown in Figure D.8a with 100 MHz passband and the interference level is given in Figure D.8b. As can be seen, the Gaussian pulse decays very quickly in the TD compared with other filters and the passband can be controlled easily in the FD, which proves the Gaussian filter can be a good candidate for the RC TD analysis.

By sweeping the centre frequency of the BPF and applying the least-squares fit to the PDP [1] in the range of 500–5000 ns, the chamber decay constant can be extracted and is shown in Figure D.9. As expected, all BPFs have very similar results and no significant difference is observed since all the interference levels are quite small. Even for the rectangular filter, it is smaller than −30 dB.

In the TD signal analysis of the RC, different BPFs have been studied to quantify their effects. It has been found that for an RC with a long decay constant (τ_{RC}), the extracted τ_{RC} is not sensitive to the type of filters selected. A simple rectangular filter is good enough for most applications, but in some applications the effect of BPF could be significant, such as investigating the early time behaviour of the RC [9], scattering damping time measurement, and total scattering cross section (TSCS) measurement in Chapter 6. In these cases, the Gaussian filter can be a good candidate and the interference level from the input signal needs to be checked to ensure the TD response is dominated by the RC but not the input signal.

(a)

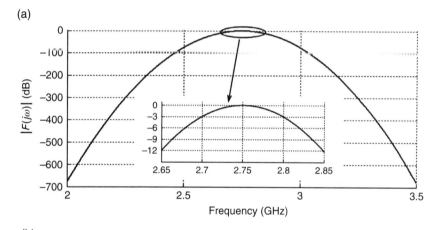

(b)

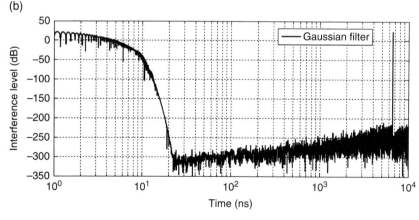

Figure D.8 (a) Transfer function of a Gaussian filter with 100 MHz passband (−3 dB). (b) Interference level from the input signal with a Gaussian filter.

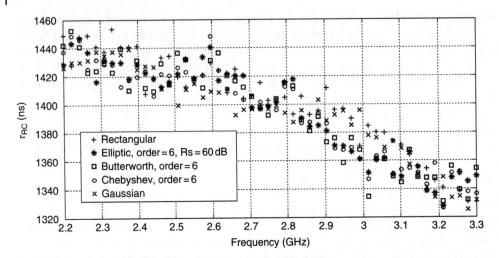

Figure D.9 Extracted chamber decay constant by using different BPFs.

References

1 Xu, Q., Huang, Y., Xing, L., and Tian, Z. (2016). Extract the decay constant of a reverberation chamber without satisfying Nyquist criterion. *IEEE Microwave and Wireless Components Letters* 26 (3): 153–155.

2 Holloway, C. L., Shah, H. A., Pirkl, R. J. et al. (2012). Reverberation chamber techniques for determining the radiation and total efficiency of antennas. *IEEE Transactions on Antennas and Propagation* 60 (4): 1758–1770.

3 Tian, Z., Huang, Y., Shen, Y., and Xu, Q. (2016). Efficient and accurate measurement of absorption cross section of a lossy object in reverberation chamber using two one-antenna methods. *IEEE Transactions on Electromagnetic Compatibility* 58 (3): 686–693.

4 Xu, Q., Huang, Y., Xing, L. et al. (2016). B-scan in a reverberation chamber. *IEEE Transactions on Antennas and Propagation* 64 (5): 1740–1750.

5 Xu, Q., Huang, Y., Xing, L. et al. (2016). The limit of the total scattering cross section of electrically large stirrers in a reverberation chamber. *IEEE Transactions on Electromagnetic Compatibility* 58 (2): 623–626.

6 Li, C., Loh, T., and Tian, Z. et al.(2015). A comparison of antenna efficiency measurements performed in two reverberation chambers using non-reference antenna methods. Loughborough Antennas and Propagation Conference (LAPC). LoughboroughK. pp. 1–5.

7 Xu, Q., Huang, Y., and Zhao, Y. et al. (2017). Investigation of bandpass filters in the time domain signal analysis of reverberation chamber. 32nd International Union of Radio Science General Assembly & Scientific Symposium (URSI GASS). Montreal, 19–26 August.

8 Taflove, A. (2005). *Computational Electrodynamics: The Finite-Difference Time-Domain Method*, 3e. Artech House.

9 Holloway, C. L., Shah, H. A., Pirkl, R. J. et al. (2012). Early time behavior in reverberation chambers and its effect on the relationships between coherence bandwidth, chamber decay time, RMS delay spread, and the chamber buildup time. *IEEE Transactions on Electromagnetic Compatibility* 54 (4): 714–725.

Appendix E

Compact Reverberation Chamber at NUAA

As an example, an in-house developed compact electromagnetic reverberation chamber (RC) at the Nanjing University of Aeronautics and Astronautics (NUAA) is reported in this appendix. The dimensions of the RC are 0.8 m × 1.2 m × 1.2 m. The field uniformity (FU), the ratio of the standard deviation and the mean value of the received power, the Q factor, and the enhanced backscatter coefficient have been measured. Results show that the lowest usable frequency (LUF) is about 1 GHz when the standard deviation of the average received power is less than 2 dB, which agrees well with the designed value.

E.1 Design Guidelines

As we know, the resonant frequencies of a cavity can be scaled up or down with physical dimensions, and we can predict the LUF from existing measurement results. Theoretical calculation from mode number (density) is also useful. A rough evaluation is that an enclosure has at least 60–100 resonant modes and the mode density should be larger than 1.5 modes/MHz [1]. From the Wely's law [1], we have:

$$\frac{8\pi V f^3}{3} \frac{f^3}{c^3} \geq 60 \sim 100 \tag{E.1}$$

and

$$8\pi V \frac{f^2}{c^3} \geq 1.5 \, \text{modes/MHz} \tag{E.2}$$

where $c = 3 \times 10^8$ m/s is the speed of light in free space, f is the frequency, and V is the volume of the RC.

A good way to estimate the LUF of a given RC is to scale the volume according to a reference RC with known LUF. If the RC at the University of Liverpool (UoL) is used as a reference [2], the dimensions are 3.6 m × 5.8 m × 4 m, the volume is 83.52 m^3. From (E. 1), the predicted LUF is about 132 MHz for 60 modes, and 157 MHz for 100 modes; from (E. 2), the predicted LUF is about 139 MHz. Measurement results have confirmed that the LUF is about 177 MHz (stirrers without slots) [3] according to the International Electrotechnical Commission (IEC) standard, which is of course dependent on a number of things such as the equipment under test (EUT) area and the stirrer design. Suppose the dimensions are scaled with a factor of 1/5, to be 0.72 m × 1.16 m × 0.8 m. The LUF should be scaled up to a factor of 5 which is about 177 × 5 = 885 MHz. Considering the available volume in practice, we choose the dimensions as 0.8 m × 1.2 m × 1.2 m. Obviously, the stirrers must be large enough to stir the field when the frequency is close to the LUF [3], and the performance limit of a stirrer depends on the stirring surface (Section 6.4–6.6). The diameter of the stirrer is also scaled, 0.4 m is used.

The side view and the top view of the schematic plot are illustrated in Figure E.1a,b, a XY platform is used to sweep the antenna in a 2D plane. To realise an irregular stirrer, one side of the z-shaped stirrer is cut randomly

Anechoic and Reverberation Chambers: Theory, Design, and Measurements, First Edition. Qian Xu and Yi Huang.
© 2019 John Wiley & Sons Ltd. Published 2019 by John Wiley & Sons Ltd.

(a)

(b)

(c)

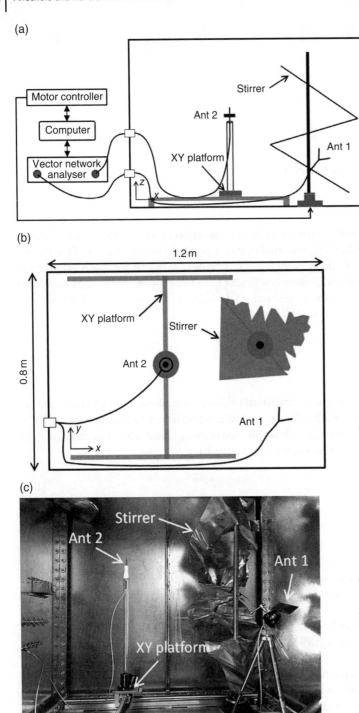

Figure E.1 Reverberation chamber at NUAA: (a) side view of the schematic plot, (b) top view of the schematic plot, and (c) field uniformity measurement. A plastic tube is used to hold Ant 2.

to have a zigzag edge. The computer controls the triggering of the vector network analyser (VNA) and the rotation of the stirrer, which also records measurement data (*S*-parameters in Figure E.1a).

E.2 Measurements

In this section, measurements are performed to validate the performance of the RC. FU, Q factor, the ratio of the standard deviation and the mean value of the received power, and the enhanced backscatter coefficient are measured.

E.2.1 Field Uniformity

According to the standard IEC 61000-4-21 [3], only eight points (four points in a plane) are enough to evaluate the FU. As an XY platform is available, a better resolution can be obtained, thus we use $6 \times 6 = 36$ sample points in the 2D plane to measure the FU, as shown in Figure E.2. In the *x*-direction, the step size is 8 cm and in the *y*-direction, the step size is 8.8 cm. The coordinates of the point at the left corner are (20, 21.5), as a little space is occupied by the motors of the XY platform. At each sample point, the monopole antenna (Ant 2 in Figure E.1c) is positioned in the *x*, *y*, and *z* directions; for each sample point and each direction, 360 stirrer rotations are used (1°/step). S_{21} with 1601 points in the frequency range of 10 MHz–3 GHz are recorded for each stirrer rotation. We have also measured the FU with three different heights to the ground of the RC: H_1 = 30 cm, H_2 = 62 cm, and H_3 = 98 cm. A summary of measurement scenarios is given in Table E.1.

An in-house developed automatic control system is used to synchronise the trigger of the VNA, the rotation of the stirrer, and the movement of Ant 2 in the XY platform. A workflow is illustrated in Figure E.3. Before the trigger of the VNA, we need to wait a few seconds until the rotation of the stirrer is finished. Continuous rotation is also acceptable [3], but if the time domain response needs to be inverted, we cannot use continuous rotation as the VNA needs time to complete the frequency sweep. After the trigger of the VNA, a few seconds may be necessary until the VNA finishes the frequency sweep. A control branch will determine if the next step is the movement of Ant 2, the rotation of the stirrer, or the end of the measurement.

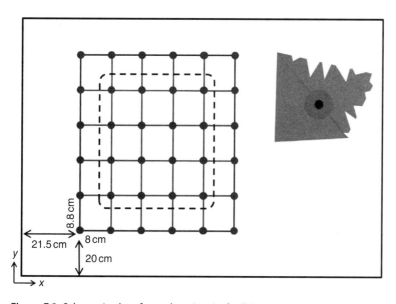

Figure E.2 Schematic plot of sample points in the FU measurement. 36 point are used in the whole 2D plane and 16 points are used in the area within the dashed line.

Table E.1 FU measurement scenarios.

Height	Direction	Sample points	Stirrer rotations
$H_1 = 36$ cm	x		
	y		
	z		
$H_1 = 62$ cm	x		
	y	36	360
	z		
$H_1 = 98$ cm	x		
	y		
	z		

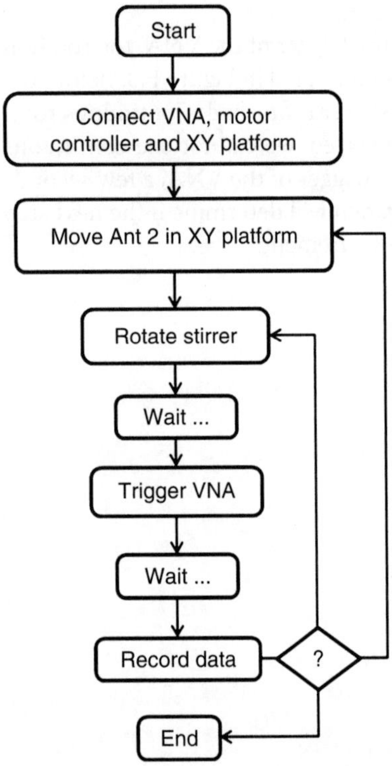

Figure E.3 The flowchart of the measurement control system.

A typical S_{21} curve at one stirrer position is illustrated in Figure E.4a, the maximum received power at 36 sample points at $H_2 = 62$ cm, z direction is illustrated in Figure E.4b, and the averaged received power at 36 sample points at $H_2 = 62$ cm, z direction is illustrated in Figure E.4c. To obtain a direct illustration of FU, we plot the stirred part of S_{21}. The average stirred part is defined as $\langle |S_{21,s}|^2 \rangle = \langle |S_{21} - \langle S_{21} \rangle|^2 \rangle$, where $\langle \cdot \rangle$ means the average over all stirrer positions. At frequencies 1 GHz, 2 GHz, and 3 GHz, $\langle |S_{21,s}|^2 \rangle$ are illustrated in Figure E.4d–f. As can be seen, the field close to the boundaries of the RC is not as uniform as that in the centre region; also when the frequency increases, a better FU is obtained.

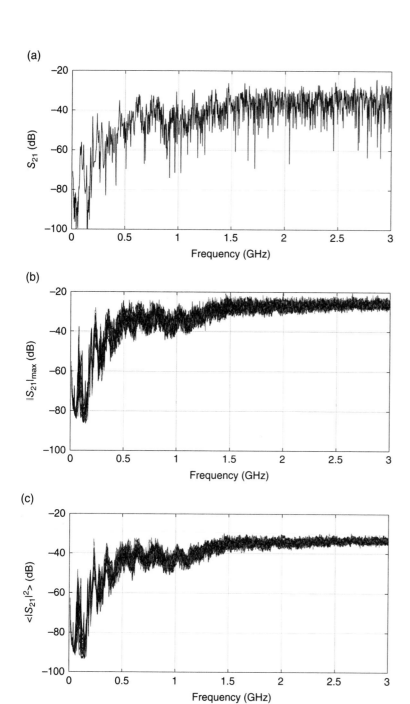

Figure E.4 H$_2$ = 62 cm, *z* direction: (a) typical measured S_{21} at one stirrer position, (b) maximum received power at 36 sample points, 36 curves are overlaid, (c) averaged received power at 36 sample points, (d) $\langle |S_{21,s}|^2 \rangle$ at 1 GHz, (e) $\langle |S_{21,s}|^2 \rangle$ at 2 GHz, and (f) $\langle |S_{21,s}|^2 \rangle$ at 3 GHz. Linear interpolation is used for points not at the sample positions, units are in dB.

(d)

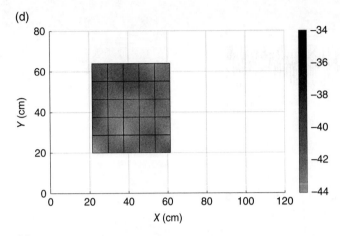

(e)

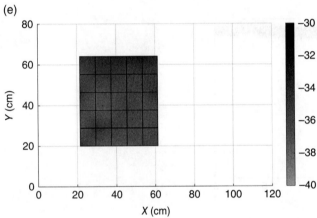

(f)

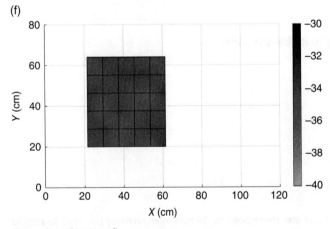

Figure E.4 (Continued)

The FU can be quantified using the standard deviation [3, 4]

$$\text{Std}_{\text{max}} = \sqrt{\frac{\sum_{p=1}^{N}\left(|S_{21}|^2_{\text{max},p} - \left\langle |S_{21}|^2_{\text{max}}\right\rangle\right)^2}{N-1}} \tag{E.3}$$

$$\text{Std}_{\text{aver}} = \sqrt{\frac{\sum_{p=1}^{N}\left(|S_{21}|^2_{\text{aver},p} - \left\langle |S_{21}|^2_{\text{aver}}\right\rangle\right)^2}{N-1}} \tag{E.4}$$

where N is the sample point number, $\langle \cdot \rangle$ means the average value over N sample points, and max (aver) means the maximum (averaged) value over all stirrer positions. If the RC is used mainly for electromagnetic compatibility (EMC) purposes, or the maximum E-field or received power is the main concern, (E.3) is used; if the averaged chamber transfer function is the main concern, such as antenna efficiency measurement [1, 2], absorption cross section measurement [5, 6], etc., (E.4) can be used. In dB form, we have

$$\text{FU or Std (dB)} = 10\log_{10}\frac{\text{Std (linear)} + \text{Mean (linear)}}{\text{Mean (linear)}} \tag{E.5}$$

'10log' is used when we use the received power. If the E-field magnitude is used, '20log' is required.

The measured FU of the maximum received power and average received power are illustrated in Figures E.5–E.7. 'All max' means all directions are used, thus there are three times more samples. As can be seen, when the antenna is swept in the plane of $H_1 = 30$ cm from the ground, the FU is not as good as that in the plane of $H_2 = 62$ cm. This is because the XY platform is metallic and has a height of 23 cm, thus Ant 2 is only 7 cm from the support of the XY platform. It is well known that when the receiving antenna is too close to a scatterer, the boundary condition on the scatterer will affect the received power [1], leading to a nonuniform statistical distribution. At low frequencies, the FU in the dashed line area (N = 16) shows better performance, but overall the FUs are comparable because we already have 20 cm from the boundaries.

As expected, when the height increases, a better FU is achieved which is less than 2 dB (average value) when the frequency is higher than 1 GHz. This is comparable to the measured results in the RC at UoL [2]. At UoL the FU is smaller than 2 dB at 200 MHz, while we have scaled the RC with a factor of 1/5, thus the LUF is scaled with a factor of 5, which is around 1 GHz. Obviously, the LUF depends on the acceptable threshold; if a 4 dB threshold is acceptable in Figure E.6, the LUF can be considered as 500 MHz. In practical measurement, to improve the FU, one can use or combine frequency stir, source stir (platform stir), and polarisation stir techniques to reduce the uncertainty.

E.2.2 The Ratio of Standard Deviation and Mean Value

It is well known that for a random variable with exponential distribution, the standard deviation equals the mean value. This provides a quick way to assess the performance of the RC [7, 8]. By comparing the standard deviation against the mean value of the received power, we can evaluate how close it is to an ideal exponential distribution. Although it is not rigorous in mathematics (hypothesis test is more rigorous), the result is easy to obtain for all frequencies. We calculated 10lg(Std/Mean) for all 36 sample points by using 360 stirrer positions, and the curves are shown in Figure E.8. Note that when the frequency is higher than 1 GHz, the difference between the standard deviation and the mean value of the received power is smaller than 1 dB, which means the distribution of the received power is very close to the ideal exponential distribution.

E.2.3 *Q* Factor

To measure the Q factor in the RC, we use the time domain method [9], as the insertion loss of the cables and antennas does not affect the Q factor measured in the time domain. Two broadband Vivaldi antennas are used

(a)

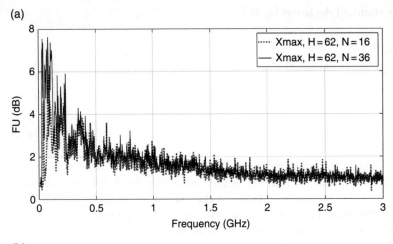

(b)

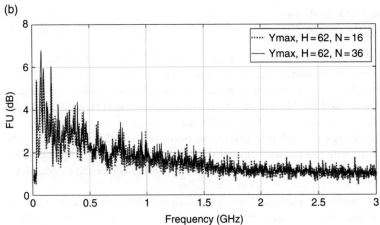

(c)

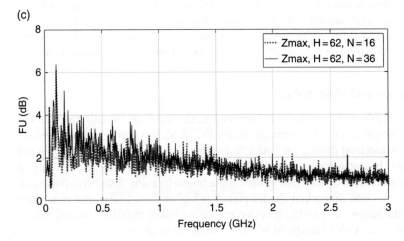

Figure E.5 The FU of the maximum received power in the plane of $H_2 = 62$ cm: (a) *x* direction, (b) *y* direction, and (c) *z* direction. N = 16 means the FU is measured within the dashed line area in Figure E.2a, and N = 36 means all the sample points in the plane are used.

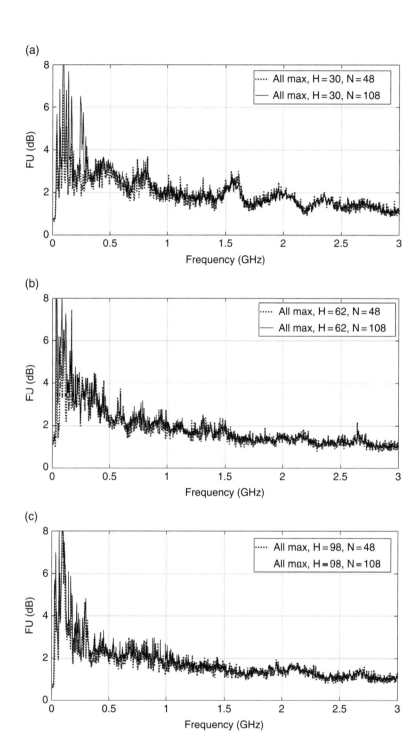

Figure E.6 The total FU of the maximum received power (including *x*, *y*, and *z* directions) in the plane of (a) H_1 = 30 cm, (b) H_2 = 62 cm, and (c) H_3 = 98 cm. *N* = 48 means the FU is measured within the dashed line area in Figure E.2a, and *N* = 108 means all the sample points in the plane are used.

(a)

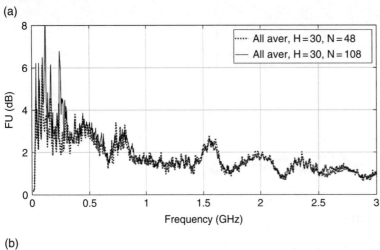

(b)

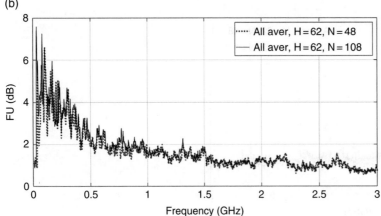

(c)

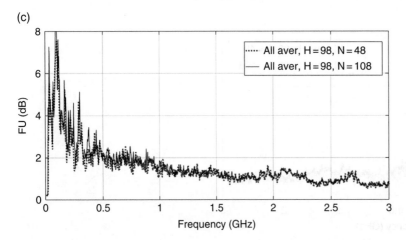

Figure E.7 The total FU of the average received power (including *x*, *y*, and *z* directions) in the plane of (a) H_1 = 30 cm, (b) H_2 = 62 cm, and (c) H_3 = 98 cm. N = 48 means the FU is measured within the dashed line area in Figure E.2a, and N = 108 means all the sample points in the plane are used.

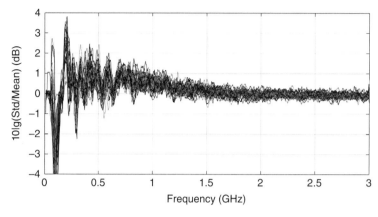

Figure E.8 The ratio of the standard deviation and the mean value of the received power. 36 curves are overlaid, the results are smoothed by using the nearest 17 frequency points (before converting to dB).

to measure the S-parameters as shown in Figure E.9; 10 001 frequency points are used in the frequency range of 400 MHz–6 GHz. By applying a bandpass filter with 100 MHz passband to the measured S_{21}, applying the inverse Fourier transform and using the least-squares method [9, 10], the chamber decay constant (τ_{RC}) can be obtained. 360 stirrer positions are used to obtain the power delay profile (PDP), and the extracted τ_{RC} is given in Figure E.10. Finally, the Q factor can be obtained from $Q = \omega\tau_{RC} = 2\pi f\tau_{RC}$.

We have also measured the RC with no XY platform. As expected, by removing unnecessary objects (e.g. wires, motors, and tracks) in the RC, the loss is reduced thus τ_{RC} is increased. Correspondingly, the Q factors are shown in Figure E.11.

Although the least-squares fit is used to fit the PDP to extract τ_{RC} [9], at low frequencies the PDP no long decays exponentially [11] and the extracted Q factor is not accurate. The PDP at centre frequencies of 500 MHz, 1 GHz, and 2 GHz are illustrated in Figure E.12. As can be seen when the frequency is lower than 1 GHz, the PDP does not decay exponentially and a curved PDP decay is observed when plot in dB unit.

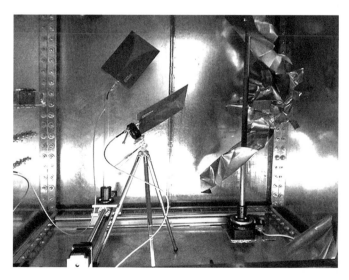

Figure E.9 Q factor measurement scenario using broadband antennas.

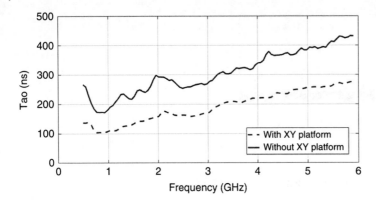

Figure E.10 Measured chamber decay time.

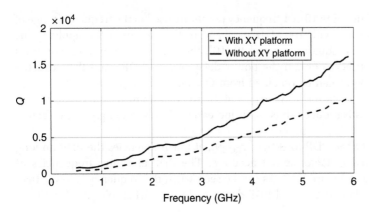

Figure E.11 Measured Q factors.

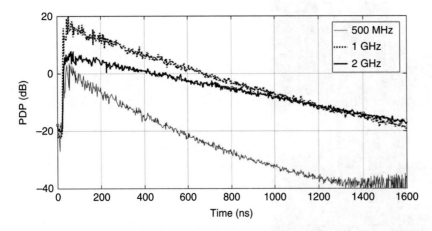

Figure E.12 Measured PDP at centre frequencies of 500 MHz, 1 GHz, and 2 GHz. The peak values have not been normalised to 0 dB (not necessary).

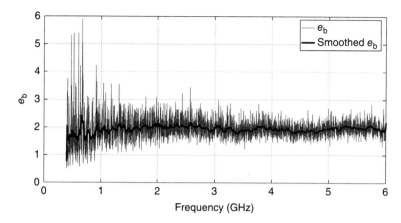

Figure E.13 Measured e_b. The smoothed result is obtained by averaging the nearest 17 points (1/100 of the full bandwidth).

Also note it takes about 50 ns to achieve the peak value of the PDP because the reference planes of two ports are calibrated to the output ports of the VNA (the very beginning of cables). This does not affect our measurement and it is very convenient if the insertion loss of cables does not affect the results. There is no need to recalibrate the measurement system every time the cables are moved.

E.2.4 Enhanced Backscatter Coefficient

It has been found that, for a well-stirred RC, the enhanced backscatter coefficient (e_b) equals 2 [1, 12] and e_b can be used to validate the performance of the RC. The definition of e_b has been given in [12]:

$$e_b = \frac{\sqrt{\langle|S_{11,s}|^2\rangle\langle|S_{22,s}|^2\rangle}}{\langle|S_{21,s}|^2\rangle} \tag{E.6}$$

where $S_{*,s}$ is the stirred part of the measured S_* and $\langle\cdot\rangle$ means the average operation over different stirrer positions. The measured e_b is illustrated in Figure E.13. When the frequency is lower than 1 GHz, e_b starts to deviate from 2.

E.3 Summary

A compact RC (0.8 m × 1.2 m × 1.2 m) measurement system has been briefly reported and evaluated. By applying the in-house developed automatic measurement system, the FU, standard/mean ratio, Q factor, PDP, and enhanced backscatter coefficient have been measured. The results have shown that the LUF of the RC is about 1 GHz. Note that we have used very dense sample points to evaluate the FU, and it took weeks to complete all the measurements. For a quick evaluation, only eight sample points are enough [3]. This appendix could provide a good reference for the RC design and the size of the RC can be scaled up or down according to the required LUF.

References

1 Hill, D. A. (2009). *Electromagnetic Fields in Cavities: Deterministic and Statistical Theories.* Wiley-IEEE Press.
2 Boyes, S. and Huang, Y. (2016). *Reverberation Chambers: Theory and Applications to EMC and Antenna Measurements.* Wiley.

3 IEC 61000-4-21 (2011). Electromagnetic compatibility (EMC) – Part 4–21: Testing and measurement techniques – Reverberation chamber test methods. 2.0e. International Electrotechnical Commission.

4 Montgomery, D. C. and Runger, G. C. (2013). *Applied Statistics and Probability for Engineers*, 6e. Wiley.

5 Melia, G. C. R., Robinson, M. P., Flintoft, I. D. et al. (2013). Broadband measurement of absorption cross section of the human body in a reverberation chamber. *IEEE Transactions on Electromagnetic Compatibility* 55 (6): 1043–1050.

6 Bamba, A., Gaillot, D. P., Tanghe, E. et al. (2015). Assessing whole-body absorption cross section for diffuse exposure from reverberation chamber measurements. *IEEE Transactions on Electromagnetic Compatibility* 57 (1): 27–34.

7 Ladbury, J., Koepke, G., and Camell, D. (1999). Evaluation of the NASA Langley research center mode stirred chamber facility. US National Institute of Standards and Technology Technical Note 1508.

8 Demoulin, B. and Besnier, P. (2011). *Electromagnetic Reverberation Chambers*. Wiley.

9 Xu, Q., Huang, Y., Xing, L., and Tian, Z. (2016). Extract the decay constant of a reverberation chamber without satisfying Nyquist criterion. *IEEE Microwave and Wireless Components Letters* 26 (3): 153–155.

10 Xu, Q., Huang, Y., and Zhao, Y. (2017). Investigation of bandpass filters in the time domain signal analysis of reverberation chamber. *32nd International Union of Radio Science General Assembly & Scientific Symposium (URSI GASS)*. Montreal, 19–26 Aug, 2017.

11 Dunlap, C. (2013). Reverberation Chamber Characterization Using Enhanced Backscatter Coefficient Measurements. PhD Thesis, Department of Electrical, Computer and Energy Engineering, University of Colorado.

12 Holloway, C. L., Shah, H. A., Pirkl, R. J. et al. (2012). Reverberation chamber techniques for determining the radiation and total efficiency of antennas. *IEEE Transactions on Antennas and Propagation* 60 (4): 1758–1770.

Appendix F

Relevant Statistics

Many mathematical techniques can be used for the interpretation of the measured data in the reverberation chamber (RC). This appendix presents frequently used definitions, facts, theorems, and methods in probability theory and statistics.

F.1 Propagation of Uncertainty

The uncertainty analysis of the RC measurement is very important, as one needs to justify the significance or uncertainty of the measurement result. There are many references for the metrology and uncertainty analysis, and more details can be found in references [1–3].

Generally, the uncertainty can be calculated using the first-order expansion of the measurand equation. The uncertainty or error (although the use of 'error' is unpopular in some books, engineers accept the use of it) in y, where y is a function of variables x_i [4], is

$$u(y)^2 = \sum_{i=1}^{N} \left(\frac{\partial y}{\partial x_i} \right)^2 u(x_i)^2 \tag{F.1}$$

where $u(x_i)$ is the uncertainty/error in x_i, and the partials are evaluated at the central values of x_i.

When correlations are included, the formula also involves the covariances $u(x_i, x_j)$ between the quantities-with-error [4]:

$$u(y)^2 = \sum_{i=1}^{N} \left(\frac{\partial y}{\partial x_i} \right)^2 u(x_i)^2 + 2 \sum_{i=1}^{N-1} \sum_{j=i+1}^{N} \frac{\partial y}{\partial x_i} \frac{\partial y}{\partial x_j} u(x_i, x_j) \tag{F.2}$$

The covariance $u(x_i, x_j)$ can be expressed in terms of the correlation $r(x_i, x_j)$ and the uncertainties/errors $u(x_i)$, $u(x_j)$ as:

$$u(x_i, x_j) = r(x_i, x_j) u(x_i) u(x_j) \tag{F.3}$$

A useful package called ScientificErrorAnalysis in Maple software can be used for uncertainty/error assignment and computations. A typical example in Maple is given below which calculates the uncertainty/error of a combined variable $y = f(x_1, x_2) = (x_1^3 - x_2)/x_2$, where the correlation coefficient between x_1 and x_2 has also been set as 0.2.

```
> with(ScientificErrorAnalysis):
> x1:=Quantity(2.3,0.1,relative);
> x2:=Quantity(9.5,0.2,relative);
```

Anechoic and Reverberation Chambers: Theory, Design, and Measurements, First Edition. Qian Xu and Yi Huang.
© 2019 John Wiley & Sons Ltd. Published 2019 by John Wiley & Sons Ltd.

```
> SetCorrelation(x1,x2,0.2);
> combine((x1^3-x2)/x2,errors);
```

The result is

```
Quantity(0.2807368421, 0.4169777056)
```

which means the central value is 0.28 and the uncertainty is 0.42 (rounded values).

Generally, the Monte Carlo simulation can be used to perform a general analysis, and the probability density function (PDF) and the cumulative density function (CDF) of the combined variable y can be obtained [3]. For the correlated variables, generally if we needs to generate n correlated Gaussian distributed random variables $\mathbf{Y} = (Y_1, Y_2,..., Y_n)$ with a given covariance matrix $\mathbf{\Sigma}$ and mean value $\mathbf{\mu} = (\mu_1, \mu_2,..., \mu_n)$, first we need to simulate a vector of uncorrelated Gaussian random variables $\mathbf{Z}$, then find a matrix $\mathbf{C}$ such that $\mathbf{CC}^T = \mathbf{\Sigma}$ (Cholesky decomposition), then the target vector can be obtained as:

$$\mathbf{Y} = \mathbf{\mu} + \mathbf{CZ} \tag{F.4}$$

If two random variables are used and the correlation coefficient is ρ, $\mathbf{C}$ can be chosen as

$$\mathbf{C} = \begin{bmatrix} 1 & 0 \\ \rho & \sqrt{1-\rho^2} \end{bmatrix} \tag{F.5}$$

A typical example in Matlab is

```
X1=randn(1,10000);
X2=randn(1,10000);
rho=0.6;
C=[1 0;rho sqrt(1-rho^2)];
Y=C*[X1;X2];
Y1=Y(1,:);Y2=Y(2,:);
corrcoef(Y1,Y2)
```

The result is

$$\begin{bmatrix} 1.0000 & 0.5946 \\ 0.5946 & 1.0000 \end{bmatrix}$$

which is very close to the given value 0.6. Note the results can have a slight difference after each run.

Finally, the type B uncertainty (evaluation of uncertainty by non-statistical methods) needs to be included for the total uncertainty of the result.

F.2 Parameter Estimation

The parameters in the PDF can be estimated by using methods such as the method of moments, maximum likelihood estimation (MLE) or Bayesian approach [5]. MLE is generally preferable because it has a better efficiency, but the method of moments is easy to compute.

Method of Moments: Let X_1, X_2, ..., X_n be a random sample from the PDF $f(x; \theta_1, \theta_2,..., \theta_m)$, the kth population moment (or distribution moment) is $E(X^k)$, $k = 1, 2, ..., m$, and the corresponding kth sample moment is $1/n\sum_{i=1}^{n}X_i^k, k = 1,2,...,m$. The estimated values of the unknown parameters $\theta_1, \theta_2,..., \theta_m$ can be found by equating the first m population moments to the first m sample moments and solving the resulting equations:

$$\int_{-\infty}^{+\infty} x^k f(x;\theta_1,\theta_2,...,\theta_m)dx = \frac{1}{n}\sum_{i=1}^{n} X_i^k, \ k=1,2,...,m \tag{F.6}$$

It is interesting to note that the method of moments actually estimates the Fourier transform of the PDF using the Taylor series [6]:

$$\begin{aligned}
\text{FT}[f(x)] &= \int_{-\infty}^{+\infty} e^{-j\omega x} f(x)dx = \int_{-\infty}^{+\infty} \sum_{k=0}^{\infty} \frac{(-j\omega x)^k}{k!} f(x)dx \\
&= \sum_{k=0}^{\infty} \frac{(-j\omega)^k}{k!} \int_{-\infty}^{+\infty} x^k f(x)dx = \sum_{k=0}^{\infty} \frac{(-j\omega)^k}{k!} \langle X^k \rangle
\end{aligned} \tag{F.7}$$

If the kth moments are known, the Fourier transform of the PDF of the first kth order is known. Thus the method of moments estimates the PDF by approximating the Fourier transform of it.

Maximum Likelihood Estimation: Suppose X is a random variable with PDF $f(x;\theta)$ where θ is a single unknown parameter. Let $x_1, x_2, ..., x_n$ be the observed values in a random sample of size n. Then the likelihood function of the sample is

$$L(\theta) = f(x_1;\theta) \cdot f(x_2;\theta) \cdot ... \cdot f(x_n;\theta) = \prod_{i=1}^{n} f(x_i,\theta) \tag{F.8}$$

The maximum likelihood estimator of θ is the value of θ that maximises the likelihood function $L(\theta)$. The method can be generalised to multiple parameters $\theta_1, \theta_2, ..., \theta_n$, and the estimated values can be found by solving the system of equations

$$\frac{\partial L(\theta_1,\theta_2,...,\theta_m)}{\partial \theta_k} = 0, \ k=1,2,...,m \tag{F.9}$$

It is interesting to note that in the regime of a non-ideal random field, the PDF of the magnitude of the electric field can be described by using the Weibull distribution with two parameters [7–10]:

$$f(x;a,b) = abx^{b-1} \exp\left(-ax^b\right) \tag{F.10}$$

The Rayleigh distribution in (5.60) is a special case of the Weibull distribution. When $a = 1/(2\sigma^2)$, $b = 2$, (F.10) becomes

$$f(x;\sigma) = \frac{x}{\sigma^2} \exp\left(-\frac{x}{2\sigma^2}\right) \tag{F.11}$$

which is the Rayleigh distribution. The maximum likelihood function of the Weibull distribution is

$$L(a,b) = (ab)^N \left(\prod_{i=1}^{N} x_i^{b-1}\right) \exp\left(-a\sum_{i=1}^{N} x_i^b\right) \tag{F.12}$$

The estimated values for a and b can be obtained by solving the system of equations [7–10]:

$$\begin{cases} \dfrac{\partial}{\partial a}\ln[L(a,b)] = N - a\sum_{i=1}^{N} x_i^b = 0 \\[2mm] \dfrac{\partial}{\partial b}\ln[L(a,b)] = N + b\sum_{i=1}^{N}\ln x_i - ab\sum_{i=1}^{N} x_i^b \ln x_i = 0 \end{cases} \tag{F.13}$$

It has been shown in [7] that when the frequency increases, b is close to 2 (Rayleigh distribution). Thus the Weibull distribution can be used to characterise how close a practical RC is to an ideal RC.

F.3 Useful Laws and Theorems

There are some laws and theorems used frequently in the application of the RC. We have introduced the transformation of variable in (5.153) and the order statistics in (5.135) and (5.136), but the well-known central limit theorem (CLT) and the generalised extreme value (GEV) distribution (not that well known) have not been given. They are listed here.

The Lyapunov CLT: Suppose $X_1, X_2, ..., X_n$ is a sequence of independent random variables, each with finite expected value μ_i and variance σ_i^2, define

$$s_n^2 = \sum_{i=1}^n \sigma_i^2 \tag{F.14}$$

If for some $\delta > 0$, Lyapunov's condition

$$\lim_{n \to \infty} \frac{1}{s_n^{2+\delta}} \sum_{i=1}^n E\left[|X_i - \mu_i|^{2+\delta}\right] = 0 \tag{F.15}$$

is satisfied, then a sum of X_i converges in a normal random variable as n goes to infinity: $\lim_{n \to \infty} \sum_{i=1}^n X_i$ has a distribution of normal form with mean value of $\sum_{i=1}^n \mu_i$ and variance of s_n^2. The average value $\lim_{n \to \infty} \frac{1}{n} \sum_{i=1}^n X_i$ has a distribution of normal form with mean value of $\frac{1}{n} \sum_{i=1}^n \mu_i$ and variance of s_n^2/n^2.

Specifically, when $\mu_1 = \mu_2 = ... = \mu_n = \mu$ and $\sigma_1^2 = \sigma_2^2 = ... = \sigma_n^2 = \sigma^2$, the average value $\lim_{n \to \infty} \frac{1}{n} \sum_{i=1}^n X_i$ has a distribution of normal form with mean value of μ and variance of σ^2/n (standard deviation of $\sigma/\sqrt{n}$). Lyapunov condition can be replaced by a weaker condition known as the Lindeberg CLT. The CLT has been used in many places to derive the PDF of unknown variables. It has also been generalised to multidimensional CLT [11].

Fisher–Tippett–Gnedenko Theorem: Like the CLT, the Fisher–Tippett–Gnedenko theorem gives the distribution of extreme order statistics. Suppose $X_1, X_2, ..., X_n$ is a sequence of independent and identically distributed random variables, and $M_n = \max \{X_1, X_2, ..., X_n\}$. If a sequence of pairs of real numbers (a_n, b_n) exists such that each $a_n > 0$ then

$$\lim_{n \to \infty} P\left(\frac{M_n - b_n}{a_n} \le x\right) = F(x) \tag{F.16}$$

where F belongs to either the Gumbel, the Fréchet, or the Weibull family. The CDF is

$$F(x)_{\text{GEV}} = \begin{cases} \exp\left[-\left(1 + k\frac{x-m}{s}\right)^{-1/k}\right], & k \ne 0 \\ \exp\left[-\exp\left(1 - \frac{x-m}{s}\right)\right], & k = 0 \end{cases} \tag{F.17}$$

where k is the shape parameter, m is the location parameter, and s is the scale parameter. For the application in the RC (maximum E-field magnitude or received power), the CDF is the reverse Weibull type ($k < 0$) and $x < m - s/k$, which is bounded [12–14]. This theorem is very useful as it does not need the parent PDF. It can not only be applied in the overmoded region but also in the undermoded region where the PDF of the parent distribution is not necessarily known. Care should be taken in the undermoded region, as the independent sample should be large enough to use the GEV distribution. This is not easy as when the RC works in the under-moded region the independent sample number is small. It can also be used to estimate the upper bound of the electric field, as in practice the magnitude of the electric field is always finite but the Rayleigh distribution is unbounded.

References

1 Bell, S. (2011). *The Beginner's Guide to Uncertainty of Measurement.* National Physical Laboratory.

2 Birch, K. (2003). *Estimating Uncertainties in Testing.* British Measurement and Testing Association.

3 Willink, R. (2013). *Measurement Uncertainty and Probability.* Cambridge University Press.

4 Scientific Error Analysis packages. Available online: http://www.maplesoft.com.

5 Montgomery, D. C. and Runger, G. C. (2014). *Applied Statistics and Probability for Engineers.* Wiley.

6 Scientific Error Analysis packages. Available online: http://mathworld.wolfram.com/CentralLimitTheorem.html.

7 Demoulin, B. and Besnier, P. (2011). *Electromagnetic Reverberation Chambers.* Wiley.

8 Amador, E., Lemonine, C., and Besnier, P. (2011). An empirical statistical detection of non-ideal field distribution in a reverberation chamber confirmed by a simple numerical model based image theory. *Annals of Telecommunications* 66: 445–455.

9 Lemoine, C., Besnier, P., and Drissi, M. (2007). Investigation of reverberation chamber measurements through high-power goodness-of-fit tests. *IEEE Transactions on Electromagnetic Compatibility* 49 (4): 745–755.

10 Orjubin, G., Richalot, E., Mengue, S., and Picon, O. (2006). Statistical model of an undermoded reverberation chamber. *IEEE Transactions on Electromagnetic Compatibility* 48 (1): 248–251.

11 van der Vaart, A. W. (2000). *Asymptotic Statistics.* Cambridge University Press.

12 Orjubin, G. and Wang, M.–. F. (2011). Experimental determination of the higher electric field level inside an overmoded reverberation chamber using the generalized extreme value distribution. *Annals of Telecommunications* 66: 457–464.

13 Orjubin, G. (2007). Maximum field inside a reverberation chamber modeled by the generalized extreme value distribution. *IEEE Transactions on Electromagnetic Compatibility* 49 (1): 104–113.

14 Gradoni, G. and Arnaut, L. R. (2010). Generalized extreme-value distributions of power near a boundary inside electromagnetic reverberation chambers. *IEEE Transactions on Electromagnetic Compatibility* 52 (3): 506–515.

Index

Anechoic and Reverberation Chambers: Theory, Design, and Measurements, First Edition. Qian Xu and Yi Huang.
© 2019 John Wiley & Sons Ltd. Published 2019 by John Wiley & Sons Ltd.